T0231265

FOOD TECHNOLOGY
Applied Research and
Production Techniques

Innovations in Agricultural and Biological Engineering

FOOD TECHNOLOGY

Applied Research and Production Techniques

Edited by
Murlidhar Meghwal, PhD
Megh R. Goyal, PhD, PE
Mital J. Kaneria, PhD

APPLE ACADEMIC PRESS

Apple Academic Press Inc.
3333 Mistwell Crescent
Oakville, ON L6L 0A2 Canada

Apple Academic Press Inc.
9 Spinnaker Way
Waretown, NJ 08758 USA

© 2018 by Apple Academic Press, Inc.

First issued in paperback 2021

Exclusive worldwide distribution by CRC Press, a member of Taylor & Francis Group
No claim to original U.S. Government works

ISBN 13: 978-1-77-463685-5 (pbk)
ISBN 13: 978-1-77-188509-6 (hbk)

Library and Archives Canada Cataloguing in Publication

Food technology : applied research and production techniques / edited by Murlidhar Meghwal, PhD, Megh R. Goyal, PhD, PE, Mital J. Kaneria, PhD.
(Innovations in agricultural and biological engineering)
Includes bibliographical references and index.
Issued in print and electronic formats.
ISBN 978-1-77188-509-6 (hardcover).--ISBN 978-1-315-36565-7 (PDF)
1. Food industry and trade--Technological innovations. I. Goyal, Megh Raj, editor II. Meghwal, Murlidhar, editor III. Kaneria, Mital J., editor IV. Series: Innovations in agricultural and biological engineering
TP370.F65 2017   664   C2017-902306-3   C2017-902307-1

Library of Congress Cataloging-in-Publication Data

Names: Meghwal, Murlidhar, editor. | Goyal, Megh Raj, editor. | Kaneria, Mital J., editor.
Title: Food technology : applied research and production techniques / editors, Murlidhar Meghwal, PhD, Megh R. Goyal, PhD, PE, Mital J. Kaneria, PhD.
Other titles: Food technology (Apple Academic Press)
Description: Toronto ; Waretown, New Jersey : Apple Academic Press, 2017. |
Includes bibliographical references and index.
Identifiers: LCCN 2017015045 (print) | LCCN 2017026428 (ebook) | ISBN 9781315365657 (ebook) | ISBN 9781771885096 (hardcover : acid-free paper)
Subjects: LCSH: Food--Biotechnology.
Classification: LCC TP248.65.F66 (ebook) | LCC TP248.65.F66 F6495 2017 (print) | DDC 664/.024--dc23
LC record available at https://lccn.loc.gov/2017015045

Apple Academic Press also publishes its books in a variety of electronic formats. Some content that appears in print may not be available in electronic format. For information about Apple Academic Press products, visit our website at **www.appleacademicpress.com** and the CRC Press website at **www.crcpress.com**

CONTENTS

LIST OF CONTRIBUTORS

Soumitra Banerjee, PhD
Assistant Professor, Food Technology, Centre for Emerging Technologies, Jain University, Jakkasandra – 562112, Ramanagara, Karnataka; Mobile: +91-9480443846; E-mail: soumitra.banerjee7@gmail.com

Yogesh Baravalia, PhD
Assistant Professor, Department of Biochemistry, Saurashtra University, Rajkot – 360005, Gujarat, India. Mobile: +91-9725127393; E-mail: yogesh_baravalia@yahoo.co.in

Ewelina Basiak, PhD
Research Scientist, Department of Food Engineering and Process Management, Faculty of Food Sciences, Warsaw University of Life Sciences-SGGW (WULS-SGGW), Warsaw, Poland; UMR PAM A 02–102, Food Processing and Physical – Chemistry Lab Université Bourgogne, Franche-Comté – AgroSup, Dijon, France; Department of Horticultural Engineering, Leibniz – Institute for Agricultural Engineering E.V., Potsdam, Germany. Tel.: +48-607966644, +49-1748147736; E-mail: ewelina_basiak@sggw.pl

Khyati C. Bhojani, MSc
Student, Department of Biochemistry, Saurashtra University, Rajkot – 360–005, Gujarat, India. E-mail: amitbhojani3620@yahoo.com

Ravi Kumar Biradar, MTech
Research Scholar, Food Technology, Centre for Emerging Technologies, Jain University, Jakkasandra – 562112, Ramanagara, Karnataka, Mobile: +91-9739204027; E-mail: ravikumar.aer@gmail.com

Sumitra Chanda, PhD
Professor, Department of Biosciences (UGC-CAS), Saurashtra University, Rajkot – 360005, Gujarat, India. Mobile: +91-9426247893; E-mail: svchanda@gmail.com.

D. V. Chidanand, PhD (Pursuing)
Assistant Professor, Indian Institute of Crop Processing Technology, Pudukkottai Road, Thanjavur – 613005, Tamil Nadu. Mobile: +91-9750968417; E-mail: chidanand@iicpt.edu.in

Mahuya Hom Choudhury, PhD
Scientist – C, Patent Information Centre, West Bengal State Council of Science and Technology, Department of Science and Technology, Govt. of West Bengal, Vigyan Chetana Bhavan, DD Block Salt Lake, Kolkata-700091, India; Mobile: +91-9007780898; E-mail: mhc123ster@gmail.com, mhc_123@rediffmail.com

Ofeoriste D. Esiegbuya, PhD
Senior Research Officer (Mycologist/Phytopathologist), Plant Pathology Division, Nigerian Institute for Oil Palm Research (NIFOR), P. M. B. 1030, Benin City, Edo State, Nigeria. Mobile: +2347054662459; E-mail: esiegbuya@gmail.com

Rajesh A. Dave, PhD
Research Associate, Food Testing Laboratory, Department of Biotechnology and Biochemistry, Junagadh Agricultural University, Junagadh – 362001, Gujarat, India. Mobile: +91-98980-99064; E-mail: drrajeshdave12@gmail.com

B. L. Dhananjaya, PhD

Associate Professor, Toxicology and Drug Discovery Unit, Centre for Emerging Technologies, Jain Global Campus, Jain University, Jakksandra Post, Kanakapura Taluk, Ramanagara – 562112, Karnataka, India; Mobile: +91-8197324276 and +91-9597031796; E-mail: chandu_greeshma@rediffmail.com, chandudhananjaya@gmail.com

Shailesh B. Gondaliya, PhD

Associate Professor, Central Instrumentation Lab, Directorate of Research, SD Agricultural University, Sardarkrushi Nagar – 385506, Gujarat, India. Mobile: +91-9879584774; E-mail: drgondaliya@yahoo.com

T. K. Goswami, PhD

Professor, Agricultural and Food Engineering Department, Indian Institute of Technology, Kharagpur – 721302, West Bengal, India. Tel.: +91-3222-283123 (R), 283122 (O); Fax: +91-3222-282244. E-mail: tkg@agfe.iitkgp.ernet.in

Megh R. Goyal, PhD, PE

Retired Faculty in Agricultural and Biomedical Engineering from General Engineering Department, University of Puerto Rico – Mayaguez Campus; and Senior Technical Editor-in-Chief in Agriculture Sciences and Biomedical Engineering, Apple Academic Press Inc., New Jersey, USA. E-mail: goyalmegh@gmail.com

Uday Heddurshetti, MTech

Research Scholar, Food Technology, Centre for Emerging Technologies, Jain University, Jakkasandra – 562112, Ramanagara, Karnataka, Mobile: +91-9739204027; E-mail: udayheddurshetti88@gmail.com

F. I. Okungbowa, PhD

Professor, Senior Lecturer, Department of Plant Biology and Biotechnology, University of Benin, Benin City, Edo State, Nigeria. Mobile: 234–7065150189; E-mail: fiokun2002@yahoo.com

Mital J. Kaneria, PhD

Assistant Professor, Department of Biosciences (UGC-CAS), Saurashtra University, Rajkot – 360005, Gujarat, India. Mobile: +91-9879272607; E-mail: mitalkaneria@gmail.com

Murlidhar Meghwal, PhD

Assistant Professor, Department of Food Science and Technology, National Institute of Food Technology Entrepreneurship & Management, Kundli – 131028, Sonepat, Haryana, India; Mobile: +91 9739204027; Email: murli.murthi@gmail.com

Pooja Moteriya, MPhil

Research Scholar, Department of Biosciences (UGC-CAS), Saurashtra University, Rajkot – 360005, Gujarat, India. Mobile: +91-9624240142; E-mail: poojamoteriya@gmail.com

Hemali Padalia, MPhil

Research Scholar, Department of Biosciences (UGC-CAS), Saurashtra University, Rajkot – 360005, Gujarat, India, Tel.: +91-96248-17140; E-mail: hemalipadalia@gmail.com

Harsha Patel, PhD

Assistant Professor, Gujarat Biotechnology Agricultural Institute, Navsari Agricultural University, Surat, Gujarat, India. Mobile: +91-9228188371; E-mail: vaghasiyah27@gmail.com

Jinisha T. Patel, BTech

Senior Research Assistant, Main Dry Farming Research Station, Junagadh Agricultural University, Targhadia (Rajkot) – 360003, Gujarat, India; E-mail: patel.jinishaa@gmail.com

Rohit M. Patel, PhD

Scientist, Gujarat Institute of Desert Ecology (GUIDE), Bhuj-Kachchh – 370001, Gujarat, India, Mobile: +91-97243–37687; E-mail: rmpecology@gmail.com

Komal V. Pokar, MSc
Graduate Student, Department of Biochemistry, Saurashtra University, Rajkot – 360005, Gujarat, India; E-mail: komalpokar1710@gmail.com

Anu Rachit, MTech
Researcher, Food Technology, Centre for Emerging Technologies, Jain University, Jakkasandra – 562112, Ramanagara, Karnataka; Mobile: +91-9739204027; E-mail: rachit.anu@gmail.com

Kalpna D. Rakholiya, PhD
Senior Research Fellow, Main Dry Farming Research Station, Junagadh Agricultural University, Targhadia (Rajkot) – 360003, Gujarat, India. Mobile: +91-9726599451; E-mail: kalpna.rakholiya@gmail.com

Jalpa Ram, MPhil
Graduate Student, Department of Biosciences (UGC-CAS), Saurashtra University, Rajkot – 360005, Gujarat, India. Mobile: +919624271800; E-mail: ramjalpa@gmail.com.

Ashish Rawson, PhD
Assistant Professor, Indian Institute of Crop Processing Technology, Thanjavur – 613005, Tamil Nadu, India. Mobile: +91-7373068426; E-mail: ashish.rawson@iicpt.edu.in

Kirubanandan Shanmugam, MSc
Former Graduate Research Student, Laboratory of Multiphase Process Engineering, Chemical Engineering Division, Department of Process Engineering and Applied Science, Dalhousie University, Halifax, NS, Canada. Mobile: +91-9444682247; E-mail: S.Kirubanandan@dal.ca

Jagruti M. Sonagara, MSc
Graduate Student, Department of Biosciences (UGC-CAS), Saurashtra University, Rajkot – 360005, India. Mobile: +91-8511633844; E-mail: jsonagara@gmail.com

C. K. Sunil, PhD (Pursuing)
Assistant Professor, Indian Institute of Crop Processing Technology, Thanjavur – 613005, India. Mobile: +91-9750968423; E-mail: sunil.ck@iicpt.edu.in

G. S. Sutaria, PhD
Research Scientist, Main Dry Farming Research Station, Junagadh Agricultural University, Targhadia (Rajkot) – 360003, Gujarat, India. E-mail: gssutaria@jau.in

Yogesh Vaghasiya, PhD
Assistant Professor, Department of Biochemistry, Saurashtra University, Rajkot – 360005, Gujarat, India. Mobile: +91-9227797424. E-mail: vaghasiyay@gmail.com

V. D. Vora, MSc
Assistant Research Scientist, Main Dry Farming Research Station, Junagadh Agricultural University, Targhadia (Rajkot) – 360003, Gujarat, India. E-mail: vdvora@jau.in

LIST OF ABBREVIATIONS

AA	ascorbic acid
AACC	American Association of Cereal Chemists
ABTS	2,2-azino-bis-(3-ethylbenzothiazoline-6-sulfonic acid)
AC	acetone
AICTE	All India Council for Technical Education
ALL	*Aerva lanata* leaf
ALS	*Aerva lanata* stem
AOAC	Association of Official Analytical Chemists
AQ	aqueous
ASAE	American Society of Agricultural Engineers
ASBI	American Shea Butter Institute
AYUSH	Ayurveda, Unani, Siddha and Homeopathy System of Medicines
BARD	Binational Agricultural Research and Development Fund
BC	*Bacillus cereus* ATCC29737
BDR	*Boerhaavia diffusa* root
Bo	bond number
BS	*Bacillus subtilis* ATCC6833
CA	*Candida albicans* ATCC2091
CCl_4	carbon tetrachloride
CE	*Candida epicola* NCIM3102
CF	*Citrobacter freundii* ATCC10787
CF	*Citrobacter freundii* NCIM2489
Cfu	colony forming unit
CG	*Candida glabrata* NCIM3448
CH	chloramphenicol
CN	*Cryptococcus neoformans* ATCC34664
CR	*Corynebacterium rubrum* ATCC14898
CS	cefotaxime sodium
CSIR	Council for Scientific and Industrial Research
DAE	Department of Atomic Energy

DBT	Department of Biotechnology
DFPI	Department of Food Processing Industries
DHA	docosahexaenoic acids
DHA-Et	docosahexaenoic acids–ethyl ester
DMSO	dimethyl sulfoxide
DNA	deoxyribonucleic acid
DOE	Department of Education
DPPH	2,2-diphenyl-1-picrylhydrazyl
DRDO	Defense Research and Development Organization
DSC	differential scanning calorimeter
DST	Department of Science and Technology
DTA	differential thermal analysis
DTG	derivative thermo gravimetry
EA	*Enterobacter aerogenes* ATCC13048
EA	ethyl acetate
EC	*Escherichia coli* ATCC25922
EC	*Escherichia coli* NCIM2931
EPA	eicosapentaenoic acids
EPA-Et	eicosapentaenoic acid-ethyl ester
FAO	Food and Agriculture Organization
FRAP	ferric reducing antioxidant power
GA	gallic acid
GEN	gentamycin
GMPs	Good Manufacturing Practices
HACCP	Hazard Analysis and Critical Control Points
HCl	hydrochloric acid
HPLC	high performance liquid chromatography
HPTLC	high performance thin layer chromatography
ICAR	Indian Council of Agricultural Research
ICMR	Indian Council of Medical Research
IEEE	Institute of Electrical and Electronics Engineers
INSA	Indian National Science Academy
ISO	International Organization for Standardization
KA	*Klebsiella aerogenes* NCIM2098
KP	*Klebsiella pneumonia* NCIM2719
LC-MS	liquid chromatography-mass spectrometry

LDL	low density lipoprotein
LLE	liquid–liquid extraction
LM	*Listeria monocytogenes* ATCC19112
MBC	minimum bactericidal concentrations
ME	methanol
MF	*Micrococcus flavus* ATCC10240
MFCS	Ministry of Food and Civil Supplies
MIC	minimum inhibitory concentrations
ML	*Micrococcus luteus* ATCC10240
MNRE	Ministry of New and Renewable Energy
MRSA	methicillin-resistant *Staphylococcus aureus*
NA	not applicable
NAAS	National Academy of Agricultural Sciences
NADH	nicotinamide adenine dinucleotide reduced
NBT	nitroblue tetrazolium
NDDB	National Dairy Development Board
NOSVODB	National Oil seeds and Vegetable Oils Development Board
OCH	old corn hair
OH	hydroxyl radical scavenging activity
OPEC	Organization of the Petroleum Exporting Countries
OX	oxacillin
PA	*Pseudomonas aeruginosa* ATCC27853
PA	*Pseudomonas aeruginosa* ATCC9027
PAE	acetone extract
PCE	chloroform extract
PCR	crude powder
PE	petroleum ether
PHE	hexane extract
PME	methanol extract
PMS	phenazine methosulfate
PQSS	Product Quality, Safety and Standards
PS	*P. syrigae* NCIM5102
PT	*P. testosterone* NCIM5098
PTE	toluene extract
QMS	Quality Management System
ROS	reactive oxygen species

RSM	ripe seed methanol extract
RTE	ready-to-eat
SA	*Staphylococcus aureus* ATCC29737
SA-1	*Staphylococcus aureus* ATCC25923-1
SA-2	*Staphylococcus aureus* ATCC29737-2
SE	*S. epidermidis* ATCC12228
SE	*Staphylococcus epidermidis* NCIM2493
SEM	scanning electron microscope
SEM	standard error of mean
SERC	Science and Engineering Research Council
SI	self inspection
SICE	The Society of Instrument and Control Engineers
SO	superoxide
SOP	Safety Operation Process
SSOP	Sanitation Standard Operating Procedures
ST	*Salmonella typhimurium* ATCC23564
TBF	*T. bellerica* fruit rind
TBL	*Terminalia bellirica* leaf
TBS	*Terminalia bellirica* stem
TCF	*T. chebula* fruit rind
TCL	*Terminalia catappa* leaf
TCS	*Terminalia chebula* stem
TG	thermo gravimetry
TGA	thermogravimetric analysis
TLC	thin layer chromatography
TPC	total phenol content
TPTZ	2,4,6-tri(2-pyridyl)-s-triazine
TTF	*Tribulus terrestris* fruit
UGC	University Grants Commission
USAID	United States Agriculture International Aid
USDA	United states department of Agriculture
UV	ultra violet
VAN	Vancomycin
WHO	World Health Organization
WS	woody stem
YCH	Young corn hair
YS	young stem

LIST OF SYMBOLS

Aqua	aqueous phase
Bo	Bond Number
Ca	Capillary Number
CCl_4	carbon tetrachloride
Cfu	colony forming unit
$FeCl_3$	ferric chloride
$FeSO_4$	ferrous sulfate
Fish Oil EE	fish oil ethyl ester
H_2SO_4	sulfuric acid
IC_{50}	50% inhibitory concentration
K_2CO_3	potassium carbonate
$K_2S_2O_8$	potassium persulfate
NaCl	sodium chloride
Omega 3 PUFA	omega 3 poly unsaturated fatty acids
Orga.	organic phase
PMi	*Proteus mirabilis* NCIM2241
PMo	*P. morganii* NCIM2040
PPi	*P. pictorum* NCIB9152
PPu	*P. putida* NCIM2872
Re	Reynolds Number
R_f	refractive index
Sal	*Staphylococcus albus* NCIM2178
Q	volumetric flow rate (ml/min)
We	Weber Number
M	fluid dynamic viscosity (Pa.s)
P	fluid density (Kg/m^3)
U	velocity of fluids in mini-channel (m/sec)
ρH	density of heavy phase (Kg/m^3)
ρL	density of light phase (Kg/m^3)
σAB	interfacial tension (mN/m)

FOREWORD BY T. K. GOSWAMI

I feel very delighted and honored to write this foreword for the book on *Food Technology: Applied Research and Production Techniques* under the book series *Innovations in Agricultural and Biological Engineering*. This book is edited by Murlidhar Meghwal, Megh R. Goyal, and Mital J. Kaneria.

Food technology is the applied science dedicated to the study of food, edible oils, herbs and spices, nutrition, their health effect, and various processing parameter and changes. It is a discipline in which the engineering, biological, and chemical principles of food processing and physical sciences are used to study the nature of foods, the causes of deterioration, packaging, storage, the principles underlying food processing, and the improvement of foods for the consuming public.

In this book, first four sections cover important topics on "Food Technology and Processing and Food Science." They are namely principles and practical applications in good manufacturing practices for food processing industries, research funding agencies around the globe in the food engineering, use of plastics in the twenty-first century food industry, latest trends in thermal processing in food technology, nondestructive technique of soft X-ray for evaluation of internal quality of agricultural produce, *in vitro* antioxidant efficacy of selected medicinal plants, antioxidant activities of some marine algae, omega-3 PUFA from fish oil of silver-based solvent extraction, antioxidant and antibacterial properties of extracts, *in vitro* antimicrobial activity, and antimicrobial properties of leaf extract.

The fifth section covers isolation, validation and characterization of major bioactive constituents from mango ripe seed, isolation and characterization of lycopene from tomato and its biological activity, and food processing using microbial control system.

I congratulate the editors for their timely decision of bringing out this book for use by scientists, engineers, professionals, and students. I am sure that it will be a very useful reference book for professionals working in food technology, food science, food processing, and nutrition.

Prof. Tridib Kumar Goswami, PhD,
Agricultural and Food Engineering Department,
Indian Institute of Technology,
Kharagpur – 721302, India
Tel.: +91 (03222) 283122 (off), (03222) 283123 (Res);
Fax: (03222) 255303
E-mail: tkg@agfe.iitkgp.ernet.in

FOREWORD BY G. S. DAVE

I take this auspicious opportunity to congratulate the editorial team of Dr. Murlidhar Meghwal, Dr. Megh R. Goyal, and Dr. Mital J. Kaneria for their extensive input in this book volume. I have been personally enriched while glancing through this compendium.

We are living in the era of global warming—climate change, food, water and natural resource crises—along with advancements in food, environmental and agricultural technologies. Current advancements in technology have made both pros and cons for humanity and the environment. Recent developments and sustainable technologies are combined in this book. I personally feel that this book will be a great resource in the updating and development of agricultural and food technologies in near future.

Advanced topics on food processing, preservation, nutritional analysis, quality checks and maintenance as well as good manufacturing practices in food industries are covered in this book. The editors and the contributing authors have generated highly focused reports to direct development of food and agriculture based on current knowledge into promising technologies.

Readers and stakeholders in agricultural technologies will gain a tremendous amount of information on (i) gaps of interdisciplinary approaches, (ii) food science and technology, and (iii) possible research groups for collaboration. Moreover, this book targets audiences from academia, a wide range of researchers, undergraduate/graduate and postgraduate students, postdoctoral researchers, medical staff, food/pharmaceutical companies, dieticians, private producers, and farmer-innovators. Institutes of higher learning and universities are the main academic sector contributing in teaching and research on various subjects, which are covered in this book.

I give my best compliments to editors, authors, and readers of this book.

Prof. Gaurav S. Dave, PhD (Biochemistry)
Department of Biochemistry,
Saurashtra University,
Rajkot – 360005, Gujarat, India

PREFACE 1

The food technologies and industries include various activities, such as good manufacturing practices (GMP), research, isolation and characterization, extraction, expression, antimicrobial activity, thermal processing, food production, transportation, packaging and distribution. This book volume provides information on the technology and suggests devices, standardization, packaging, ingredients, laws and regulatory guidelines and information on infrastructure to transform technology into highly value- added products.

The targeted audience for this book is food technologists, practicing food engineers, researchers, lecturers, teachers, professors, food professionals, those in the dairy industry, and food industries, students of these fields and all those who have inclination for food science and processing sector.

Part I on "*Good Manufacturing Practices and Research in Food Technology*" covers chapters on good manufacturing practices for food processing industries: principles and practical applications, and food engineering research funding agencies around the globe. Part II is focused on "*Latest Food Technologies,*" which includes have chapters on use of plastics in the twenty-first in the food industry, latest trends on thermal processing in food technology, and nondestructive technique of soft X-ray for evaluation of internal quality of agricultural produce. Part III covers "*Role of Antioxidants in Foods*", such as *in vitro* antioxidant efficacy of selected medicinal plants of Gujarat, antioxidant activities of some Marine algae as a case study from India, omega-3 PUFA from fish oil by silver-based solvent extraction, and antioxidant and antibacterial properties of extracts *Terminalia chebula* and *Terminalia bellerica*. Part IV focuses on "*Antimicrobials Activities in Food*" and presents an *in vitro* antimicrobial activity study, and antimicrobial properties of leaf extract: *polyalthialongi foliavar pendula* under *in-vitro* conditions. The last section on active constituents of foods provides details about isolation, validation and characterization of major bioactive constituents from mango ripe seed isolation and characterization of lycopene from tomato and its biological activity food processing using microbial control system.

The coverage of each topic is comprehensive and can serve as an overview of the most recent and relevant research and technology. Numerous references are included at the end of each chapter.

My own training and work experience as a dairy and food process engineer and teacher was crucial in conceiving this book, *Food Technology: Applied Research and Production Techniques*. I wish to thank the contributors, who did the real great work, for their time and energy to create scholarly and practical chapters. Their professionalism is appreciable, and they have my utmost appreciation and admiration.

My thanks also to Almighty God, whose love and blessings help us immensely.

—Murlidhar Meghwal, PhD
March 31, 2017

PREFACE 2

Deep in our refrigerator,
there's a special place
for food that's been around awhile…
we keep it, just in case.
'It's probably too old to eat,'
my mother likes to say.
'But I don't think it's old enough
for me to throw away.'

It stays there for a month or more
to ripen in the cold,
and soon we notice fuzzy clumps
of multicolored mold.
The clumps are larger every day,
we notice this as well,
but mostly what we notice
is a certain special smell.

When finally it all becomes
a nasty mass of slime,
my mother takes it out, and says,
'Apparently, it's time.'
She dumps it in the garbage can,
though not without regret,
then fills the space with other food
that's not so ancient yet

—Deep In Our Refrigerator by Jack Prelutsky
http://poemhunter.com/poems/food/page-1/37365112/

We all know food is essential for our survival. The increasing world population and the continuous climate change result in reduction of agricultural lands for food production. Subsequently this urges modern food science and technology to develop sustainable food production systems and improve nutritional value of food products, while keeping the cost as low as possible. Quality and nutritional value of foods are highly dependent on environment, agricultural practices, production conditions, and consumer preferences, which all may provide different effects for human health. One of the main challenges of food science and technology is to optimize food production to have minimum environmental footprint, lower production costs, and improving quality and nutritional value.

Analysis of foods is continuously requesting the development of more robust, efficient, sensitive, and cost-effective analytical methodologies to guarantee the safety, quality, and traceability of foods in compliance with legislation and consumers' demands. A large number of works have directly focused on the analytical technique used in food technology, while others have focused on the types of food, compound, or process investigated. Regarding specific analytical techniques applied to solve different problems in food analysis, one of the more active areas is the development of food processing techniques, in good agreement with the complex nature of foods. Food processing is one of the key steps for the development of any new analytical methodology; as a result, research on new procedures is one of the most active areas in food technology.

Therefore, we introduce this book volume on *Food Technology: Applied Research and Production Techniques* under book series *Innovations in Agricultural and Biological Engineering* by http://www. appleacademicpress.com. This book covers mainly current scenario of the research on food technology, food quality, emerging technologies of food processing, antioxidant and antimicrobial potentials, isolation and characterization of bioactive compounds, etc. This book volume sheds light on different technological aspects of Food Science and Technology; and it contributes to the ocean of knowledge on Food Science. We hope that this compendium will be useful for the students and researchers of academia as well as the persons working with the food, nutraceutical and herbal industries.

We like to share the views by our cooperating authors on this book. Dr. Kalpna Rakholiya comments: *This book provides exhaustive guidance for*

Agricultural and Biosciences researchers, carrying out research in this direction and useful for society. I know that it has taken a lot of hard work by all authors to get your book to the stage it is at now. According to Dr. Rajesh A. Dave, *food problem is the most vexing problem throughout the world. It is the most insulting problem too. The Governments are trying their best to increase the food-production through block development projects, national extension projects, community projects, package programs, and grow more food campaigns. In that way, our book will be very useful for scientific community as reference, academic, professional, and guidebook. "This book provides novel and thought-provoking insights into the fundamental issues involved in food sciences and technology. This book includes the informative chapters regarding funding agencies for research in food technology, principles and practical applications of good manufacturing practices for food processing industries, modern technologies, medicinal properties and characterization of pharmaceutically important active constituents of food,"* comments Dr. Yogesh K. Baravalia.

The contribution by all cooperating authors to this book volume has been most valuable in the compilation. Their names are mentioned in each chapter and in the list of contributors. We appreciate you all for having patience with our editorial skills. This book would not have been written without the valuable cooperation of these investigators, many of them are renowned scientists who have worked in the field of food engineering throughout their professional careers. We are glad to introduce Dr. Murlidhar Meghwal (Lead Editor of this book), who is an Assistant Professor in the Food Technology, Center for Emerging Technologies at Jain University – Jain Global Campus in District Karnataka, India. With several awards and recognitions including from President of India, Dr. Meghwal brings his expertise and innovative ideas in this book series. Without his support, leadership qualities as editors of the book volume and extraordinary work on food technology applications, readers will not have this quality publication.

We will like to thank editorial staff, Sandy Jones Sickels, Vice President, and Ashish Kumar, Publisher and President at Apple Academic Press, Inc., for making every effort to publish the book when the diminishing water resources are a major issue worldwide. Special thanks are due to the AAP

Production Staff for typesetting the entire manuscript and for the quality production of this book.

I request readers to offer their constructive suggestions that may help to improve the next edition.

We express our admiration to our families and colleagues for their understanding and collaboration during the preparation of this book volume. As an educator, there is a piece of advice to one and all in the world: *Permit that our almighty God, our Creator, provider of all and excellent Teacher, feed our life with Healthy Food Products and His Grace; and Get married to your profession.*

—Megh R. Goyal, PhD, PE
Mital J. Kaneria, PhD
February 1, 2017

WARNING/DISCLAIMER

PLEASE READ CAREFULLY

The goal of this book volume is to guide the world community on how to manage efficiently for technology available for different processes in food science and technology. The reader must be aware that dedication, commitment, honesty, and sincerity are important factors for success. This is not a one-time reading of this compendium.

The editors, the contributing authors, the publisher and the printer have made every effort to make this book as complete and as accurate as possible. However, there still may be grammatical errors or mistakes in the content or typography. Therefore, the content in this book should be considered as a general guide and not a complete solution to address any specific situation in food engineering. For example, one type of food process technology does not fit all cases in engineering/science/technology.

The editors, the contributing authors, the publisher and the printer shall have neither liability nor responsibility to any person, any organization or entity with respect to any loss or damage caused, or alleged to have caused, directly or indirectly, by information or advice contained in this book. Therefore, the purchaser/reader must assume full responsibility for the use of the book or the information therein.

The mention of commercial brands and trade names are only for technical purpose. No particular product is endorsed over another product or equipment not mentioned. The author, cooperating authors, educational institutions, and the publisher, Apple Academic Press Inc., do not have any preference for a particular product.

All web-links that are mentioned in this book were active on December 31, 2016. The editors, the contributing authors, the publisher and the printing company shall have neither liability nor responsibility, if any of the web-links are inactive at the time of reading of this book.

ABOUT THE EDITOR

Murlidhar Meghwal, PhD
*Assistant Professor, Department of Food Science
and Technology, National Institute of Food
Technology Entrepreneurship & Management,
Kundli – 131028, Sonepat, Haryana, India*

Murlidhar Meghwal, PhD, is a distinguished researcher, engineer, teacher, and consultant. He is currently Assistant Professor in the Department of Food Science and Technology at the National Institute of Food Technology Entrepreneurship and Management, Kundli, Haryana, India. He was formerly a professor at the Centre for Incubation, Innovation, Research and Consultancy (CIIRC) at the Jyothy Institute of Technology, Bengaluru, India. Dr. Meghwal is currently working on the characterization of instant coffee powder; development of nutrient and fiber rich functional cookies from fruit fibers residues; developing inexpensive, disposable, and biodegradable food containers using agricultural wastes; and quality improvement, quality attribute optimization, and freeze-drying of milk. He has authored or edited six books in the field of food technology, food science, food engineering, dairy technology, and food process engineering and has published many research papers. He is a reviewer and an editorial board member of several journals. He has attended many national and international workshops, seminars, and conferences. Dr. Meghwal received his BTech degree in Agricultural Engineering from the College of Agricultural Engineering Bapatla, Acharya N. G. Ranga Agricultural University, Hyderabad, India. He received his MTech. degree in Dairy and Food Engineering from the Indian Institute of Technology Kharagpur, West Bengal, India; and his PhD degree in Food Process Engineering from the Indian Institute of Technology Kharagpur. Dr. Meghwal was actively involved in establishing the Food Technology Program (MTech) at the Centre for Emerging

Technologies at Jain University, Bengaluru, and he acted as course coordinator, placement in-charge, and head. He earlier worked as a research associate at INDUS Kolkata (a rice parboiling, milling, and processing company) in the eastern part of the India on the development of a quicker and industrial-level parboiling system for paddy and rice milling. He is recipient of several scholarship, fellowships, and an award from the President of India.

ABOUT SENIOR EDITOR-IN-CHIEF

Megh R. Goyal, PhD
*Retired Professor in Agricultural and Biomedical
Engineering, University of Puerto Rico,
Mayaguez Campus Senior Acquisitions Editor,
Biomedical Engineering and Agricultural Science,
Apple Academic Press, Inc.
E-mail: goyalmegh@gmail.com*

Megh R. Goyal, PhD, PE, is a Retired Professor in Agricultural and Biomedical Engineering from the General Engineering Department in the College of Engineering at University of Puerto Rico–Mayaguez Campus; and Senior Acquisitions Editor and Senior Technical Editor-in-Chief in Agriculture and Biomedical Engineering for Apple Academic Press Inc. He has worked as a Soil Conservation Inspector and as a Research Assistant at Haryana Agricultural University and Ohio State University. He was the first agricultural engineer to receive the professional license in Agricultural Engineering in 1986 from the College of Engineers and Surveyors of Puerto Rico. On September 16, 2005, he was proclaimed as "Father of Irrigation Engineering in Puerto Rico for the twentieth century" by the ASABE, Puerto Rico Section, for his pioneering work on micro irrigation, evapotranspiration, agroclimatology, and soil and water engineering. During his professional career of 45 years, he has received many prestigious awards. A prolific author and editor, he has written more than 200 journal articles and textbooks and has edited over 48 books. He received his BSc degree in engineering from Punjab Agricultural University, Ludhiana, India; his MSc and PhD degrees from Ohio State University, Columbus; and his Master of Divinity degree from Puerto Rico Evangelical Seminary, Hato Rey, Puerto Rico, USA.

ABOUT CO-EDITOR

Mital J. Kaneria, PhD
*Assistant Professor, UGC-CAS Department of
Biosciences, Saurashtra University, Rajkot, India*

Mital J. Kaneria, PhD, is presently working as Assistant Professor in the Department of
Biosciences, Saurashtra University, Rajkot, India.
He formerly worked as a research associate at
GUIDE (Gujarat Institute of Desert Ecology), Bhuj-Kachchh, India, in the
terrestrial ecology division on the study of floral diversity, mangrove monitoring, herbarium preparation, and soil-water analytical parameters. He
has published more than 40 research papers in national and international
journals and has also written book chapters and books. He is a reviewer
and editorial board member of many journals, has attended and presented
several papers in several national and international seminars, and conferences and has received best paper awards. His current research involves
isolation and characterization of bioactive phyto-constituents focused on
in vitro and *in vivo* antimicrobial, antioxidant, and pharmacological activities of medicinal plants, particularly in relation to safety profiling, ageing,
and various acute and chronic diseases and disorders.

Dr. Kaneria received his BSc degree in botany from M.D. Science
College, Porbandar, Saurashtra University, Gujarat, India, and his MSc
degree in botany from Department of Biosciences, Saurashtra University,
Rajkot, Gujarat, India. He earned his PhD degree in botany under the
guidance of Prof. Sumitra Chanda, from the same Institute. During his
PhD training, he was awarded a BSR Fellowship from UGC, New Delhi,
India, for three years. His doctoral research is based on the phytochemical
and pharmacological potency of a selected medicinal plant from Gujarat
region.

BOOK ENDORSEMENTS

A highly informative, value added, well researched interpretation will excite students and researchers and showcase recent advances in major areas in food technology. Insights of highly experienced scientists and experts in this field create diversity in the chapters. "Food Technology: Applied Research and Production Techniques" will have high impact in the universities and research institutions and act as a guideline for the food processing units.

—Arpita Das, PhD
Visiting Faculty
Dept. of Pharmacutical Technology
Jadavpur University

This book provides a comprehensive coverage of the various aspects of food technology. Topics will be very useful to students and professionals. The Increasing awareness of consumers on food technology, food processing and preservation and the growing processed food market make this book an excellent source for reference in these areas.

—Narendra Reddy, PhD
Professor and Ramalingaswami Fellow
Centre for Emerging Technologies,
Jain University, Jain Global Campus, Bangalore, India

OTHER BOOKS BY
APPLE ACADEMIC PRESS, INC.

Management of Drip/Trickle or Micro Irrigation
Megh R. Goyal, PhD, PE, Senior Editor-in-Chief

Evapotranspiration: Principles and Applications for Water Management
Megh R. Goyal, PhD, PE, and Eric W. Harmsen, Editors

Book Series: Research Advances in Sustainable Micro Irrigation
Senior Editor-in-Chief: Megh R. Goyal, PhD, PE
  Volume 1: Sustainable Micro Irrigation: Principles and Practices
  Volume 2: Sustainable Practices in Surface and Subsurface Micro Irrigation
  Volume 3: Sustainable Micro Irrigation Management for Trees and Vines
  Volume 4: Management, Performance, and Applications of Micro Irrigation Systems
  Volume 5: Applications of Furrow and Micro Irrigation in Arid and Semi-Arid Regions
  Volume 6: Best Management Practices for Drip Irrigated Crops
  Volume 7: Closed Circuit Micro Irrigation Design: Theory and Applications
  Volume 8: Wastewater Management for Irrigation: Principles and Practices
  Volume 9: Water and Fertigation Management in Micro Irrigation
  Volume 10: Innovation in Micro Irrigation Technology

Book Series: Innovations and Challenges in Micro Irrigation
Senior Editor-in-Chief: Megh R. Goyal, PhD, PE
Volume 1: Principles and Management of Clogging in Micro Irrigation

Volume 2: Sustainable Micro Irrigation Design Systems for Agricultural
          Crops: Methods and Practices
Volume 3: Performance Evaluation of Micro Irrigation Management:
          Principles and Practices
Volume 4: Potential Use of Solar Energy and Emerging Technologies in
          Micro Irrigation
Volume 5: Micro Irrigation Management: Technological Advances and
          Their Applications
Volume 6: Micro Irrigation Engineering for Horticultural Crops: Policy
          Options, Scheduling, and Design
Volume 7: Micro Irrigation Scheduling and Practices
Volume 8: Engineering Interventions in Sustainable Trickle Irrigation:
          Water Requirements, Uniformity, Fertigation, and Crop
          Performance

Book Series: Innovations in Agricultural and Biological Engineering
Senior Editor-in-Chief: Megh R. Goyal, PhD, PE
- Dairy Engineering: Advanced Technologies and their Applications
- Developing Technologies in Food Science: Status, Applications, and
  Challenges
- Emerging Technologies in Agricultural Engineering
- Engineering Interventions in Agricultural Processing
- Engineering Practices for Agricultural Production and Water
  Conservation: An Interdisciplinary Approach
- Flood Assessment: Modeling and Parameterization
- Food Engineering: Modeling, Emerging Issues and Applications.
- Food Process Engineering: Emerging Trends in Research and Their
  Applications
- Food Technology: Applied Research and Production Techniques
- Modeling Methods and Practices in Soil and Water Engineering
- Processing Technologies for Milk and Milk Products: Methods,
  Applications, and Energy Usage
- Soil and Water Engineering: Principles and Applications of Modeling
- Soil Salinity Management in Agriculture: Technological Advances
  and Applications

- Technological Interventions in Management of Irrigated Agriculture
- Technological Interventions in the Processing of Fruits and Vegetables
- State-of-the-Art Technologies in Food Science
- Sustainable Biological Systems for Agriculture
- Novel Dairy Processing Technologies: Techniques, Management, and Energy Conservation
- Technological Interventions in Dairy Science: Innovative Approaches in Processing, Preservation, and Analysis of Milk Products
- Engineering Interventions in Foods and Plants

EDITORIAL

Apple Academic Press, Inc., (AAP) is publishing book volumes in the specialty areas as part of *Innovations in Agricultural and Biological Engineering book series*, over a span of 8 to 10 years. These specialty areas have been defined by *American Society of Agricultural and Biological Engineers* (http://asabe.org).

The mission of this series is to provide knowledge and techniques for Agricultural and Biological Engineers (ABEs). The series aims to offer high-quality reference and academic content in Agricultural and Biological Engineering (ABE) that is accessible to academicians, researchers, scientists, university faculty, and university-level students and professionals around the world. The following material has been edited/modified and reproduced below *"Goyal, Megh R., 2006. Agricultural and biomedical engineering: Scope and opportunities. Paper Edu_47 at the Fourth LACCEI International Latin American and Caribbean Conference for Engineering and Technology (LACCEI' 2006): Breaking Frontiers and Barriers in Engineering: Education and Research by LACCEI University of Puerto Rico – Mayaguez Campus, Mayaguez, Puerto Rico, June 21–23."*

WHAT IS AGRICULTURAL AND BIOLOGICAL ENGINEERING (ABE)?

"Agricultural Engineering (AE) involves application of engineering to production, processing, preservation and handling of food, fiber, and shelter. It also includes transfer of technology for the development and welfare of rural communities," according to http://isae.in." *ABE is the discipline of engineering that applies engineering principles and the fundamental concepts of biology to agricultural and biological systems and tools, for the safe, efficient and environmentally sensitive production, processing, and management of agricultural, biological, food, and natural resources systems,"* according to http://asabe.org.

"*AE is the branch of engineering involved with the design of farm machinery, with soil management, land development, and mechanization and automation of livestock farming, and with the efficient planting, harvesting, storage, and processing of farm commodities,*" definition by: http://dictionary.reference.com/browse/agricultural+engineering.

"*AE incorporates many science disciplines and technology practices to the efficient production and processing of food, feed, fiber and fuels. It involves disciplines like mechanical engineering (agricultural machinery and automated machine systems), soil science (crop nutrient and fertilization, etc.), environmental sciences (drainage and irrigation), plant biology (seeding and plant growth management), animal science (farm animals and housing) etc.,*" by: http://www.ABE.ncsu.edu/academic/agricultural-engineering.php.

According to https://en.wikipedia.org/wiki/Biological_engineering: "*BE (Biological engineering) is a science-based discipline that applies concepts and methods of biology to solve real-world problems related to the life sciences or the application thereof. In this context, while traditional engineering applies physical and mathematical sciences to analyze, design and manufacture inanimate tools, structures and processes, biological engineering uses biology to study and advance applications of living systems.*"

SPECIALTY AREAS OF ABE

Agricultural and Biological Engineers (ABEs) ensure that the world has the necessities of life including safe and plentiful food, clean air and water, renewable fuel and energy, safe working conditions, and a healthy environment by employing knowledge and expertise of sciences, both pure and applied, and engineering principles. Biological engineering applies engineering practices to problems and opportunities presented by living things and the natural environment in agriculture. BA engineers understand the interrelationships between technology and living systems, have available a wide variety of employment options. "*ABE embraces a variety of following specialty areas,*" http://asabe.org. As new technology and information emerge, specialty areas are created, and many overlap with one or more other areas.

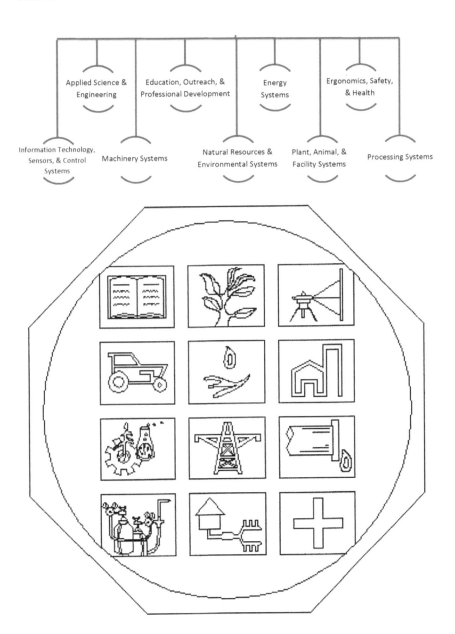

1. **Aquacultural Engineering**: ABEs help design farm systems for raising fish and shellfish, as well as ornamental and bait fish. They specialize in water quality, biotechnology, machinery, natural resources, feeding and ventilation systems, and sanitation. They

seek ways to reduce pollution from aquacultural discharges, to reduce excess water use, and to improve farm systems. They also work with aquatic animal harvesting, sorting, and processing.

2. **Biological Engineering** applies engineering practices to problems and opportunities presented by living things and the natural environment.

3. **Energy:** ABEs identify and develop viable energy sources – biomass, methane, and vegetable oil, to name a few – and to make these and other systems cleaner and more efficient. These specialists also develop energy conservation strategies to reduce costs and protect the environment, and they design traditional and alternative energy systems to meet the needs of agricultural operations.

4. **Farm Machinery and Power Engineering**: ABEs in this specialty focus on designing advanced equipment, making it more efficient and less demanding of our natural resources. They develop equipment for food processing, highly precise crop spraying, agricultural commodity and waste transport, and turf and landscape maintenance, as well as equipment for such specialized tasks as removing seaweed from beaches. This is in addition to the tractors, tillage equipment, irrigation equipment, and harvest equipment that have done so much to reduce the drudgery of farming.

5. **Food and Process Engineering:** Food and process engineers combine design expertise with manufacturing methods to develop economical and responsible processing solutions for industry. Also food and process engineers look for ways to reduce waste by devising alternatives for treatment, disposal and utilization.

6. **Forest Engineering**: ABEs apply engineering to solve natural resource and environment problems in forest production systems and related manufacturing industries. Engineering skills and expertise are needed to address problems related to equipment design and manufacturing, forest access systems design and construction; machine-soil interaction and erosion control; forest operations analysis and improvement; decision modeling; and wood product design and manufacturing.

7. **Information and Electrical Technologies Engineering** is one of the most versatile areas of the ABE specialty areas, because it is

applied to virtually all the others, from machinery design to soil testing to food quality and safety control. Geographic information systems, global positioning systems, machine instrumentation and controls, electromagnetics, bioinformatics, biorobotics, machine vision, sensors, spectroscopy: These are some of the exciting information and electrical technologies being used today and being developed for the future.

8. **Natural Resources:** ABEs with environmental expertise work to better understand the complex mechanics of these resources, so that they can be used efficiently and without degradation. ABEs determine crop water requirements and design irrigation systems. They are experts in agricultural hydrology principles, such as controlling drainage, and they implement ways to control soil erosion and study the environmental effects of sediment on stream quality. Natural resources engineers design, build, operate and maintain water control structures for reservoirs, floodways and channels. They also work on water treatment systems, wetlands protection, and other water issues.

9. **Nursery and Greenhouse Engineering**: In many ways, nursery and greenhouse operations are microcosms of large-scale production agriculture, with many similar needs – irrigation, mechanization, disease and pest control, and nutrient application. However, other engineering needs also present themselves in nursery and greenhouse operations: equipment for transplantation; control systems for temperature, humidity, and ventilation; and plant biology issues, such as hydroponics, tissue culture, and seedling propagation methods. And sometimes the challenges are extraterrestrial: ABEs at NASA are designing greenhouse systems to support a manned expedition to Mars!

10. **Safety and Health:** ABEs analyze health and injury data, the use and possible misuse of machines, and equipment compliance with standards and regulation. They constantly look for ways in which the safety of equipment, materials and agricultural practices can be improved and for ways in which safety and health issues can be communicated to the public.

11. **Structures and Environment:** ABEs with expertise in structures and environment design animal housing, storage structures, and

greenhouses, with ventilation systems, temperature and humidity controls, and structural strength appropriate for their climate and purpose. They also devise better practices and systems for storing, recovering, reusing, and transporting waste products.

CAREERS IN AGRICULTURAL AND BIOLOGICAL ENGINEERING

One will find that university ABE programs have many names, such as biological systems engineering, bioresource engineering, environmental engineering, forest engineering, or food and process engineering. Whatever the title, the typical curriculum begins with courses in writing, social sciences, and economics, along with mathematics (calculus and statistics), chemistry, physics, and biology. Student gains a fundamental knowledge of the life sciences and how biological systems interact with their environment. One also takes engineering courses, such as thermodynamics, mechanics, instrumentation and controls, electronics and electrical circuits, and engineering design. Then student adds courses related to particular interests, perhaps including mechanization, soil and water resource management, food and process engineering, industrial microbiology, biological engineering or pest management. As seniors, engineering students team up to design, build, and test new processes or products.

For more information on this series, readers may contact:

| Ashish Kumar, Publisher and President Sandy Sickels, Vice President Apple Academic Press, Inc. Fax: 866-222-9549 E-mail: ashish@appleacademicpress.com http://www.appleacademicpress.com/ publishwithus.php | Megh R. Goyal, PhD, PE Book Series Senior Editor-in-Chief *Innovations in Agricultural and Biological Engineering* E-mail: goyalmegh@gmail. com |

PART I

GOOD MANUFACTURING PRACTICES AND RESEARCH IN FOOD TECHNOLOGY

CHAPTER 1

GOOD MANUFACTURING PRACTICES FOR FOOD PROCESSING INDUSTRIES: PRINCIPLES AND PRACTICAL APPLICATIONS

MURLIDHAR MEGHWAL, UDAY HEDDURSHETTI, and RAVIKUMAR BIRADAR

CONTENTS

1.1   INTRODUCTION

Food manufacturing and producing industries should follow and adopt *Good Manufacturing Practices* (GMPs) [12] to make sure that all products are manufactured under a safe and healthy environment, which ensures the safety and quality of their products to fulfill requirement of standards regulations [4–6, 13, 15]. GMPs are regulations given to ensure effective hazard free overall practices to ensure product quality, safety and standards (PQSS) [6, 12, 13]. GMPs must follow for various practices of product testing, manufacturing, storage, handling, and distribution. GMPs should fulfill the standards of Safety, Integrity, Purity, Quality, and Composition (SISPQC) [12, 13]. The practices of Hazard Analysis and Critical Control Points (HACCP) and GMP programs give confidence and faith to consumers that proper testing consistency and safety and quality checks have been maintained throughout manufacturing, packaging, and distribution of products [13]. GMP is risk assessment trail and it is in currently adopted by industries to ensure the PQSS [9, 12, 15, 16]. GMP has science and technology based rules, regulations and standards. It also has integrated systems approach for quality, facilities and equipment, materials, production, packaging, labeling and laboratory control. It keeps proper records for proposed amendments regarding validation and cross-contamination [3, 4, 6].

This chapter presents principles and practical applications of good manufacturing practices for food processing industries.

1.1.1 WHY GMP/HACCP IS REQUIRED?

GMP or HACCP establishes minimum GMP for methods to be used and the facilities or controls to be used for, the manufacture, processing, packing or holding of a food or drug to assure that the food or drug is safe, has the appropriate identity, purity and strength, meets quality and purity characteristics [6, 9]. Food safety and quality control is essential in ensuring that food aid supplies are safe, of good quality and available in adequate amounts, in time, at affordable prices to ensure an acceptable nutritional and health status for all population groups [13]. HACCP, GMP, ISO, WHO [5] and Codex Alimentarius are major food safety and quality systems [11, 13]. GMP is a system to ensure that products meet food PQSS and legal requirements. HACCP can be part of GMP and is a systematic program to assure food safety [6, 16, 10]. GMP provides a high level assurance that food items, drugs or medicines are manufactured in a way that ensures their safety, efficacy and quality and also gives marketing authority with stability [9, 12–15].

1.1.2 WHO SHOULD FOLLOW GMP?

All food processer, medicinal drug manufacturer, food product manufacturers, packagers, labelers, and distributors, warehouse/storage facilities keepers should strictly follow GMP and given different regulations of food safety and standards [7, 12, 13, 15].

1.1.3 PURPOSE OF GMP

Drinking and other food preparation purposes used water, facilities for personal hygiene, air quality and ventilation lighting storage operation controls time and temperature control cross contamination raw materials packaging product information traceability pest control personal hygiene transportation training food marketing food services, verification. The brief outline of the GMP structure for a company is given below [9, 10, 12, 15, 19].

a. Quality assurance: Every industry should have its quality management department.

b. GMP for food, medicine and pharmaceutical products [14].

c. Sanitation and hygiene: This very important and crucial part of GMP.

d. Qualification and validation.

e. Complaints.

f. Product recalls.

g. Contract production and analysis: It can be subcategorized as general, contract giver, contract accepter and the contract.

h. Self-inspection (SI) and quality audits: (i) Items for SI (ii) SI team, (iii) Repetition of SI, (iv) SI report, (v) Follow-up action, (vi) Audit for quality assurance, and (vii) Suppliers' audits and approval.

i. Personnel: In any industry there are two types of personal (i) General Worker and Operator, and (ii) Key personnel.

j. Training.

k. Personal hygiene.

l. Premises: It may include general area, ancillary areas, storage areas, weighing areas, production areas and quality control area.

m. Equipment.

n. Materials can be general materials, starting materials, packaging materials, intermediate and bulk products, finished products, rejected, recovered, reprocessed and reworked materials, recalled products, returned goods, reagents and culture media, reference standards, waste materials and miscellaneous.

o. Documentation can include general records, documents on labels, testing procedures records, specifications for starting and packaging materials, for intermediate and bulk products and for finished products records, files on master formulae and Batch Processing Records, Packaging instructions and Batch Packaging Records, Standard Operating procedures (SOP's) and records and finally Logbooks.

p. GMP in production are general neat and clean practices, GMP in prevention of cross-contamination and bacterial contamination during production, GMP in processing operations and packaging operations.

q. GMP in quality control are control of starting materials and intermediate, bulk and finished products, test requirements, batch record review and stability studies [12].

1.2 FUNDAMENTALS OF GMP

• Quality Control: Product meets specifications.
• Quality Assurance: Systems ensure control and consistency; validation.
• Documentation: If it is not documented, it did not happen or it is just rumor.
• Verification and self-inspection [12].

1.3 TEN PRINCIPLES OF GMP FOR FOOD

GMP consists of following 10 principles that introduce employees to critical behaviors established by food regulations and industry leaders to maintain GMP in plants [3, 4, 15].

1. Writing procedures.
2. Following written procedures.
3. Documenting for traceability.
4. Designing facilities and equipment.
5. Maintaining facilities and equipment.
6. Validating work.
7. Job competence.
8. Cleanliness.
9. Component control.
10. Auditing for compliance.

1.4 HAZARD ANALYSIS AND CRITICAL CONTROL POINT (HACCP): APPLICATIONS IN THE FOOD PROCESSING INDUSTRY

HACCP is well recognized in the food industry as an effective approach to establishing good production, sanitation, and manufacturing practices that produce safe foods [17, 16]. HACCP is a system of process control used by the industry to prevent hazards to the food supply and as a tool in the control, reduction and prevention of pathogens in food. HACCP make GMP complete. It is very important and need to strictly observe by any food industries. First of all, HACCP program was developed in 1960s for NASA to ensure the safety of food products that were to be used by the astronauts in the space program [9, 12, 13, 16].

HACCP is a systematic process control system designed to determine potential hazards and implement control measures to reduce or eliminate the likelihood of their occurrence. The focus is on hazard prevention, rather than hazard detection. HACCP determining the step or steps that the really serious problems occur or could occur in your production process and monitoring these steps so you know there are problems and finally fixing any problems that arise. HACCP is designed to prevent food safety problems rather than catch then after they occur, and it includes seven principles. HACCP was adopted by the FDA in the 1990's [3, 7, 10, 13]. In HACCP, "hazards" refer to conditions or contaminants in foods that can cause illness or injury [16].

1.4.1 TYPES OF FOOD HAZARDS

1. **Biological Food Hazard:** Biological food hazard includes hazard in food due to Microorganisms, Yeast, Mold, Bacteria, Viruses, Protozoa and Parasitic worms.
2. **Chemical Food Hazard:** There are three types of food hazard in food due to chemicals:
   Naturally Occurring, Intentionally added and Unintentionally added.
   - **Types of Naturally Occurring Chemical Hazards:** Mycotoxins (e.g., aflatoxin), Scombrotoxin, Ciguatoxin, Shellfish toxins, Paralytic shellfish poisoning (PSP), Diarrhetic shellfish poisoning (DSP), Neurotoxic shellfish poisoning (NSP) and Amnesic shellfish poisoning (ASP)/Domoic Acid.
   - **Intentionally Added Chemicals – Food Additives:** Direct (allowable limits under GMPs), Preservatives (e.g., nitrite and sulfating agents), Nutritional additives (e.g., niacin, vitamin A) and Color additives [9, 20].
   - **Unintentionally or Incidentally Added Chemicals**
     a. Agricultural chemicals-pesticides, fungicides, herbicides, fertilizers, antibiotics, growth hormone.
     b. Prohibited substances (21 CFR, Part 21, 189).
     c. Toxic elements and compounds, e.g., lead, zinc, arsenic, mercury, cyanide.
     d. Secondary direct and indirect, e.g., lubricants, cleaning compounds, sanitizers, paint.

3. **Physical Food Hazard:** Any potentially harmful extraneous matter not normally found in food such as glass, wood, stones, metal and plastic.

1.5 HACCP LEGAL AND REGULATORY REQUIREMENTS

a. Each and every food industry hazard analysis must be performed.
b. They should develop flowchart describing steps in the process.
c. They should be clearly mention the intended consumer of the finished product.
d. All possible food safety hazards must be listed out and must be recorded.
e. Make the list of the critical control points.
f. Make the list the critical limits.
g. List the monitoring procedures and frequencies.
h. Should have list the corrective actions to be followed in response to deviation from a critical limit.
i. Should maintain a recordkeeping system.
j. Should list the verification procedures and frequencies.
k. All documents must be signed and dated by responsible establishment official.
l. It must be developed by someone who has completed a course of instruction in the application of HACCP principles to food processing [13].

1.6 HOW DOES HACCP WORK?

HACCP processors must assemble a HACCP team to design their plan and describe the product and its method of production, distribution and intended consumer. Develop and verify process flow diagrams to identify at each step of the production flow chart any hazard to food safety such as chemical, physical, bacterial or support the hazard with a decision making document and scientific data. If a CCP deviation is found then identification of the cause of deviation, description for how the critical limit was restored, how the deviation can be prevented from happening again and how the adulterated product was reconditioned or what happened to the product must take place [1, 9, 16].

1.7   HACCP: SEVEN PRINCIPLES FOR FOOD SAFETY

1. **Analyze hazards:** the hazard could be biological, such as a microbe; chemical, such as a toxin; or physical, such as ground glass or metal fragments [13].
2. **Identify critical control points:** the points in food production at which the potential hazard can be controlled or eliminated. Examples are cooking, cooling, packaging, and metal detection.
3. **Establish preventive measures with critical limits for each control point:** for example, for a cooked food setting minimum cooking temperature and time required to ensure the elimination of any harmful microbes.
4. **Establish procedures to monitor the critical control points:** it includes determining how and by whom cooking time and temperature should be monitored.
5. **Establish corrective actions to be taken, when monitoring shows that a critical limit has not been met,** such as reprocessing or disposing of food if the minimum cooking temperature is not met.
6. **Establish procedures to verify that the system is working properly:** for example, testing time-and-temperature recording devices to verify that a cooking unit is working properly.
7. **Establish effective recordkeeping to document the HACCP system:** records of hazards and their control methods, the monitoring of safety requirements and action taken to correct potential problems.

1.8   ADVANTAGES OF HACCP

- Focuses of HACCP is on identifying and preventing hazards from contaminating food.
- HACCP is based on sound science and technology.
- HACCP places responsibility for ensuring food safety on the food manufacturer or distributor [13].
- HACCP helps food companies to compete more effectively in the world market.
- HACCP reduces barriers to international trade.

- HACCP permits more efficient and effective government oversight, because the recordkeeping allows investigators to see how well a firm is complying with food safety laws over a period rather than how well it is doing on any given day [9, 13, 16].

1.9 FIVE P'S OF GMP

The *place, primary materials, people, process* and *product* are the five important factors of production and processing of food product that affect the quality and safety while following GMPs. Places and premises must be clean and well organized. All surfaces and roofs, sidewalls should be well cleaned and designed to prevent any kind of contamination. People who are working in the industry must be well qualified with the help of education, training, workshops, experience and seminars. People of Quality Assurance department are responsible to send each product for sale, release raw material and packaging components for use, approve operational procedures, master formulas and specifications, review returns prior to resale, and also investigate each complaint. In Processes Sanitation programs include documented procedures for effective cleaning of the premises, equipment, handling of substances, and the health and hygienic behavior of personnel. In spices process operation following system should be followed: (i) Review and release of all raw spices materials and packaging materials; (ii) Water that is potable and meets the Guidelines for Indian Drinking Water Quality or applicable regulations should be used [19, 18]; (iii) Complete batch records for each batch allowing for traceability; (iv) Each finished spice package should be identified with a lot number and expiry date; and (v) Labels should be secured and controlled and Quality agreements should in place with contracts [1, 2, 4].

1.9.1 FIRST 'P': PLACE/PREMISES – BUILDING AND FACILITIES

Premises must be clean and equipments should be orderly arranged; surfaces should allow for effective cleaning and should be designed to prevent contamination of food. Equipment and containers should be easy to carry and convey for adequately cleaned, disinfected and maintained to avoid the contamination of food.

- Separate well defined areas as are necessary to prevent contamination or mix-ups.
- Air filtration systems (HVAC) in production areas.
- Sanitation.
- Manufacturing area should be suitably located, designed, constructed and maintained.
- Effective measures should be taken to avoid contamination from environment and from pests.
- Employed painted line, plastic curtain, flexible barrier as rope/tape to prevent mix-up.
- Appropriate changing rooms and facilities should be provided.
- Toilets should be separated from the production areas to prevent product cross contamination.
- Defined areas should be provided for: (i) materials receiving; (ii) material sampling; (iii) incoming goods and quarantine; (iv) starting materials storage; (v) weighing and dispensing; and (vi) processing.
- Storage of bulk products.
- Defined areas should be provided for:(i) quarantine storage before final release of products (ii) storage of finished products; (iii) loading and unloading; (iv) laboratories; (v) equipment washing; and (vi) packaging.
- Wall and ceiling, where applicable should be smooth and easy to maintain.
- The floor in processing areas should have a surface that is easy to clean and sanitize.
- Drains should be of adequate size and should have trapped gullies and proper flow.
- Try to avoid but if open channels required they should be easy in cleaning and disinfection.
- Air intakes and exhausts and associated pipework and ducting should not contaminate product.
- Buildings should be ventilated appropriate and have sufficient light.
- Pipework, light fittings, ventilation points should be fixed in such a way as to avoid uncleanable process and run outside the processing areas.
- Production areas and quality control laboratories should be physically separated.
- Storage facility should be of adequate space, have suitable lighting, dry and clean.

- Such areas should be suitable for effective separation of quarantined materials and products.
- Special and separate storage facility should be provided for storage of explosive, flammable, highly toxic, rejected and recalled substances.
- For control of temperature and humidity should be provided special storage conditions.
- Storage facility should permit separation of labels and printed materials to avoid mix-up.

1.9.2 SECOND 'P': PRIMARY MATERIALS

All incoming materials should be checked, tested and recorded. Impure, adulterated, broken, toxic, etc. raw materials should e rejected and sent back.

1.9.3 THIRD 'P': PEOPLE

Working people must be in adequate in number, qualified by education, training, or experience to perform their respective functions. Quality Assurance is responsible to release each product for sale, release raw material and packaging components for use, approve all operational procedures, master formulas and specifications, review all returns prior to resale, and investigate each complaint. People known, or suspected, to be suffering from, or to be a carrier of a disease or illness likely to be transmitted through food, it leads to contaminate the food and it can transmit illness to consumers; so that all food handlers shall be medically examined by a registered medical practitioner and vaccinated according to current legislative requirements for avoiding the food contamination during the process of food products. Wear outer garments, light colored and suitable to the operation in a manner that protects against the contamination of food, food-contact surfaces, or food packaging materials. Remove all jewelry and other objects that might fall into food, equipment, or containers. Wear an effective manner, hairnets, headbands, hand gloves, caps, beard covers, or other effective hair restraints. Wash hands properly in an adequate hand-washing facility before starting work, after each absence from the workstation, immediately after using toilet and at any other time when the hands may have become contaminated. Store clothing or other personal belongings in areas other than where food is exposed or where equipment or utensils are washed. Take any other necessary precautions to protect against contamination of food. They should be in good health and capable of handling the duties assigned to them [1].

1.9.4 FOURTH 'P': PROCESS

Sanitation programs include documented procedures for effective cleaning of the premises, equipment, handling of substances, and the health and hygienic behavior of personnel. Raw materials and primary packaging materials are stored and handled in a manner which prevents their mix-up, contamination with microorganisms or other chemicals. Raw materials, in-process samples and finished products are tested or examined to verify their identity. Lot or batch identification is essential in traceability and product recall and also helps effective stock rotation. Packaged foods should be labeled with clear instructions to enable the next person in the food chain to handle, display, store and use the product safely [1].

1.9.5 FIFTH 'P': PRODUCTS

Specifications are available for every product and include purity, quantity, and identity of medicinal ingredients, potency, and test methods [8]. Every lot must be tested for conformance with its finished product specifications. Importers may follow a reduced testing program in which the first lot of each product is fully tested; and for each subsequent lot: a certificate of analysis with actual test results is reviewed. The lot is positively identified upon receipt. Transportation and storage conditions do not adversely impact the product. Full confirmatory testing is conducted on at least one lot per year dosage form, per supplier. Stability results demonstrate that the product will meet specifications under the recommended storage conditions. Samples are retained under the recommended storage conditions for one year past the expiry date of product. There are sufficient samples available to perform complete testing of the product. Records must be retained for one year past the expiration date of the product to which they refer. Recall procedure is documented and effective by performing a Mock Recall. Sterile NHPs must be manufactured and packaged in a separate enclosed area under the supervision of a trained person using scientifically proven methods to ensure sterility [1].

1.10 EQUIPMENTS REQUIREMENT TO FULFILL GMP REGULATIONS AND GUIDELINES

Equipment should be well-planed, designed and located to suit the production of the product [4].

1.10.1 DESIGN AND CONSTRUCTION OF EQUIPMENTS

The surfaces of equipment which comes in contact with any food in-process items or material should not react with or adsorb the materials which are processed. Equipment should not adversely affect the product through leaking valves, lubricant drips and through inappropriate modifications or adaptations. Equipment should be designed in such a way so that they can be easily cleaned. Equipment used for flammable substances should be explosion proof.

1.10.2 INSTALLATION AND LOCATION FOR EQUIPMENTS IN INDUSTRY

Equipment should be located to avoid congestion and should be easily identified. Wherever required water, steam and pressure or vacuum lines should be installed required all phases of operation. They should be easily cleaned, handled and clearly identified. Support systems such as heating, ventilation, air conditioning, water (such as potable, purified, distilled), steam, compressed air and gases (example nitrogen) should function as designed and identifiable.

1.10.3 MAINTENANCE OF EQUIPMENTS

Weighing, measuring, testing and recording equipment should be serviced and calibrated regularly. All records should be maintained.

1.11 QUALITY MANAGEMENT SYSTEM (QMS) OF FOOD INDUSTRY

A QMS should be developed, established and implemented by each and every food industry to achieve objectives and implement the GMP/HACCP policies. QMS should give the organizational structure, functions, responsibilities, procedures, instructions, processes and resources for implementing the GMP/HACCP and quality management. The QMS should make ensure that samples of raw materials, starting materials, intermediate, and finished products are collected and tested, examined to decide their release or rejection based on their results and other available evidence of quality parameter.

The personal involved in production sections a quality control sections should not be same person and should be guided and headed by different persons to avoid the partiality in decisions. The senior and other employee of production should be properly/adequately trained, experienced and well equipped with knowledge in drug/food/cosmetic manufacturing and production.

Head of production and manufacture should be given responsibilities and authority to regulate the production and manufacturing process. The head of QMS should have full responsibility and authority in all quality control duties (*establishment, verification* and *implementation*). The QMS head should designate person, to approve starting materials, intermediates, bulk and finished products. Role, responsibilities, functions and authority of each worker, employee and key personnel should be clearly defined and also should be mentioned in written record. Sufficient number of trained officers, trainee, and personnel should be appointed to carry out direct inspections in production line and quality control unit.

1.11.1 TRAINING

All the workers and people involved directly in the food manufacturing and production process must be appropriately trained and given fist hand knowledge in manufacturing and production operations as per given guidelines of GMP/HACCP principles. Specific care must be given to training and educating the personnel working and handling any hazardous materials. There should be regular and continuous training in GMP and all working people, officers, managerial staff, quality control people must attend it and keep the records. Proper record of all training activities should be maintained and effectively examined periodically.

1.12 BENEFITS OF GMP/HACCP

1. Operating cost drops as rework and penalties due to noncompliance reduces and efficiencies increases.
2. Increases the customer satisfaction, employees, stockholders, regulators.
3. Respect for an organization that demonstrates commitment to NHP safety.
4. GMP/HACCP covers all safety and written procedures (SOP), which makes employees more efficient and reduces errors during the manufacturing process [16].

1.13 MAINTENANCE AND SANITATION IN THE FOOD INDUSTRY

In food industries all the equipments, devices and utensils should be cleaned and maintained in accordance with an established GMP/HACCP procedure and program [16]. The cleaning operation program should have description of cleaning methods, level of cleanliness required, and frequency of cleaning activities, list of all cleaning and sanitizing agents used in the cleaning process including their concentrations and uses must be in written document and should be maintained.

1.14 TRANSPORTATION OF RAW MATERIAL FROM COLLECTION POINT TO FOOD INDUSTRY AND FINISHED PRODUCTS

Raw materials and finished food products must be sufficiently saved and protected in transportation. Based the type of raw materials and food products the type of conveyances or containers may vary for example while transporting ice cream, truck should be refrigerated where for transporting dry food such as biscuits, spices it is not required. The food conveyances and bulk containers should not contaminate foods or packaging, allow effective separation of different foods or foods from nonfood items where ever required during transport; should be able to effectively maintain the temperature, humidity, atmosphere and other conditions necessary to protect food from harmful microorganism [2].

1.15   FOOD PRODUCT INFORMATION: LABELING

Insufficient product information, and/or inadequate knowledge of general food hygiene, can lead to products being mishandled at later stages in the food chain. Such mishandling can result in illness, or products becoming unsuitable for consumption, even where adequate hygiene control measures have been taken earlier in the food chain. Products should bear appropriate information to ensure that adequate and accessible information is available to handle, store, process, prepare and display the product safely and correctly. Prepackaged food items must be labeled with clear instructions to enable the next person in the food chain to handle, display, store and use the product safely and shall meet the requirements of the relevant legislations. Lot number or batch identification number is essential in tracing and during product recall and also helps effective stock rotation. Each container of food should be permanently marked to identify the producer and the lot or batch.

1.16   SANITATION AND HYGIENE IN THE FOOD PROCESSING INDUSTRY

Sanitation and hygiene of personnel, premises, equipment/apparatus and production materials and containers should be practiced to avoid contamination of the manufacturing of products.

1.16.1   PERSONNEL HYGIENE

Personnel should be healthy to perform their assigned works. Regular medical examination must be conducted for all production personnel involved with manufacturing. People known, or suspected, to be suffering from, or to be a carrier of a disease or illness likely to be transmitted through food, should not be allowed to enter any food handling area if there is a likelihood of their contaminating food. Any person so affected should immediately report illness or symptoms of illness to the management. All food handlers shall be medically examined by a registered medical practitioner and vaccinated according to current legislative requirements. Any personnel shown at any time to have an apparent illness or open lesions that may adversely affect the quality of products should not be allowed to handle raw materials, packaging materials, in-process materials, and finished products. Personnel should wear protective and clean attire appropriate to the duties they perform. Smoking, eating, drinking and chewing, food, drinks and smoking materials and other materials that might contaminate are not permitted in production, laboratory, storage or other areas where they might adversely affect product quality [19]. All authorized personnel entering the production areas should practice personal hygiene including proper attire. All persons working in direct contact with food, food-contact surfaces, and food-packaging materials shall conform to hygienic practices while on duty to the extent necessary to protect against contamination of food. Wear outer garments of light colored and it should be suitable to the operation in a manner that protects against the contamination of food, food-contact surfaces, or food packaging materials. Maintain adequate personal cleanliness such as short and clean hair, fingernails. Wash hands thoroughly (and sanitizing to protect against contamination with undesirable microorganisms) in an adequate hand-washing facility before starting work, after each absence from the workstation, immediately after using toilet and at any other time when the hands may have become soiled or contaminated. Remove protective clothing such as overall, head cover, apron, feet covering, etc., before visiting the toilet. During operation and working in industry, take of all the jewelry and other objects that might fall into food, equipment, or containers. The gloves should be of an impermeable material and should be maintained in an intact, clean, and sanitary condition. Do not keep or store clothing and other personal belongings in areas where food materials where exposed and or where equipment and utensils are washed.

1.16.2 PREMISES HYGIENE

1. Every food industry should have adequate washing and well ventilated toilet facilities for all employees, workers and security.

2. The toilet, drinking water and food canteen for employees should be provided and separated from the production area to avoid any contaminations.
3. To storage and keep the clothing and personal belongings suitable locker facilities should be provided at appropriate location to all employees.
4. Waste material should be regularly collected in suitable receptacles for removal to collection points outside the production area.
5. The use of rodenticides, insecticides, fumigating agents and sanitizing materials in cleaning and killing rodents, insects, fungus and microbes must not contaminate equipment, raw materials, packaging materials, in-process materials or finished products.

1.16.3 EQUIPMENT AND APPARATUS HYGIENE

Equipment and utensils should be kept clean and hygienic condition. Now-a-days, vacuum or wet cleaning methods are preferred. In cleaning process compressed air and brushes should be used with care and avoided if possible, as they increase the risk of product contamination. Always SOP must be followed for cleaning and sanitizing of major machines. Documented cleaning procedures shall be kept and it should include responsibility for cleaning, item/area to be cleaned, frequency of cleaning, method of cleaning, including dismantling equipment when required, cleaning chemicals and concentrations, cleaning materials to be used and cleaning records and responsibility for verification.

1.16.4 SANITATION STANDARD OPERATING PROCEDURES (SSOP)

According to food regulations and orders the SSOP should include the general maintenance, list of substances used in cleaning and sanitizing, proper storage facility for toxic materials, good pest control system, proper sanitation of food-contact surfaces, sstorage and handling of clean portable equipment and utensils and facility for disposal of rubbish materials [4].

1.16.5 TIME AND TEMPERATURE CONTROL

Time and temperature controls include of recording of time and temperature parameter of cooking, cooling, processing and storage.

1.17 PRODUCTION AND MANUFACTURE IN THE FOOD PROCESSING INDUSTRY

1.17.1 WATER

Water systems should be sanitized according to well-established procedures given by GMP, FDA, WHO, HACCP, CODEX aliment Arius [5, 10, 11]. The chemical and microbiological quality of water used in production should be monitored regularly as per written guidelines and any deviation in quality should be followed by corrective action. Selection of method for water sanitizing treatment (deionization, distillation or filtration) depends on product requirement. Storage of water and its supply system should be properly maintained.

1.17.2 VERIFICATION OF MATERIALS IN THE FOOD PROCESSING INDUSTRY

Raw material samples should be manually tested for quality conformity to specifications before it is permitted for production and manufacture use. The raw material and packaging material should be examined and cross checked for their conformity to specifications. Raw materials as well finished products should be clearly labeled as per food laws and orders given by government and food regulating bodies. Packaging must be clean, appropriate, protective, free from leakage, perforation or exposure.

1.17.3 REJECTED MATERIALS

- Deliveries of raw materials that do not comply with specification should be segregated and disposed according to standard operating procedures.
- Defected and damaged final products also should be removed and recorded

1.17.4 BATCH NUMBERING SYSTEM FOR FINAL PRODUCTS

- Every finished product should bear a production identification number which enables the history of the product to be traced.
- A batch numbering system should be specific for the product and a particular batch number should not be repeated for the same product in order to avoid confusion.
- Whenever possible, the batch number should be printed on the immediate and outer container of the product.
- Records of batch number should be maintained.

1.17.5 MEASUREMENT AND WEIGHING OF INCOMING MATERIALS AND OUTGOING PRODUCTS

- Weighing of incoming raw materials should be carried out in the defined areas from these material get into the industry using calibrated equipment.
- Weighing and measurement of product should be done in processing line only before packaging.
- All weighing and measurement should be recorded and where applicable, counter checked.

1.17.6 PROCESSING PROCEDURE IN THE FOOD PROCESSING INDUSTRY

All starting incoming raw materials should be verified as per the specifications guidelines. Each and every manufacturing operation must be performed in line with written procedures, instruction and guidelines. Every needed in-process controls must be carried out and recorded. Wherever required all bulk products should be properly labeled until approved by QMS. Special care and attention must be given to overcome the cross-contamination issues in all stages of processing, productions and manufactures.

1.17.6.1 Ready-to-Eat Food

Ready-to-Eat (RTE) food products are important, easy to carry, easy to pack, easy to store and convenient food products. Now days, RTE food products are at large marketed and consumed. RTE can be seen as snack foods, sandwich fillings, picnic items, deli buffet foods, namkeen, kurkure, popcorn, samosa, khakhre, etc. Such item doesn't need much further processing. All over the globe millions of people consume these RTE due their convenient and generally consumers presume them to be safe to eat as purchased. A microorganism known as *Listeria* exists widely in the environment and it is pathogenic to both human and animal population. *Listeria* spp. is very resistant to adverse conditions and continues to grow under refrigerated conditions also. Therefore, heat processing of RTE products is required to destroy such pathogenic bacteria and also specific packaging and handling are recommended to avoid contamination of *Listeria* spp. due to its resistance nature it continue to grow even under vacuum packing and refrigerated conditions. The identification of pathogenic microbes at very low levels on product is always difficult and it becomes complicated when the distribution and occurrence is sporadic. A systematic, very well planned and broadways strategies good to assure the safety of RTE products.

1.18 CONSEQUENCES OF GMP/HACCP VIOLATIONS

This is very damaging and severe consequences for not following proper GMP/HACCP guidelines given for governmental bodies: ISO, WHO [5], FDA [10], FSSAI etc. Some of them are given below:

a. Product will be declared as "adulterated" if not fulfill safety and quality requirement.
b. Court by law and order force to shut down the manufacturing facility.
c. Product will be black listed in social media and will be ceased.
d. There can be recall for the product.
e. There will be front page press coverage of the product and company.
f. Consumer losses faith in company and their products, loss of business.
g. In such situation competitor will get better opportunity.
h. There can be suspension or hold on product applications.
i. There can be criminal inspection, investigations and indictments.
j. There can be lawsuits against such industry which don't follow GMP/HACCP, etc.

1.19   COMPLAINTS CORRECTIONS AND PROCEDURE FOR COMPLAINTS APPLICATION

In every industry there should be designated responsible person to take care of complaints and grievances of the consumers. Such designated person should be able to handle complaint, take decision on measure to be taken, should be authorized person if not then must advise authorized person of results, should have sufficient support staff and access to records.

- Written procedure (SOP): There should be written records describing action to be taken and also should include need to consider a recall (e.g., possible product defect).
- Thorough investigation: (i) quality control should be involved; (ii) with special attention to establish whether "counterfeiting" may have been the cause; and (iii) fully recorded investigation – reflect all the details.
- Due to product defect (discovered or suspected): (i) consider checking other batches; and (ii) batches containing reprocessed product.
- Investigation and evaluation should result in appropriate follow-up actions: May include a "recall".
- All decisions and measures taken should be recorded.
- Referenced in batch records.
- Records reviewed – trends and recurring problems.
- Inform competent authorities in case of serious quality problems, such as faulty manufacture, product deterioration, and counterfeiting.
- Have a thorough recall procedure that is consistent with the complaints handling procedure.
- Trend complaints, their investigations and results.

1.20   SELF-INSPECTION OF INDUSTRY GMP APPLICATION FOR IMPROVEMENT AND PROPER FUNCTIONING

The main objective and focus of self-investigation is to evaluate that is company's operations are remain in line with the GMP and HACCP guidelines or not. It can be very useful to avoid any violation of GMP/HACCP and to avoid unnecessary trouble. The self-inspection program should include following items:

- cover all aspects of production and quality control.

- be designed to detect shortcomings in the implementation of GMP.
- recommend corrective actions.
- set a timetable for corrective action to be completed.

The self-inspection should be carried out on regular and routine basis. Also, on some special occasions such as recalls and repeated rejections. This SI operation should be done with the help of a team appointed by quality management system, with:

a. authority.
b. sufficient experience, expertise in their own field and knowledge of GMP.
c. may be from inside or outside the company.

Frequency of SI should generally at least once in a year depending up on company requirements and the number of the company activities and the size of the company. The report prepared at completion of inspection, must include: (i) results; (ii) evaluation; (iii) conclusions; and (iv) recommended corrective measures.

Time to time follow-up action should be done for the smooth and effective follow-up program and to ensure that company management will evaluate both the report and corrective actions properly on time.

1.21 SUMMARY

GMP compliance is not an option but it should be followed. Quality should be built into the food product. GMP cover all aspects of manufacturing activities prior to supply. It is very important to understand the requirement and importance of documentation practices to supply safe food products. In considering microbiological criteria for spices, the microbiological safety of food is principally achieved through the implementation of control measures throughout the production and processing chain [2, 13]. The role and involvement of the senior management is crucial in applying GMP. GMP can be a great help to the food industry to fulfilling legal requirement. HACCP was initially developed for astronaut food but now it is common to be every food industry to apply and practice it. If GMP are adequately followed, practiced and applied, then the industry can avoid many severe bad consequences.

DISCLAIMER

The recommended principles and practices contained herein in this chapter are intended as guidelines, not standards or requirements. Every company must determine which practices are most effective and appropriate for the products that it manufactures and/or distributes. This guidance document is not intended as, and should not substitute for, legal advice. Companies should consult their own legal counsel to ensure compliance with applicable laws and regulations. The companies and organizations that have developed these guidelines do not warrant that they will ensure that the products they or their members manufacture and/or distribute are safe, wholesome, or correctly labeled, and they expressly disclaim any such warranties.

KEYWORDS

- complaint
- consumer
- GMP compliance
- good manufacturing practices
- hazard
- health
- health
- hygiene
- industries
- labeling
- legal
- management
- manufacturing
- practices
- regulations
- standards
- storage
- testing
- training

REFERENCES

1. Bagchi, Debasis. *Nutraceutical and Functional Food Regulations in the United States and Around the World*. Second edition, Department of Pharmacological and Pharmaceutical Sciences, College of Pharmacy, University of Houston, Houston – USA.
2. CAC (1995). Code of hygienic practice for spices and dried aromatic plants. CAC/RCP42-1995.
3. CAC (2003). Recommended international code of practice – general principles of food hygiene. CAC/RCP1-1969, Revised April 2003.
4. Chahar, Digambar (2014). *Nutraceutical and Functional food Regulations in the United States and Around the World*. In: A Volume in Food Science and Technology. Canada. Second edition, pp. 55–61.
5. FAO (2014). *The Joint FAO/WHO Expert Meeting on Microbiological Hazards in Spices and Dried Aromatic Herbs*, 7–10 October. FAO.
6. FDA (1997). *Good Manufacturing Practice (GMP): Guidelines/Inspection Checklist*. February 12, 1997; Updated April 24, 2008
7. FDA (2010). *Guidance for Industry: Current Good Manufacturing Practice in Manufacturing, Packaging, Labeling, or Holding Operations for Dietary Supplements; Small Entity Compliance Guide*. Available at: (http://www.fda.gov/Cosmetics/GuidanceComplianceRegulatory Information/
8. FDA (2010). *New Dietary Ingredients in Dietary Supplements – Background for Industry, What Information Must the Notification Contain?* Available at: http://www.fda.gov/Food/DietarySupplements/ucm109764.htm#what info
9. FDA (2011). *Proposed Rule Current Good Manufacturing Practice and Hazard Analysis and Risk-Based Preventive Controls for Human Food*. Report #N-0921 by FDA.
10. FDA (2012). *FDA Foods Program, The Reportable Food Registry: A New Approach to Targeting Inspection Resources and Identifying Patterns of Adulteration*. Second Annual Report: September 8, 2010–September 7, 2011, by FDA.
11. *Food Chemicals Codex*. http://online.foodchemicalscodex.org/online/login
12. Health Canada, (2009). *Good Manufacturing Practices (GMP): Guidelines*. Version GUI-0001, 21 CFR part 210 and 211, Health Canada.
13. Hoffmann, S. (2011). US food safety policy enters a new era. *Amber Waves, 9*, 24.
14. Mukherjee, P. K., Venkatesh, M., & Kumar, V. (2007). Overview on the development in regulation and control of medicinal and aromatic plants in the Indian system of medicine. *Bol. Latinoam Caribe Plant Med. Aromat., 6*, 129–136.
15. NHPD (2006). *Good Manufacturing Practices: Guidance Document*. Natural Health Products Directorate (NHPD), Health Canada.
16. Pierson, M. D., & Corlett, D. A. (Eds.) (1992). Principles and Guidelines for HACCP. Van Nostrand Reinhold, New York.
17. Sikora, T. (2015). Good manufacturing practice (GMP) in the production of dietary supplements, Dietary Supplements, Safety, Efficacy and Quality. In: *A Volume in Woodhead Publishing Series in Food Science, Technology and Nutrition*, pp. 25–36.
18. WHO (2007). Quality assurance of pharmaceuticals: a compendium of guidelines and related materials. In: *Good Manufacturing Practices and Inspection, Volume 2*. World Health Organization (WHO).

19. WHO (2008). *Guidelines for Drinking Water Quality, Volume 1*. Third edition. World Health Organization (WHO). pp. 668.
20. WHO (2010). *Compendium of Food Additive Specifications*. 74th Annual Meeting by WHO.

CHAPTER 2

RESEARCH PLANNING AND FUNDING AGENCIES: FOCUS ON FOOD ENGINEERING

MURLIDHAR MEGHWAL, SOUMITRA BANERJEE,
B. L. DHANANJAYA, ANU RACHIT, UDAY HEDDURSHETTI,
and RAVI KUMAR BIRADAR

CONTENTS

2.1   INTRODUCTION

The chapter on comprehensive list of all research and innovation funding agencies can be very useful and motivating for getting funding for search for those who are in need and are searching for the funding to start their research. Especially this chapter will be useful for the researcher, scientist, post-doctoral students and even for under graduate and post graduate students.

This briefly describes about funding organizations and agencies, their existence, rules and policies that may change with time. For full details, person can directly contact them.

2.1.1   RESEARCH AND ITS IMPORTANCE

The research is a diligent and systematic investigation process employed to increase, discover or revise current knowledge by discovering new facts, theories and application. The research can be categorized into two general groups: (i) basic research – it is focused on increasing scientific knowledge; and (ii) applied research – it is effort aimed at using basic research knowledge for solving practical problems or developing new processes, products and techniques. Moreover, the explosion of information technology has enabled clinicians and researchers to easily transcend geographical distances so that they may collaborate to answer important questions.

Research has a lot of impact on our life. Consider the example of clinical research, because it links the clinical experience and evidence base to the laboratory which lends itself to statistical and trend analyzes, collaboration, discovery and ultimately to positive impacts on patient diagnosis, care and safety, therefore it has a significant role. Ultimately, research funding enhances the expertise and clinical skills of our clinicians and provide a rich environment for teaching and developing trainees, the public and other medical professionals in the areas of diagnostic imaging, interventional radiology and radiation oncology.

2.1.2   WHO IS A RESEARCHER?

Researcher is the person who is engaged in an organized and systematic way to search and enrich self and society by finding outcomes that would increase human knowledge domain or would be used for the betterment of society. Organized and systematic approach should have either scientific or logical methodology. Researchers work in vast domain of subjects from engineering to medical for betterment of human life. Henceforth with active involvement of researchers and scientists in various domains of science and technologies, there is continuous development and enrichment of human life style, comfort level, health and many countless other aspects in day to day life. One cannot forget the effort of researcher working day and night to develop a life-saving drug or a security technology to make precious human life more safe and secured.

Question may come to students who wants to involve themselves in serious research that where to start or how to start a research. A general view in the world would give an impression that every aspect and development has already been done and there is no scope or need for any new research and development. What makes a researcher different from the crowd is their level of knowledge, understanding and courage to ask question. The most important criteria of a researcher should be curiosity to explore unknown aspects.

The following flow diagram is provided, which would be helpful to a new researcher to start scientific research in their specialization domain (Figure 2.1). Before starting of research, one should have sound understanding in the field of work. The researcher should study existing conventional facility or process or knowledge available and identify the problems with the conventional system or research gap in that particular area. This is called problem identification. After problem identification, researcher must do an exhaustive study of past and on-going research on that particular problem. This step is called literature reviewing. After thorough review of literature, the researcher gets an idea that what are the approaches which had been tried till date and as the problem still exists, means the problem has not been solved yet. Therefore, the researcher sets own plan of work to solve the problem or research gap.

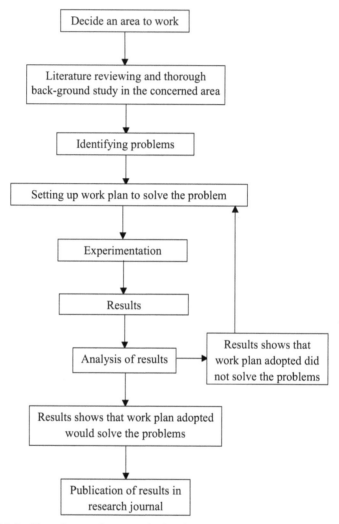

FIGURE 2.1   Flow diagram for research planning.

Work planning involves dividing total work into small phases and experimentation planning. Statistical experimental planning is given more importance since better logical inferences could be drawn out from data analysis of the obtained results. Work plan should be followed by experimentations. Before conducting experiments to avoid any confusion or chances of errors, one should be sure enough to understand the experimental procedures and methodology from standard references. Confusion during experiments causes error chances and unnecessary delay of work, costing time, effort and experimental consumables. Once experiment is over, the obtained data is to be analyzed to draw valuable conclusion that is obtained from the experimental data. Drawn conclusion would guide the researcher that whether the hypothesis adopted for solving the identified problem is correct or not. If it is correct, the researcher must repeat the experimentation again to confirm the results. Once the researcher becomes confident about the repeatability of the obtained data from the experiment, then one should proceed for publication of the findings of research, so that the research finding is available to the betterment of society.

2.1.3   STEPS FOR RESEARCH FUNDING

To choose a career of research is a good decision, but major hurdle faced as one thinks of starting the work is the availability of funds. Funding is another challenge faced by the researcher, as only having an idea would not be valid unless supported by some experimental results. For conducting research work, first thing that would be required is funding. Step by step reader would be made aware how to cross this hurdle.

Before beginning for the process of fund requests, one should realize the fact that the fund to be released on the competence of oneself. Before asking for fund, the researcher must be having strong academic candidature. Academic candidature is measured by number of valued publication in the domain of work. Researcher with good publication would be a better candidate for applying for research funds.

First step is to identify all potential funding agencies that provide funding for research. Internet surfing along with taking guidance from the experienced professionals from that field might help the researcher to get idea about who are the potential funding agencies in particular domain. Seeking for fund from a agency which does not fund for that field, may be a waste of researcher's valuable time. For this reason, preparing funding agency list should be done with utmost care. After matching the domain, the researcher should also check whether the researcher is matching the eligibility criteria or not. Some finding agencies have strict guidelines that the university from where the researcher work should be associated with a reputed organization or the researcher should have previous project handling experiences.

Second step is to plan what are the research facilities available with the researcher's institute and what facilities would be necessary for carrying out the work. Facilities may be in terms of equipments, materials, consumables or other research requirements. Facilitates not available with the researcher, to complete the project may be asked from the funding agency in term of monetary budget. A proper justification regarding the purchase should be prepared to justify the monetary request. If the request be made un-justified, possibilities stays of rejection and even drawing a negative view over the submitted project. Rational monetary requests with justified purchase budgets are more entertained by the finding agencies.

Fourth step is to consider the need of any living entity testing, e.g., human or animal. It may require clearance from ethical committees and other legal authorities depending upon the nature of trials to be conducted. Before applying for project, one should be confirm that the work to be conducted would not be legally challenged when started. After fund approval, once the project starts, at that point, if the work be forced to stop due to legal issue, then both the approved fund and time of researcher would be wasted. This may draw a low light on researcher's profile, affecting the future project's financial approvals.

Fifth step is to calculate the project cost including human resource required, office maintenance cost, money to be paid of the institute for handling infra-structure, etc. This thing includes many factors which to be prepared with the help of financial officer. Financial officer help would be vitally required for completing the financial part.

Sixth step involves the writing the application to funding agencies. After completing all five steps and having a justification of the research work to be carried out, then only the application writing part should be initiated. One should write in short and compress application, including justified answers of all questionnaires in the application form. While preparing the application, support of research officers and faculties would be helpful [1, 2, 4, 10].

12.2.4   OBJECTIVES OF THIS STUDY

Funding is essential part for initiating any research project. A researcher having sound knowledge may with time come up with an idea that would be solving problem or providing valuable answer to an unsolved question or may reduce the research gap. But when time comes to check whether the idea or hypothesis would fill the research gap, it has to be tested in terms of conducting experimental trials. Conducting experiments requires a research facility, where the work could be conducted at a level where all required facilities are present to conduct the experimental run. To arrange this facility, there would be requirement of funding. Institutes are places where defined research facilities are present but problem comes while accessing certain specific research instrument, process or other specified entities. Research laboratories are equipped with research facilities, but all required facilities would not be available. As discussed earlier, facilities would not be in terms of only scientific equipments or facilities, but also includes human resources.

For getting what facilities are not available, there would be requirement of funds. The basic objective of this chapter is to make researcher aware about what are the steps for getting financial funding, what the sources are from where funding could be received and other related and significant topics. Objective of this writing would be completed, if after going through the chapter, the reader gets complete idea about how to start the process of making fund requests, what are the major funders, etc.

2.2   TYPES OF RESEARCH: BASIC, APPLIED, FIELD, LABORATORY, DEMONSTRATION, ETC.

Research has been classified in different types by different groups of specialists and committees. United Kingdom's Medical Research Council has classified research in two broad groups, e.g., basic and applied research.

Basic research is carried out for the advancement of knowledge which may or may not have significant application or economic value. Increase in knowledge level is the primary goal of the work. Output of the research work, may have potential application for its further application in development of processes or products. But as such the finding may not have any significant application in day-to-day life.

Applied research mainly targets a specific application on particular necessities, problems, and marketable opportunities. These researches mainly focus on particular area and try to develop ready to apply solutions to bridge the present requirement with the user [4].

Depending on where experiment to be performed, the research may be classified as *field based* or *laboratory based*. General tread of any research is laboratory based, which is performed in a laboratory condition. Laboratory is usually equipped with all the required facilities that would be required for conducting the process, under controlled conditions with manipulative variables. This helps the researcher to study the effects of variables on estimated responses [12]. But sometimes research studies may involve facilities that become difficult to accommodate in one roof laboratory or investigation needs to be carried out on site/field. For those reasons on-field research needed to be carried out. This is very common in the field of agricultural and biological engineering, where for conducting experiments need for field is very essential. Beside this, engineering projects are carried out at industrial levels (field) when it becomes difficult to conduct the same in laboratory. *Field research* is quite popular because researcher understands the problem better by interacting with the people and understanding the problem first hand. Researcher also gets better view to analyze the problem and understand its social impact [5]. But some drawbacks of field research are observation documentation may be erroneous, too much factors affecting the subject of study, time taking, affected by natural conditions like day/night, weather conditions, etc.

2.3   FOCUS AREAS OF FOOD ENGINEERING

Food engineering has a wide range of focus areas and huge potential in future also. Some of the major focus areas are:

Food preservation, nondestructive methods for food preservation, hurdle-preserved food products, minimal processing in food, high pressure processing, Ohmic heating, mathematical modeling in food, food freezing technologies, frozen processing, active packaging, intelligent or "smart" packaging, tamper evident food packaging, thermal processing, drying process: solar, novel, infrared, industrial, grain and seed drying, fruits and vegetable drying, nonthermal process: pulsed electric fields, pulsed light, ionizing radiation, high hydrostatic pressure, ultrasound processing, food regulations, hazard analyzes and critical control point program, organic food processing, food safety engineering, novel nonthermal intervention technologies, biosensors in food engineering, grain harvesting machinery, grain storage systems, dairy engineering, milk processing, milk products, dairy processing equipment, heat and mass transfer in food processing, food rheology, unit operations in food engineering, food process modeling, simulation and optimization, food process control systems, energy efficiency and conservation in emerging food processing systems, ascetic food processing, functional food and nutraceuticals, fortification, food and nutrition, encapsulation, algae foods: principles, practices and challenges, food additives- overall quality, safety, nutritive value, appeal, convenience, and economy of foods, food biotechnology, food fermentation, food waste and its management, aromatic and medicinal plant and food engineering, food milling, grinding, mixing and filtration, entrepreneurship and management of food plant, food analysis and quality control, health and medical foods, food sustainability, seafood professing: principles, practices and challenges, innovative food products development, plant based bioactive and their effects, food antioxidants agent and aging effect, dietary fibers and health, functional lipids related to food, food allergy and food poisoning, self-heating food packaging, emerging ways for safety and quality in food processing, food fortification and enrichment, food additives, their properties and effects, bioactive peptides, food-borne illness, carbohydrate as food source, common vegetables and their food functionalities, traditional therapeutic foods for type II diabetes, functional food, functionality of fermented foods, fruits as a functional food, carotenoids and polyphenolic compounds in foods, antioxidant and prooxidants in foods, food and cardiac health, health benefits of tea, functional food and degenerative diseases, functional food for chronic diseases, functional food for hypertension, emerging technologies in food processing, progresses in the transport and storage of liquid and solid foods, food enzyme kinetics, microbiological and chemical testing of food items, advanced mass transfer in foods, food dietary supplements, nutrition for seniors, functional beverages and food products, use of safe, nutritious, and wholesome food, impact of genetically modified food on health, molecular nutrition, disease prevention and treatment, decaffeination of coffee and tea, food labeling, nutritional effects of food processing, enzyme application in dairy industry, mal absorption syndromes, raw food diets, food supplements, body and masculinity building food.

2.4   RESEARCH PRIORITIES IN FOOD ENGINEERING AROUND THE GLOBE

Food engineering holds key to the solution of global problems related with food processing and preservation. The most challenging problem is food security, which may be addressed by this modern discipline of study. Food security which stands on three pillars, e.g., food availability, food accessibility and food usage, is a complicated issue that is linked with our health, economic growth and business/trade. Processing and producing affordable health food is another problem especially in under-developed and also in developing countries. Human population has been growing in much rapid rate. To provide every human being nutritional healthy food at affordable cost is a big challenging aspect. Current area of work is involved with genetically modified food, whether it could address all these problems or not. Genetically modified foods are those which are not naturally occurring but are developed by modifying genetic material of the source. They have been trapped in debates, whether consumption of modified food have any effects on consumer health. These are some broad areas of priorities in research for food engineers. Beside these, some recommended research priorities are listed below [6, 14, 15]:

- Rapid microbial/chemical/foreign contamination identification in food.
- Rapid identification of adulteration/pesticides/toxic levels.
- Effective use of emerging trends for efficient processing, product development and value addition of existing products.
- Real time contamination and identification before processing.
- Effective food industries pollution control technique.
- Efficiently using available resources for producing safer and healthier food.

Rapid development in the field of technology and arising modern day problems are making research priorities change. Problems related with food processing that was there in past has been solved or research ongoing to solve the problems. But problems and difficulties arising today and that would arise tomorrow also needs to be thought about. Modern day's innovations in interdisciplinary studies would make it easier to address the priorities and tackle the problems with safe, healthy and economical solutions.

2.5   ETHICAL ISSUES/CANNONS IN RESEARCH

2.5.1   GUIDELINES FOR SPECIFIC STUDY TYPES

2.5.1.1   Human Subject's Research

All research involving human objects must have been approved by the researchers' Institutional Review Board (IRB) or by equivalent ethics committee(s), and must have been conducted according to the principles expressed in the *Declaration of Helsinki*. Subjects must have been properly instructed and have indicated that they consent to participate in the study by signing the appropriate informed consent paperwork. If consent is of verbal instead of written, or if consent could not be obtained, the researchers must explain the reason, and the use of verbal consent or the lack of consent must have been approved by the IRB or ethics committee.

For studies involving humans categorized by race/ethnicity, age, disease/disabilities, religion, sex/gender, sexual orientation, or other socially constructed groupings, researchers should:

- Explicitly describe their methods of categorizing human populations.
- Define categories in as much detail as the study protocol allows.
- Justify their choices of definitions and categories, including for example whether any rules of human categorization were required by their funding agency.
- Explain whether (and if so, how) they controlled for confounding variables such as socioeconomic status, nutrition, environmental exposures, or similar factors in their analysis.

2.5.1.2   Clinical Trials

Clinical trials should be subject to all policies regarding human research *and should follow* the *World Health Organization's* (WHO) definition of a clinical trial, *"A clinical trial is any research study that prospectively assigns human participants or groups of humans to one or more health-related interventions to evaluate the effects on health outcomes [...] Interventions include but are not restricted to drugs, cells and other biological products, surgical procedures, radiologic procedures, devices, behavioral treatments, process-of-care changes, preventive care, etc."*

Further, all clinical trials must be registered in one of the publicly accessible registries approved by the WHO.

2.5.1.3   Animal Research

All research involving vertebrates or cephalopods must have been approved from the researchers Institutional Animal Care and Use Committee (IACUC) or equivalent ethics committee(s), and must have been conducted according to applicable national and international guidelines.

2.5.1.4   Non-Human Primates

Research involving nonhuman primates must include details of animal welfare, including information about housing, feeding, and environmental enrichment, and steps taken to minimize suffering, including use of anesthesia and method of sacrifice if appropriate, in accordance with the recommendations.

2.5.1.5   Observational and Field Studies

Research on observational and field studies should include:

- Should obtain permits and approvals for the work, specifying the authority that approves the study; if none were required, researchers should explain why.
- Whether the land accessed is privately owned or protected.

- Whether any protected species were sampled.
- Full details of animal husbandry, experimentation, and care/welfare, where relevant.

2.5.1.6 Paleontology and Archaeology Research

Specimen numbers and complete repository information, including museum name and geographic location, are required to be document by the researchers.

2.5.1.7 Cell Lines

Researchers using cell lines should give information when and where they obtained the cells, giving the date and the name of the researcher, cell line repository, or commercial source (company) who provided the cells, as appropriate.

For *de novo* (new) cell lines, including those given to the researchers a gift:

- Details of institutional review board or ethics committee approval;
- For human cells, confirmation of written informed consent from the donor, guardian, or next of kin.

For established cell lines: A reference to the published article that first described the cell line; and/or

- The cell line repository or company the cell line was obtained from, the catalog number, and whether the cell line was obtained directly from the repository/company or from another laboratory.

2.6 GENERAL REQUIREMENTS FOR FUNDING

2.6.1 *GENERAL DETAILS REQUIRED FOR MOST FUNDING PROPOSALS*

Most funding agencies have several components to be made available by an application, many of which are common across agencies. A general overview of details to be furnished is provided here as a reference. However, an applicant shall check the guidelines of prospective funding agency before applying.

2.6.2 PROFORMA FOR SUBMISSION (IN GENERAL WHILE APPLYING)

Part 1

- General information
- Name of the Institute/University/Organization
- Project title
- Category
- Specific area
- Single or multiple institutional
- Project duration and cost
- Project summary

Part 2 (Technical details)

- Origin of the proposal
  - Rationale of the study
  - Hypothesis
  - Key questions
- Present status of research and development in the subject
- Relevance
- Outcome of the proposed study
- Preliminary work done so far
- Scope of application
- Specific objectives and work plan
- Start and end months of the research work

Part 3 (Budget particulars)

- Non-recurring (equipment/instrument, accessories)
- Justification
- Human resource details (Technical Assistant/JRF/SRF)
- Consumables (justify)
- Travel (to attend training programs and present papers in conference)

- Contingency (future expenses which cannot be predicted):
  – Approximate amount per annum
  – Overhead (5–10% of the total sum or as specified)
  – Accounting fees (audit)
  – Legal fees
  – Rent, repairs
- Research incentive to Principal Investigator (PI)
- Account holder details (name of the college and email id, and bank details)

Part 4 (Existing facilities)

- Human resource, equipment, other resources

Part 5 (Details of Investigators)

- PI and Co-PI

Part 6

- Declaration

2.6.3 DOCUMENTS TO BE SUBMITTED AFTER THE PROPOSAL IS RECOMMENDED

- List of existing laboratory equipment
- Proposal forwarded through HOI
- No Objection Certificate (NOC)
- Quotation of the equipment
- Copy of Bye laws
- Three years of audited account statement
- Registration certificate

2.6.4 DOCUMENTS TO BE SUBMITTED AFTER PROPOSAL IS AWARDED FOR FUNDING

- Submission of documents
- Memorandum of Agreement between the University and the respective funding agency
- Copy of sanction order

- Copy of terms and conditions
- Release of funds
- Advertisement
- Purchase

2.6.5 DOCUMENTS TO BE SUBMITTED AFTER COMPLETION OF PROJECT

- Audited utilization certificate
- Project completion report
- Project evaluation committee
- Account statement
- Details of publications or Patent applied

2.7 SPECIFIC REQUIREMENTS FOR FUNDING

- Industrial partner

2.8 RESEARCH FUNDING IN INDIA

Several government funding agencies in India offer grant or funds such as: All India Council for Technical Education (AICTE), Department of Science and Technology (DST), Department of Biotechnology (DBT), Vision Group of Science and Technology (VGST), Department of AYUSH, Indian Council of Medical Research (ICMR), University Grants Commission (UGC), Council for Scientific and Industrial Research (CSIR), Department of Atomic Energy (DAE), Science and Engineering Research Council (SERC), Defense Research and Development Organization (DRDO), Indian National Science Academy (INSA) are funding research proposals in India.

2.8.1 VARIOUS GOVERNMENT FUNDING AGENCIES IN INDIA

- AICTE (All India Council for Technical Education)
- CSIR (Council for Scientific and Industrial Research)

- DAE (Department of Atomic Energy)
- DBT (Department of Biotechnology)
- Dept. of AYUSH (Department of Ayurveda, Unani, Siddha and Homeopathy System of Medicines)
- DFPI (Department of Food Processing Industries)
- DOE (Department of Education)
- DRDO (Defense Research and Development Organization)
- DST (Department of Science and Technology)
- ICAR (Indian Council of Agricultural Research)
- ICMR (Indian Council of Medical Research)
- INSA (Indian National Science Academy)
- MFCS (Ministry of Food and Civil Supplies)
- MNRE (Ministry of New and Renewable Energy)
- NAAS (National Academy of Agricultural Sciences)
- NDDB (National Dairy Development Board)
- NOSVODB (National Oil seeds and Vegetable Oils Development Board)
- SERC (Science and Engineering Research Council)
- SERC (Science and Engineering Research Council)
- UGC (University Grants Commission)

From the above listed government funding agencies, researchers and academicians can choose a funding agency that suits the requirements of proposed research question.

2.8.2 VARIOUS STATE GOVERNMENT FUNDING AGENCIES IN INDIA

- Haryana State Council for Science and Technology
- VGST (Vision Group of Science and Technology) (For Karnataka State only)

2.8.3 FUNDING FROM PRIVATE AGENCIES/ ORGANIZATIONS/FOUNDATIONS

- Haryana State Council for Science and Technology
- Lady Tata Memorial Fellowship

- Rajiv Gandhi Foundation (RGF)
- Sitaram Jindal Research Fellowship Scheme
- VGST (Vision Group of Science and Technology): For Karnataka State only

2.9 RESEARCH FUNDING AROUND THE GLOBE

- ACI Fellowship Award (Material Science)
- Afton and Dorothy O. Allen Memorial Endowed Scholarship Fund
- Albert and Mary Kay Bolles Scholarship in Food Science
- Alfred E. Knobler Scholarship – Material Science/African American
- Alfred E. Knobler Scholarship – Material Science/Female
- Alfred E. Knobler Scholarship – Material Science/Hispanic American
- Alfred E. Knobler Scholarship – Material Science/Native American Indian
- Allison Family Endowment
- America's Farmers Grow Ag Leaders
- Annette Rachman Scholarship
- ASM International Scholarship
- AZ Vegetable Growers Association/M.O. Best Scholarship
- Beth and Holy Fryer Endowed Scholarship
- Bill Griffin Memorial Scholarship
- Blackmore Food Science Scholarship
- Bob and Debbie Macy Endowment for Excellence in Animal Husbandry
- Bob and Debbie Macy Scholarship
- Borden Foods Scholarship in Food Science and Technology
- Boyd-Schools Scholarship- Food Technology
- Bradford-Adams Endowed Scholarship For CANR
- Bradford-Adams Endowed Scholarship For Natural Science
- Carl and Viola Smith Endowed Scholarship
- Cecil and Eloise Robinson Food, Agricultural and Biological Engineering Endowment Fund
- Charles H. (Chuck) and Marjorie (Marge) Bagans Memorial Endowed
- Charles M. Stine Memorial Scholarship
- Constance Parks Hagelshaw and Doretha June Hagelshaw
- Dairy Farmers of America Scholarship
- Daniel and Anne Guyer Sparrtan Cornerstone Scholarship Challenge
- Daniel Brackeen Food Technology Scholarship

- Daughters of Ralph J. Stolle Food Science and Technology Support Fund
- David and Mary Jessup Dietetic Internship
- Delton and Dianne Parks Scholarship
- Dorothy E. Dugan Memorial Scholarship
- Dorothy Z. Bergman Scholarship-Food Technology
- Dr. Richard L. Hall Scholarship in Flavor Science
- Eva Patchen Memorial Scholarship-Food Technology
- Evan Turek Memorial Scholarship (Mondelez International)
- Feeding Tomorrow General Education Scholarships
- FNA Senior Professional Development Award
- FNA Undergraduate Scholarship
- Frozen Food Industry Scholarship
- Future Leaders Mentoring Scholarship
- Malcolm Trout Undergraduate Scholarship
- Gerber 'Carl G. Smith' Award for Excellence
- Gerber Foundation Endowed Scholarship in Pediatric Nutrition and Food Science
- Herman B Blum Scholarship Fund in Food Science and Human Nutrition
- IFT Food Engineering Division Scholarship
- IFT Food Laws and Regulations Division Scholarship
- IFT Nutraceutical and Functional Foods Division Scholarship
- Institute of Food Technologists (IFT)
- Institute of Food Technologists Scholarships – Food Technology
- Intermountian Institute of Food Technology Award
- Lois Gordon Ridley Endowed Scholarship/Fellowship
- Margeret A. Eaegle Endowed Scholarship
- Marilyn Mook Endowed Scholarship
- Michigan Soybean Promotion Committee Scholarship
- Society of Flavor Chemists Graduate Scholarship
- The Andrew R. Jackson Dairy Food Science Scholarship
- The Chenoweth Award in Dietetics and Human Nutrition
- Tropicana Diversity Scholarship
- United Negro College Fund (UNCF)
- Verlene and Dale K. Weber Endowed Scholarship
- Virginia Dare Award
- W.K. Kellogg Institute For Food And Nutrition Research Scholarship in Food Science

2.10   SUMMARY

An overview on research funding and their agencies is very useful and relevant topic emerging and advanced age research in science and technology. This chapter can be useful for beginners, who want to start their carrier in teaching, research and development. The summarized outline for the research proposal will be very useful. The chapter briefly discusses the steps of application for research funding. The list of topics those can be taken research in food processing will be useful for researcher to take up a new research. This update provides the list of most relevant and major funding agencies. The reader is advised to consult the web sites of these agencies for more details.

KEYWORDS

- agencies
- board
- council
- engineering
- fellowship
- food
- foundation
- funding
- grant
- organization
- proposals
- research
- science
- technology

REFERENCES

1. Anonymous (2015). Step by step guide to applying for funding, Last accessed on Oct. URL: https://www.admin.ox.ac.uk/researchsupport/applying/guide/.

2. Anonymous (2015). Strategy for Getting an NIH Grant, National Institute of Neurological Disorders and Stroke. Last accessed on Oct.URL: http://www.ninds.nih.gov/funding/write_grant_doc.htm#strategy.

3. Anonymous (2015). Food Technology Scholarships. Last accessed on: Oct.URL: http://www.schoolsoup.com/scholarship-directory/academic-major/food-technology/.

4. Berry, D. C. (2010). *Gaining Funding for Research: A Guide for Academics and Institutions.* Open University Press, McGraw Hill Company.

5. Blackstone, A. (2016). *Principles of Sociological Inquiry: Qualitative and Quantitative Methods, Pros and Cons of Field Research.* Accessed on Dec.URL: http://catalog.flatworldknowledge.com/bookhub/reader/3585?e=blackstone_1.0-ch10_s02

6. Buchmann, N. (2015). Research priorities for a sustainable food system. ETH Zurith report. Last accessed on Oct. URL: https://www.ethz.ch/en/news-and-events/eth-news/news/2015/06/research-priorities-for-a-sustainable-food-system.html

7. Coralie, G., Amélie, P., Marie-Louise, K., Marion, L., & Karine, C. (2015). Evolution of public and nonprofit funding for mental health research in France between 2007 and 2011. *Original Research Article, European Neuropsychopharmacology, 25*(12), 2339–2348.

8. Holger, S., Xu, S., Genovefa, K., Tim, C., Rupert, M., James, N., & Derek, M. (2016). Creativity Greenhouse: At-a-distance collaboration and competition over research funding. *Original Research Article, International Journal of Human-Computer Studies, 87,* 1–19.

9. Fedderke, J. W., & Goldschmidt, M. (2015). Does massive funding support of researchers work?: Evaluating the impact of the South African research chair funding initiative. *Original Research Article, Research Policy, 44*(2), 467–482.

10. Lane, J. (2015). *Top 10 Tips on How to Get Funding,* NSF Program Director. Last accessed on Oct. URL: http://advance.cornell.edu/documents/top10_list_How_to_Get_Funding.pdf.

11. Michael G. Head, Joseph R. Fitchett, Vaitehi Nageshwaran, Nina Kumari, Andrew Hayward, & Rifat Atun (2015). Research investments in global health: A systematic analysis of UK infectious disease research funding and global health metrics, 1997–2013. Original Research Article, EBioMedicine, Available online 17 December 2015.

12. Psychwiki (2015). What is laboratory research? Accessed on Dec. URL: http://www.psychwiki.com/wiki/What_is_laboratory_research%3F.

13. Sita Naik (2015). Funding opportunities for research in India. *Indian Journal of Rheumatology, 10*(3), 152–157.

14. USDA (2015). United States Department of Agriculture, Food Safety and Inspection Service, Food Safety Research Priorities. Last accessed on: Oct. URL: http://www.fsis.usda.gov/wps/portal/fsis/topics/science/food-safety-research-priorities

15. WHO (2015). World Health Organization, Genetically Modified Food. Last accessed on: Oct. URL: http://www.who.int/topics/food_genetically_modified/en/.

PART II

LATEST FOOD TECHNOLOGIES

CHAPTER 3

FOOD INDUSTRY: USE OF PLASTICS OF THE TWENTY-FIRST CENTURY

EWELINA BASIAK

CONTENTS

3.1 FOOD PACKAGES: EARLY NINETEENTH CENTURY

The Industrial Revolution, which began in the eighteenth century in England and Scotland, was the transition to the process of technological, economic, social and cultural rights and it was associated with the transition from an economy based on agriculture and manufacturing industry or craft to relying mainly on mechanical factory production at large scale (industrial). The major reason of the Industrial Revolution was the rise of population. It has led to an increase in the number of inhabitants, the result of which also grew market needs. They could not satisfy them manufactory which of craft production differed only in the organization of the production process. Second, in order of importance, the cause of the industrial revolution was the agrarian revolution, which led to the transformation of traditional feudal agriculture in modern agriculture.

Manner of food production changed and demand on it increased. In order to adapt to market needs in nineteenth century appeared packaging for food [1, 2]. At the beginning, the main role of packages was protection of food products and the ability to convenient transportation – counted so almost exclusively due to practical and functional. However, technological developments made it appear never forms of packaging, and thus the same packaging acquired additional significance. Producers were able to lock products in attractive forms, which contributed to increasing consumer interest in product and sales growth. It is also conditioned the growth of interest in the visual side of the packaging by manufacturers, who saw sales potential in packaging, and later also consumers, who began to differentiate the package among themselves – over time began to play the role of color, typography and the general nature of the appearance and layout of the package. The excellent example can be a bottle of whiskey. In 1985, Jack Daniel trying to differentiate their product from the common then round bottles of this drink, put on a rectangular bottle, which further emphasized the character of this filtered through charcoal whiskey. This led to other manufacturers slowly started harder to see in the package a new feature – namely product differentiation. That made packaging history began to acquire momentum. A bottle of whiskey signed with the name Daniel, quite different from previously known bottles of alcohol, gave a new identity and brand [3]. Thirty years later in 1915, Coca-Cola company also provided the patented design of a battle. This event was scheduled between the growing interest in packaging and their potential sales. The appearance of polymer materials completely revolutionized the market of the twentieth century. Many advantages such as low density, corrosion resistance and ease of processing (low costs of running large series of finished products compared to other groups of materials) made from polymers have dominated the market for packaging in the last century. Moreover economic boom, the appearance of television led to dynamic growth in the packaging market. The main driving force behind this situation was the emergence of self-service stores, which appeared in the 1930s of the twentieth century in the United States and in Europe a little bit later – just after the end of World War II. Self-service stores revolutionized the current nature of purchase. Customers could choose by themselves and they had time for that.

Packaging design has become so popular and profitable that the resulting separate fields of study deal only packaging design [4]. Packaging became an integral part of the product. Compared to the end of the twentieth/beginning of twenty-first centuries, people used to buy in the bazaars (local markets). Today, even the salad and rolls are sold in the plastic foil. Many times the packaging value exceeds the price of the product alone. Consumers buy by eyes. With each year, there is an increasing amount of plastic packaging. The process of decomposition of plastic takes more than hundred years. The amount of plastic thrown away annually can circle the Earth four times. For instance, one American citizen throws away 84 kilograms of plastic. Furthermore, every year approximately 500 billion bags made from plastic are used worldwide. Thus, society uses more than 1 million bags every minute. The 50% of synthetic polymers is used only one time. The number of landfills is increasing due to the increase in the amount of waste. Only five percent of produced plastic is recovered. What's more, the production of synthetic materials uses around 8% of the world's oil production, therefore crude oil deposits will be completely exhausted due to use of plastic for several decades. Part of the waste goes to the reservoirs disrupting natural ecosystems (wastes are eaten by marine animals, marine transport causes environmental disasters, etc.). In oceans, pieces of plastic from one liter bottle are able to end up on every kilometer of beach throughout the world. Billions of kilograms of synthetic materials can be found in rotating convergences in the oceans representing approximately 40% of the world's ocean surfaces. The 80% of pollutants are flowing into the ocean off the land. Plastic accounts for about 90% of all rubbish floating on the surface of the ocean, with 46,000 pieces of plastic per square kilometer. The largest ocean garbage site in the world (the Great Pacific Garbage Patch) is located near the California coast. This is a floating mass of synthetic polymers twice the size of Texas. Every year, 1,000 of marine mammals and 1 million of sea birds are killed due to plastic in the oceans. Numerous studies have confirmed that the plastic is present in bodies of all sea turtle species, many kinds of fish, 44% of all seabird species and 22% of cetaceans. Also part of these compounds, which have been found to have influence on hormonal system of people (xenestrogens), has potential negative effects on human health. Chemicals present in plastic may be absorbed by the human body up to 93%.

The dioxide emission is environmental problem of twenty-first century due to overuse of plastic material in our daily life. According to United States Secretary of State John Kerry (during Expo 2015) comparing the waves of immigrants to emission of greenhouse gas, it is nothing. Participants to Kyoto Protocol 141 developed countries confirmed that it is a serious problem for the Earth ecosystems and societies. Only in this century more plastic was produced than during the last century. Therefore, it is estimated that the total dioxide emissions from the base year has increased approximately 49% [5].

3.2   FOOD PACKAGING IN THE TWENTY-FIRST CENTURY

In connection with numerous disadvantages of plastic producers, scientists have begun to search for packaging material, which would be able to replace it. "Plastics" of twenty-first century should fulfill diverse functions. One of the most important is technological function – a package protects against deterioration and undesirable from outside, protects against microbiological, chemical and physical damages and helps with food trade, protects against scattering and spilling. The package is also the carrier of information (product, contents, nutrition, mass, volume, quantity, method of use, date of shelf-life, brand, producer and others) for the consumer. In the free market economy period, marketing function seems to be quite important. Color, design, form of package, information and nutrition claims or healthy information can have influence on the increasing product demand. Moreover package must be safe (regulated in legislations). In this respect, the basic act in Europe is the European Union Regulation 1935/2004 of the European Parliament and European Council of 27 October 2004 for materials and articles intended to come into contact with foodstuffs. These materials under normal or reasonably foreseeable conditions of use cannot constitute a danger for human life, cause unacceptable changes in the composition of the food as well as the deterioration of its organoleptic characteristics. These risks could be the result of migration packaging ingredients to food exceeding the allowable amounts (Polish Committee for Standardization).

The perfect solution of the problem of plastic packaging seems to be biodegradable and/or edible films and coatings for food. Edible coatings are thin layer of edible material applied directly on food product and meantime films are produced separately from the foodstuff – as solid sheets and then they are used as a foil, tray, etc. Biopackaging can protect the food stuff from mechanical damage, chemical, physical, and microbiological activities. Such systems can be a carrier of nutraceuticals, antioxidants, antimicrobial substances, flavorings and color agents. They are able to improve several parameters and properties in food products as mechanical integrity, appearance quality and others [6]. Biopolymers are usually edible, but if customer does not want to eat food with package, he can wash it and the coating will solve into water, or consumer can throw it away [7]. Process of degradation takes usually few days to few months, it means times period of decomposition is a few hundred to few thousand shorter than in case of plastics. Production of films and coatings is safe for environment. Any greenhouse gas does not arise in the process. According to Kyoto Protocol from 1997, production of biopolymers is not only environmentally safe and but is eco-friendly. One can have profits by selling surplus. Both films and coatings can occur as single layers, bi-layers, multilayers or emulsions [8]. The source materials of biopolymers are plants (plant materials, waxes, fats), shellfish (chitin armors) and animal products (milk protein). Generally following classification is used (according to their structural component): saccharides, lipids, proteins and composites.

The most popular polysaccharides are: alginate, carrageenan, chitosan, cellulose with its derivatives, pectin, pullulan and gellan gums and starch [9]. Polysaccharides' films and coatings have good barrier properties to oxygen, carbon dioxide and other gases. Laufer et al. [10] worked with multilayered thin films. Bio totally renewable materials were produced from polysaccharides, such as: chitosan, carrageenan and montmorillonite clay. Layer-by-layer technique was used. Polymers characterized low oxygen permeability. After addition of one more chitosan layer (the 4th layer), the oxygen permeability was reduced of two orders of magnitude under the same conditions. Authors applied these coatings onto banana. Banana could be kept in 22°C and 55% of relative humidity without any visible changes in coating after 9 days. In the meantime, banana with three-layers coatings did not show any changes, thus the lowest amendments could be observed on four-layers coated fruit. After 13 days, the skin of uncoated banana was completely dark. Fruits coated with three-layers or four-layers still showed brightness after 9 days. Therefore, multilayer coatings are able to reduce significantly physiological processes of respiration and transpiration and to inhibit the fruit aging. Moreover, the postharvest losses can also be reduced. Therefore, this kind of packaging, which has high gas barrier and optical transparency, may be a promising replacement for foils for food packaging.

In addition, weak properties of water vapor permeability can affect the success of these package systems. After lipid addition to their matrix, this disadvantage can be significantly reduced. Basiak et al. [11] added rapeseed oil to wheat starch films. Water vapor permeability (at 75% of relative humidity) was reduced from 8.77×10^{-10} gm^{-1} s^{-1}Pa^{-1} to 3.55×10^{-10} gm^{-1} s^{-1}Pa^{-1} for 5% starch films (amount of starch powder in solution) compared to 4.6×10^{-10} gm^{-1} s^{-1}Pa^{-1} to 0.25×10^{-10} gm^{-1} s^{-1}Pa^{-1} in case of 3% starch films. Moreover, these polymers have excellent mechanical properties, especially tensile strength and elongation at break [6]. Most of polysaccharide materials are transparent, and these can be easily colored and printed [12]. Polysaccharide materials have particular application in fruits and vegetable industry. Research has confirmed positive influence of coatings on transpiration and respiration resistances, firmness, color, pH, sugar/acid ratio, and even on antioxidant and antimicrobial properties.

Arancibia et al. [13] produced polysaccharide films (made from agar and alginate) with cinnamon essential oil as bi-layer packaging to use them for microbial growth inhibition in chilled shrimps. The agar and alginate combinations were able to reduce the quantity of microorganisms as pathogenic bacteria *Listeria monocytogenes*, in peeled shrimps during the chilled storage. Besides, bi-layer films have strong antioxidant properties and are able to prolong the shelf-life of food products. However, negative impact on the organoleptic properties was observed.

The other antimicrobial effect was indicated in a study by Maftoonazad et al. [14]. Emulsions made from pectin and coated onto avocado fruit inhibited growth of *Lasiodiplodiatheobromae*. Additional boundary layer is able to decrease physical and physiological changes such as color and texture (caused by the aging of the fruit). Tavassoli et al. [6] mentioned that the carrageenan and alginate biomaterials have good mechanical and barrier properties.

However caseins, collagen, corn zein, egg white protein, gelatin, keratin, myofibrillar protein, quinoa protein, soy protein, wheat gluten, whey protein are commonly used as protein structural materials. Films produced from these proteins are flexible, usually transparent with superb aroma-, oxygen-, and oil – barrier properties at low relative humidity [15]. Thus to keep good barrier properties at high relative humidity, addition of second film-forming substance (fat for instance, to protein system) is necessary.

Janjarasskul et al. [16] added candelilla wax to whey protein matrix and observed the potential to reduce the water vapor permeability and oxygen permeability. Authors also admitted that the technique to prepare plays a significant role. Comparing to extruded sheets, the solution cast matrix was more effective barriers against water vapor transmission. Moreover the extruded sheets subsequently compressed to biofilms, at all incorporated candelilla wax. Besides, incorporation of agar to soy protein isolate matrix produces changes on the surface, such as the orientation of the polar groups and hydrogen bonding interactions between polysaccharide and protein isolate. Films obtained from agar and soy protein isolate have good functional properties due to the formation of a compact and reinforced three-dimensional network.

Lipids as animal and plant waxes (beeswax, candellila wax, carnauba wax, shellac, sugar cane wax), fatty acids (oleic acid, stearic acid), vegetable oils (corn oil, mineral oil, nut oil, olive oil, rapeseed oil, sunflower oil) are usually used as a monolayer or one of the layers in bio-coatings. Actually already in twelfth century in China, it was recorded using lipids in oranges and lemons for preventing moisture loss [17]. Liquid lipids cannot be used as self-standing films. Separately, they do not form films, but with other film-forming material (as a layer or as emulsions' component) they show positive sorption properties.

Polyols (glycerol, sorbitol, monoglycerides, oligosaccharides, polyethylene glycol, glucose, water, lipids) are commonly used plasticizers. They are added to film-forming solution to facilitate processing and/or to increase films flexibility and elasticity [18]. Jost and Langowski [19] tested several plasticizers, such as: glycerol, propylene glycol, triethyl citrate, polyethylene glycol castor oil and epoxidized soybean oil in content 5–20 wt.%. They observed that with the increase in quantity of plasticizer: crystallinity/ tensile strength and elongation at break were increased compared to decrease in melting temperature and Young's Modulus. Therefore, the oxygen and water vapor permeability were increased with increasing plasticizers content but to a different extent. Similar conclusions were suggested by Hanani et al. [20], who measured effects of plasticizer content on the functional properties of extruded gelatin-based composite films. Moreover authors admitted that the solubility in water increases with increasing glycerol content. Also, amount of plasticizer influenced the barrier properties.

3.3 SUMMARY

With increasing population, the gross domestic product and the consumption are increasing. Only during the first decade of twenty-first century, more petroleum-based polymers were produced compared to these products during the entire last century. Average, 80 million tons of plastic waste is generated annually, including approximately 500 billion nondegradable bags are used worldwide. Virtually every piece of plastic material, that was ever made, still exists in some shape or form (with the exception of the small amount that has been incinerated). On the second hand, the amount of landfills increases because the decomposition process of plastic takes more than 100 years and the Earth area still is the same. Moreover price of plastic is required by several countries belong to OPEC. The brilliant denouncement of monopoly for fossil fuels and reducing mass of 100 year nondegradable trash can be organic plant and animals sources, as polysaccharides, proteins and fats. These biomaterials are most suitable, abundant, renewable and low-cost materials. They are able to replace traditional plastic packaging as biodegradable and/or edible coatings and films. It is estimated that 1/3 of plastic waste can be replaced by biopolymers.

KEYWORDS

- barrier properties
- biodegradable films
- biopolymers
- composites

- edible coatings
- environmental problems
- food packaging
- laminates
- lipids,
- mechanical properties
- plastic
- plasticizers
- polysaccharides
- proteins
- saccharides

REFERENCES

1. Berlanstein, L. R. (1992). *The Industrial Revolution and Work in Nineteenth-Century Europe*. London and New York: Routledge.
2. Griffin, E. (2010). *Short History of the British Industrial Revolution*. Palgrave.
3. http://blog.michalgosk.com/opakowania/historia-opakowan-czesc-1/ (in Polish).
4. http://blog.michalgosk.com/opakowania/historia-opakowan-czesc-2/ (in Polish).
5. Kyoto Protocol – 2015. In: *Environmental Health News; 2015;* http://www.environmentalhealthnews.org/
6. Tavassoli-Kafrani, E., Shekarchizadeh, H., & Masoudpour-Behabadi, M. (2016). Development of edible films and coatings from alginates and carrageenans. *Carbohydrate Polymers, 137*, 360–374.
7. Debeaufort, F., Martin-Polo, M., & Voilley, A. (1993). Polarity, homogeneity and structure affect water vapor permeability of model edible films. Polarity and structure affect water. *Journal of Food Science, 58*, 426–429.
8. Debeaufort, F. A., & Voilley, A. (1995). Effect of surfactants and drying rate on barrier properties of emulsified films. *International Journal of Food Science and Technology, 30*, 183–190.
9. Han, J. H., & Gennadios, A. (2005). Edible films and coatings: a review. In: J. H. Han (Ed.), *Innovations in Food Packaging*. Elsevier Academic Press, San Diego.
10. Laufer, G., Kirkland, Ch., Cain, A. A., & Grunlan, J. C. (2013). Oxygen barrier of multilayer thin films comprised of polysaccharides and clay. *Carbohydrate Polymers, 95*, 299–302.
11. Basiak, E., Debeaufort, F., & Lenart, A. (2015). Effect of oil lamination and starch content on structural and functional properties of starch-based films. *Food Chemistry*, doj: 10.1016/j.foodchem.2015.04.098.

12. Guerrero, P., Garrido, T., Leceta, I., & de la Caba, K. (2013). Films based on proteins and polysaccharides: Preparation and physical–chemical characterization. *European Polymer Journal, 49*(11), 3713–3721.
13. Arancibia, M., Giménez, B., López-Caballero, M. E., Gómez-Guillén M. C., & Montero, P. (2014). Release of cinnamon essential oil from polysaccharide bilayer films and its use for microbial growth inhibition in chilled shrimps. *LWT – Food Science and Technology, 59*(1), 989–995.
14. Maftoonazad, N., Ramaswamy, H. S., Moalemiyan, M., & Kushalappa, A. C. (2007). Effect of pectin-based edible emulsion coating on changes in quality of avocado exposed to Lasiodiplodiatheobromae infection. *Carbohydrate Polymers, 68*, 341–349.
15. Janjarasskul, T., & Krochta, J. M. (2010). Edible packaging materials. *Annual Review of Food Science and Technology, 1*, 415–448.
16. Janjarasskul, T., Rauch, D. J., McCarthy, K. L., & Krochta, J. M. (2014). Barrier and tensile properties of whey protein–candelilla wax film/sheet. *LWT – Food Science and Technology, 56*, 377–382.
17. Allen, L., Nelson, A. I., Steinberg, M. P., & McGill, J. N. (1963). Edible corn-carbohydrate food coatings. II. Evaluation of fresh meat products. *Food Technology, 17*(11), 104–108.
18. Cerqueira, M. A., Souza, B. W. S., Teixeira, J. A., & Vicente, A. A. (2012). Effect of glycerol and corn oil on physicochemical properties of polysaccharide films – A comparative study. *Food Hydrocolloid, 27*, 175–184.
19. Jost, V., & Langowski, H. Ch. (2015). Effect of different plasticizers on the mechanical and barrier properties of extruded cast PHBV films. *European Polymer Journal, 68*, 302–312.
20. NurHanani, Z. A., McNamara, J., Roos, Y. H., & Kerry, J. P. (2013). Effect of plasticizer content on the functional properties of extruded gelatin-based composite films. *Food Hydrocolloids, 31*, 264–269.

CHAPTER 4

THERMAL PROCESSING IN FOOD TECHNOLOGY: LATEST TRENDS

MAHUYA HOM CHOUDHURY

CONTENTS

4.1   INTRODUCTION

Thermal processing is the combination of temperature and time required to eliminate desired number of microorganisms from food products. It is an important facet to be addressed in recent years in Food Technology frontier. In recent years, extrusion technology has been emerged as an important area in Food processing arena and made an impact on the availability and variety of food products. Food extrusion is one of these latest multidimensional food-processing techniques. Great possibilities are offered in food processing field by the use of extrusion technology to modify physicochemical properties of food components. The extruded food, besides its preserved and frequently even enhanced biological value, can be characterized by physicochemical properties superior to the original raw material.

Extrusion cooking is defined as a unique tool to introduce the thermal and mechanical energy to food ingredients, forcing the basic components of the ingredients, such as starch and protein, to undergo chemical and physical changes. Extrusion combines several unit operations including mixing, cooking, kneading, shearing, shaping and forming so it is a highly versatile unit operation that can be applied to a variety of food processes. Extrusion provided the means of producing new and creative foods. One major advantage of extrusion cooking is the capability to produce a wide range of finished products with minimum processing times and by using inexpensive raw material [40].

Food extrusion has a great socioeconomic impact in today's technology frontier. The technology is also useful for the recycling of industrial and restaurant food wastes with other advantages. Recycling waste food materials into useful by-products comprising the steps of drying by dehydrating for a time interval necessary to reduce the moisture content to less than 25%, extruding food materials in an oxygen-free atmosphere at an elevated temperature level sufficient to sterilize food materials, cooling food materials, and thereafter tumbling and drying food materials to reduce said food materials to particle form.

The variety of snacks being produced on food extruder continues to expand. Recent applications have gone far beyond the forming, cooking, expansion and texturizing of cereal and vegetable protein ingredients. The new applications often use the extruder's ability to convey and heat viscous material. An example would be the continuous production of candies either through the anhydrous melting of sucrose or continuous cooking of sugar solutions. In many cases, twin-screw extruders are used for these applications considering the point of energy edition and pressure profile to prevent boiling within the machine.

To increase the efficiency of extrusion system, greater attention is placed in operating them at maximum capacity while using a minimum amount of energy for the entire process. The optimal set of operating conditions is site specific due to difference in the cost of various forms of energy and money. Extreme cost found with wetter extrusion where steam addition to the feed mixture in a form of preconditioner or through the extruder barrel add most of the energy in the form of latent heat, thereby reducing the electrical energy required to turn the screw however increasing needs for drying capacity and energy. The other is a drier extrusion with increased electrical energy requirement to turn the screw because the dough is quiet viscous, making it the major source of heat to the system. Under dry extrusion condition, system drying requirements and related energy are reduced. Optimal condition falls somewhere between these two extremes.

Traditionally, food products were developed for taste, appearance, value and convenience of the consumer. The development of products to confer a health benefits is a relatively new trend, and recognize the growing acceptance of the role of diet in disease prevention and treatment. This change in motivation for product development has moved organization and companies involved in formulating foods for health benefit into new areas of understanding, like health risk, risk/benefit analysis, evaluation of efficacy, toxicity and health regulation.

Introduction of twin-screw extruder thus widened the scope of food extrusion technology for production of many cereal based products including ready to eat breakfast cereals, infant food formulations, snack food and modified starch. Extruded product exhibited better nutritional value than other traditional cooked products [5, 45].

The importance of producing new generation snacks with improved nutritional characteristics has been recognized as recent trends in food science research [1, 36, 45]. To improve nutritional values, protein, carbohydrate and medicinal antioxidant could be used in snack formulation for extrusion.

Complete and balanced diets depend on formulation by mixing various ingredients themselves. Commercial snacks can be classified into four types of diets viz. dry, semimoist, moist and snacks. Dry snacks comprise the largest segment of the amount and value of snacks sold worldwide [31]. Reducing the moisture content of snacks to low level provides good quality. In addition, the low water content provides an optimal self-stability during storage (extra moisture aids microbial development) and transportation (costly for transport of water). In moist foods, microbial growth can develop if the foods are not processed and/or stored correctly. This can result in spoilage and in development of toxins.

This chapter focuses on latest technological trends in thermal processing of foods and food products. This study tries to optimize the rice and low value shrimp extrusion process and characterize its related functionality through value addition. The protein glass transition was also analyzed to evaluate the product quality. The phase behavior of a complex shrimp protein and carbohydrate extruded product was also studied and represented on a state diagram.

4.2   EXTRUSION PROCESS FOR SNACK PRODUCTION

Principally, important indicators of extrusion processes include system variables (specific mechanical energy, torque, die pressure and product temperature) as well as product characteristics (degree of expansion, starch gelatinization, shear strength, water solubility index, color, etc.) of the extrudate [32].

Thus, extrusion systems for the production of multidimensional snacks are efficient, economical in operation and produce variety products with marketing flexibility due to long shelf life and high bulk density prior to frying or puffing.

In snack production, extrudates are prepared from raw materials such as fish, meat, cereals and other vegetable products in various shapes and sizes. The raw materials can be preconditioned in various ways before extrusion cooking (Figures 4.1 and 4.2). The first phase is to mix the various raw materials and add water in definite proportion. Grinding is conducted to achieve a uniform particle size distribution which promotes a uniform moisture uptake by all particles. The uniformity of the mixture prior to extrusion ensures that each particle will be adequately and uniformly cooked. This results in better appearance and palatability in snacks [39]. In the next phase, the material is processed through an extruder and several extrusion processes and types of devices have been developed for this purpose. After extrusion, extrudates are usually dried to obtain the desired consistency and storage properties. Extrudates are often sprayed with a coating, in order to add essence/spicy substances to the extrudate. In addition, sometimes energy yielding compounds are added like lipids, and sometimes the purpose is to improve the surface structure. In snack production, fat and oil based suspensions are often used as a coating. During the coating process, fat is applied to the hot, dried extrudates and fat must be absorbed rapidly and completely before cooling.

Physical characteristics of extrudates reflect the effectiveness of the process. One can also derive from this the suitability of ingredients for extrusion. Food/feed extrusion studies, therefore, have been carried out for a number of decades. However, methods of characterizing raw materials and evaluation of extrudates are not standardized as far as process variables are concerned. Thus the present study tries to optimize the rice and low value shrimp extrusion process and characterize its related functionality through value addition.

A food extruder consists of a screw which rotates in a tightly fitting cylindrical barrel. Raw ingredients are preground and blended before being put in the feeding system of the extruder (Figure 4.1). The action of the flights on the screw pushes the processing products forward. In this way, the constituents are mixed into a viscous dough-like mass [43]. Common components of an extrusion system are shown in Figure 4.2. As the material moves through the extruder, the pressure within the barrel increases due to a restriction at the discharge of the barrel [23]. This restriction is caused by one or more orifices or shaped openings called a die. Discharge pressures typically vary between 3 and 6 MPa [23]. When the flights on the screw of a feed extruder are filled, the product is subjected to high shear rate as it is conveyed and flowed forward by the rotation of the screw. These high shear rate areas tend to align long molecules in the product and this gives rise to cross-linking or restructuring. This gives the extruded food its typical and characteristic texture.

There are two types of extruders: single screw and twin-screw extruders [24] and each type has a specific range of application. Each type of extruder has its own unique operation conditions, along with advantages and disadvantages. Choosing the proper extruder configuration is crucial for successful extrusion. The choice depends on the type of raw material to be used, the desired product, the processing rate and other factors.

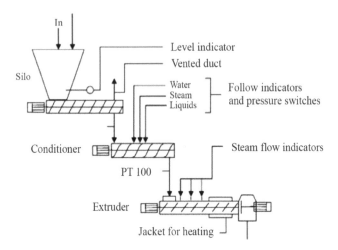

FIGURE 4.1   Extrusion cooking system.

FIGURE 4.2   Stages of Extrusion process.

With twin-screw technology, however, fat level can be higher, up to 25%. This allows twin-screw extruders to be used in processing raw animal by-products, in which the moisture content ranges from 500–900 g/kg [18], into snacks. Consistency of high fat products is easy to maintain with a twin-screw extruder.

During extrusion, mechanical shear forces and heating are applied simultaneously. This may increase the nutritional and physical quality characteristics of the product. Extrusion is used to produce a wide variety of snacks over a wide range of its processing variables such as moisture, shear, pressure, time and temperature.

Twin-screw extruders are classified according to their geometrical configuration. The main distinction is made between intermeshing and nonintermeshing extruders. Another distinguishing characteristic is the sense of rotation. The most important characteristics of the various twin-screw extruders are examined, with particular emphasis on the effect of screw geometry on the conveying characteristics. The corotating extruder appears to be best suited for melt blending operations, while the counter rotating extruder seems to be preferred in operations where solid fillers have to be dispersed in a polymer matrix. The schematic presentation of mixing mechanisms is illustrated in theoretical analysis section.

4.3 MECHANISM OF EXTRUSION PROCESS AND ITS VARIABLES

Many researchers attempted to model extrudate expansion mostly from the perspective of the influence of material and operational variables [2]. Most of the individual effects of these variables on expansion are, in general, consistent throughout the literature, discrepancies are frequently found due to the complex interactions between material and operational variables. It is therefore crucial to have a basic phenomenological understanding of the complex mechanism that governs expansion of cereal matrices, which incorporates both material properties and processing parameters.

While significant work has been done on developing quantitative mechanisms for the extrusion expansion of synthetic polymers (that is, formation of polymer foams), less work is available on the expansion of food biopolymers. This is mainly due to the complexity of such systems, which undergo continuous transformations during extrusion. A few studies have taken the challenge of modeling extrusion expansion of foods, most of them using simplifying assumptions in developing their models. Alvarez-Martinez et al. [2] developed an extrudate expansion model based on the dough viscosity model of Harper [23, 24]. Other investigators [15, 16, 29, 30, 41] have proposed more fundamental models, focusing on various aspects of expansion.

Kokini et al. [30] approached the various phases involved in extrudate expansion and illustrated the expansion mechanism on the diagram shown in Figure 4.3. The proposed expansion mechanism includes the following five major steps: Order-disorder transformations, nucleation, extrudate swell, bubble growth, and bubble collapse. First, the high shear, pressure, and temperature inside the extruder allow the transformation of cereal flours into viscoelastic melts. The degree of transformation is highly dependent on the extrusion moisture content and extruder operating variables.

Small air bubbles or impurities were entrapped during the nucleation of bubbles of the extrusion process or in the "holes" that represent the free volume of the polymer. The latter assumption could not be yet verified experimentally. These bubbles grow as the melt leaves the extruder die due to a moisture flash-off process, when the high pressure of the super-heated steam generated by moisture vaporization at nuclei overcomes the mechanical resistance of the viscoelastic melt. The bubble growth ceases upon cooling, when the viscoelastic matrix becomes glassy and no longer allows expansion to take place. These steps are discussed in the following subsections.

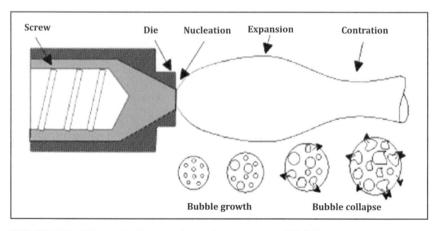

FIGURE 4.3   Schematic diagram of extrudate expansion [29, 30].

4.3.1 ORDER – DISORDER TRANSFORMATIONS DURING STARCH EXTRUSION

This step occurs inside the extruder and consists of transforming the semicrystalline starchy raw materials into a cohesive viscoelastic mass. As the predominant component of cereal flours, starch plays a major role in extrusion in general and extrudate expansion in particular. Proteins and other minor ingredients (i.e., fats or sugars) have their own contribution to expansion, which must not be overlooked.

4.3.2 MOLECULAR ORGANIZATION OF STARCH

A clear understanding of the transformations that take place during extrusion and expansion of starchy materials requires a good understanding of the macromolecular organization and properties of starch. This food biopolymer consists of two major molecular components, amylose and amylopectin, which have molecular weights in the range of 10^5 to 10^6 and 10^7 to 10^8 Da, respectively. Amylose is a long and linear polymer, while amylopectin has a highly branched molecular structure. Amylopectin is considered responsible for the structural organization of the starch granule [19] and its semicrystalline character [20]. Expansion process are the formation of microscopic pores of about 0.5 m in diameter called hilums, localized near the center of the starch granules. The hilums are considered by various researchers to be some of the nuclei at which expansion of vapor bubbles starts [9, 27, 28, 41].

The surface pores and interior channels are believed to be naturally occurring features of the starch granule structure, the pores being the external openings of the interior channels [17, 20]. The information concerning the internal organization of the starch granule is summarized in the diagram shown in Figure 4.4. It is likely that such pores could also serve as nucleation sites for expansion, although at this moment there is no experimental data to prove this hypothesis.

4.3.3 STARCH TRANSFORMATIONS DURING EXTRUSION

The severe conditions encountered during food extrusion cause various degrees of granular and molecular changes in starch [8–11, 13, 30].

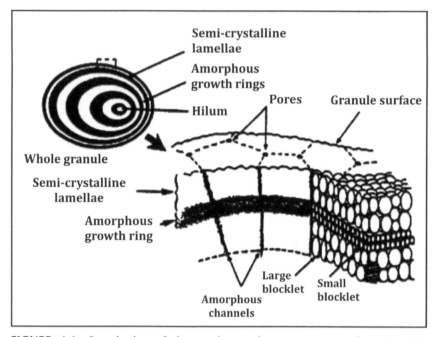

FIGURE 4.4 Organization of the starch granule structure presenting alternating crystalline (hard) and semicrystalline (soft) shells.

When starch is subjected to heat in presence of water, in a temperature range characteristic of the starch source, its native structure is disrupted and gelatinization takes place [26, 34, 38, 46]. Starch gelatinization is a sequential process, which includes the diffusion of water into the starch granule, followed by water uptake by the amorphous region, hydration and radial swelling of the granules, loss of optical bi-refringence, uptake of heat, loss of crystalline order, uncoiling and dissociation of double helices in the crystalline regions, and amylose leaching [26]. Emphasizing the importance of the loss of crystal- line order in starch during the gelatinization process, Parker et al. [38] have presented the differences in gelatinization and melting behavior of different polymorphic forms of starch. They demonstrated that the melting temperature of the amylopectin component of starch decreases with increasing water content until a critical moisture is reached; a further increase in water content does not lower further the gelatinization temperature, which instead reaches a plateau. When there is not sufficient water to complete gelatinization, the remaining crystallites simply melt upon heating, the coexistence of both gelatinized and melted starch in cooked starch at extrusion moisture contents [13, 26].

Kokini et al. [30] studied the effect of extrusion temperature and moisture content on the conversion of amylose-rich starch and amylopectin-rich starch, and found significant differences between the two starches. For instance, at 30% moisture and 200 rpm, 98% amylopectin starch had significantly higher degree of gelatinization as compared to 70% amylose starch, particularly at the lower temperatures used in the study (120°C and 135°C). Parker et al. [38] have presented the differences in gelatinization and melting behavior of different polymorphic forms of starch

4.3.4 NUCLEATION

The gas cell structure of expanded extrudates seems to be directly related to the number of bubbles nucleated in the starchy melt. Understanding the formation of extrudates' cellular matrix requires a good understanding of the nucleation phenomenon, which is responsible for the formation of individual bubbles that give rise to this structure. Shafi et al. [44] concluded that bubble nucleation and growth affect the cell size distribution in polymer foams. Despite the importance of this step in extrudate expansion, a relatively small number of studies on nucleation phenomena during extrusion of food biopolymers are available. One obvious reason is the complexity of such systems and the continuous transformations occurred during extrusion. This is why any nucleation model has to use simplifying assumptions in developing quantitative predictions of nucleation rates in extrusion of biopolymeric melts.

Kokini et al. [30] used the bubble expansion concept of Amon [3], originally developed for soap bubbles, to understand the factors that control nucleation and expansion. At saturation, the pressures in the liquid and vapor in the immediate vicinity of a vapor bubble interface are equal and at equilibrium. The force equilibrium established at the interface of the spherical vapor/gas bubble in a liquid medium can be expressed by the Laplace equation Nucleation of bubbles consists of the formation of small, thermodynamically unstable, gaseous embryos within the liquid metastable phase. Once an embryo reaches a critical size, it grows spontaneously into a stable and permanent bubble, called nucleus [9].

As superheating increases, (PV − PL) becomes larger and thus the nucleus radius (R) becomes smaller, approaching molecular dimensions at very high degrees of superheat.

Embryos can form as a result of either homogeneous or heterogeneous nucleation. Homogeneous nucleation is caused by localized thermal and density fluctuations of the metastable liquid phase, which lead to the formation of molecule clusters with vapor-like energies. Typically, the number of heterogeneous nucleation sites is less than that of homogeneous nucleation sites [12]. Guy et al. [21] noted that increases in temperature of an extruded melt did not increase the number of gas cells found in expanded extrudates, which would happen if homogeneous nucleation were the dominating mechanism.

4.3.5   EXTRUDATE SWELL

As the starchy polymer melt emerges from the die and starts cooling down, the contribution of elastic forces to the rheology of the melt increases significantly, which gives rise to the phenomenon of die swelling [6, 37]. Qualitatively speaking, elasticity affects extrudate diametric expansion. Hayter et al. [25] have shown that an increase in the die opening size results in decreased expansion and increased bulk density, which they attributed both to a reduction in the normal forces at the die (that is, reduced swelling) and to a higher viscosity caused by lower shear rate and temperature. Guy et al. [21] correlated the extrudate length/diameter ratio and the bubble number density, and suggested that *die swell* controls the overall expansion of the extrudate, especially at small number density of gas bubbles (approx. 600 g/l); as the number density of bubbles increases, the effect of die swell decreases. Della Valle et al. [15] suggested that as the number of bubbles increases, longitudinal expansion is favored and thus elastic properties may actually affect expansion through the elongational viscosity, which is the variable included in the bubble growth models. Unlike other existing model, they suggested that the elongational strain rate is actually very high due to fast expansion [15].

Die swell is relatively well characterized for synthetic polymers, but data is scarce for biopolymers. One method that can be used to measure elastic properties (that is, first normal stress difference) is slit die rheometry [14, 30]. In the case of starchy matrices, additional challenges such as water vaporization, under developed flow at the die outlet and starch modifications during the measurement may cloud the measurement of elastic properties slit die rheometry [6, 15]. Chang et al. [6] used the exit pressure method for the estimation of the first normal stress difference during starch extrusion. Since the data obtained at material temperatures >140°C was erratic, only values obtained at temperatures <120°C were used for developing first normal stress difference models.

4.3.6 BUBBLE GROWTH

Extrudate expansion is governed by the biaxial extension of individual bubbles, and the driving force for bubble growth is the pressure difference between the inside and the exterior of the matrix [16, 29, 30]. It is generally accepted that surface tension has an overall negligible effect on expansion of polymer melts [16] with little effect on initial expansion. The rheological properties of the polymeric matrix have the leading role in expansion, since they determine the resistance of the bubble wall to the pressure difference between the inside and the outside of the bubble. Fan [16] and Kokini et al. [30] considered vapor bubble growth in a viscous fluid and correlated the specific volume of the extrudate with the ratio between vapor pressure and melt viscosity.

4.3.7 BUBBLE COLLAPSE

When the bubble wall can no longer withstand the pressure inside the bubble, it collapses, especially at high moisture contents – typically above 20%. This phenomenon is also related to the extrusion temperature and material rheological properties.

An extruder is a machine that manufactures extrudate, a common form of commercial dry foods. This technology of snack processing forces the mixture of the ingredients through a spiral screw and then through the die of the extruder. During extrusion processing, the ingredients are ground, mixed and heat-treated. As a result, the extrudate is produced and dried afterwards.

4.4   ADVANTAGE OF TWIN-SCREW EXTRUSION PROCESS OVER SINGLE SCREW

Single screw extruder is like a friction pump. The plasticized food product wets the wall of the pump inner cylinder. The moving surface drags the fluid. The mechanical efficiency is low due to the dissipation of a large part of the power supplied by the shaft. Single screw extruder is thereby eminently suitable for process in which the medium being transported must be heated. The drag flow effect can be realized in a single and twin-screw extruder. The main difference between twin-screw and single screw extruder is in the mechanism of conveying. In a *single screw extruder*, the transport results from the difference in the frictional and viscous forces at the contact location of screw/product and barrel/product. In a *twin-screw extruder* with intermeshing screws the product is constrained and physically prevented from rotating with the screw. Now, the friction is less important for twin-screw extruder than in the single screw extruder, although screw geometry can have some influence.

In practice, there is considerable difference between theoretical and experimental throughput. This is caused by slip phenomenon. For food products contain a certain amount of water, there always exists the possibility that a film of water is built up upon the barrel inside or outside of the screw on which the bulk of the product will slide or slip. It is also possible that the slip layer is formed by oil or other unmixed phases.

In the case of wall slip, one may try to prevent the slip by grooving the barrel wall. This means a considerable increase of the pitch in practical circumstances for most food materials in single screw extruder in order to overcome the negative effect of the slip. This situation for twin-screw extruder is completely different, caused by the effect of intermeshing screws imply that more or less close C-shaped chambers to a certain extend filled up with product, are conveyed positively from feed zone to the die head. This means that in principle slip at the barrel wall becomes irrelevant.

4.4.1   TEMPERATURE AND HEATING EFFECTS

For a single screw extruder the temperature rise can be calculated taking into account all the restrictions made for this deviation. The overall impression that a single screw extruder causes the temperature rise by dissipation of motor power holds for a large group of materials with respect to the coefficient of internal friction (paste) or to the developed viscosity η_{app}. It is possible to make a good use of combined heating which means the use of barrel heaters extra upon the heat generated by the dissipation of mechanical power.

For a twin-screw extruder most of the heat is taken up by the product by means of the external (barrel) heating. It can be concluded that the different operational characteristics of single and twin-screw extruders lead to tailor made calculation methods for the calculation of total share, total heat input and total throughput.

Thus, in twin-screw, a single step mixing and extrusion process can make the production route more efficient and improve productivity. Narrow residence time distribution also results in uniform product quality and continuous operation coupled with the low volume of the mixing chamber lead to reduced inventories of work in progress. The 'clamshell' barrel design also allows easy access to the agitators and interior barrel surfaces for cleaning. Batch to batch variation is eliminated by continuous production. Not only that, elimination of intermediate steps such as drying reduces capital cost, space requirements and energy consumption. Elimination of organic solvents used in coating processes results in cost savings and reduces environmental impact.

4.5 MIXING MECHANISM OF TWIN-SCREW MIXERS

Mixing may be used for a number of different reasons, which include blending of ingredients, facilitation of chemical reactions, addition of energy in order to create or break molecular bonds, incorporation of air, etc. Mixing of highly viscous materials is of particular interest in food processing. Wheat flour dough is an excellent example of a highly viscous food material that requires optimal mixing. The mixing of wheat flour dough has three main purposes. The first is to distribute the ingredients homogeneously throughout the mixture. In particular, water must be thoroughly distributed and the flour particles must be hydrated and broken apart so that the protein and starch is released to form a mobile phase and allow gluten formation. The second purpose is dough development.

The current understanding focuses the application of mechanical energy to stretch the long molecules, in particular glutenin, from an unperturbed state to a more extended configuration that allows the molecules to align and form noncovalent bonds, thus imparting elasticity to dough that gives its machinability and gas retention properties. The third function of mixing is the entrainment of air. Air bubbles incorporated into the dough during mixing form the nuclei for gas cells which later expand as carbon dioxide from the fermentation process diffuses into them. The effectiveness of a dough mixer can therefore be evaluated by looking at the ability of the mixer to distribute ingredients and stretch the dough molecules, as well as disperse air bubbles and cohesive clumps.

4.5.1 MIXING IN THE INTERMESHING CO-ROTATING TWIN-SCREW EXTRUDER

In the intermeshing corotating twin-screw extruder, mixing mostly takes place in the kneading section due to shear and/or elongation. The dispersive mixing performance of the kneading elements in the intermeshing coro-tating twin-screw extruder has been studied only by a few investigators. The minor component in many blends is the dispersed phase of a blend (drops or filaments) mixed in a continuous phase of the major component. Heating, deformation and break-up of the dispersed phase occur during mixing. An elementary step in the mixing process is the deformation of a dispersed drop in a flow field. An increase in the interfacial area between the two components is accompanied by a decrease in local dimensions perpendicular to the flow direction (the striation thickness).

Velocity profiles determination in the mixers is an important phenomenon to analyze the mixing capability of mixer. Dispersive mixing is evaluated by calculating the stress and the mixing indexes [7] from the rate of deformation and vortices tensors (Figure 4.5).

In order to calculate stretching efficiencies and the distributive mixing measures, it is first necessary to follow the trajectories of a large number of material points. The material points are initially randomly distributed either in the 0.5 x 0.5 cm boxes with 100 points as shown in Figures 4.6a and 4.7a or throughout the flow domain with 1500 points in the twin-screw geometry. Figures 4.6b and 4.7b show the dispersion after 10 revolutions.

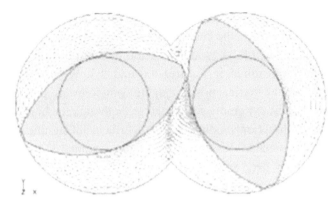

FIGURE 4.5 Velocity vectors showing direction of the flow and colored by the magnitude of the velocity (cm/s) in twin-screw mixer after the paddles have turned clockwise 67.5° from the initial position.

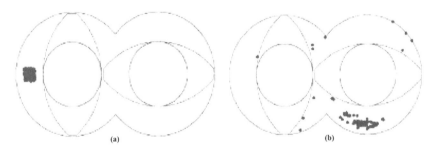

FIGURE 4.6 Particle tracking results showing the distributive mixing of a cluster (b1) of noncohesive, material points initially centered in the leftmost section of the flow field (a) initially, and (b) after 10 revolutions.

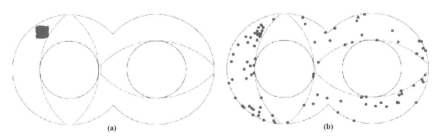

FIGURE 4.7 Particle tracking results showing the distributive mixing of a cluster (b) of noncohesive, material points initially located just behind the paddle in the leftmost section of the flow field (a) initially, and (b) after 10 revolutions.

The concentrations of particles with an initial y coordinate >0 are arbitrarily set to a value of 1, while the concentrations of the rest are set to zero.

The velocity vectors showing direction at one position in the twin-screw mixer are featured in Figures 4.6 and 4.7. When the blades are rotated 45° from the vertical position in the twin-screw mixer (Figure 4.8), the velocity field is extremely symmetric even though the blades are moving in opposite directions, indicating that inertia is not significantly affecting the flow. The material is flowing from one chamber to the other, with a small temporary dead spot where the direction of flow changes. As the blades continue to move clockwise, the dead spot gets smaller and begins to shift from in front to behind the left blade.

When the left blade is horizontal and the right blade is vertical, the material is squeezed upward through the gap between the blades due to the extreme pressure differential. The velocity in the center of the flow

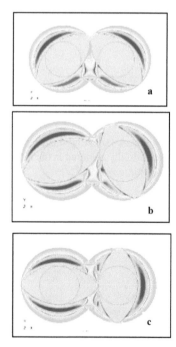

FIGURE 4.8 Mapping of the mixing index (λ_{max}) in the twin-screw mixer at positions (a) 45°, (b) 67.5°, and (c) 90° from the initial position where a value of 0 indicates pure rotation, a value of 0.5 indicates shear flow and a value of 1 indicates pure elongation.

domain both above and below the left blade is much higher than the blade speed at that radius, indicating a considerable amount of material is being forced through the gap as the lower flow chamber contracts. A mapping of the inertial reference frame mixing index is shown in Figure 4.8. Twin-screw mixers use primarily shearing mechanism. The regions of the twin-screw mixer, that are closed off from the paddle interaction region, are explained in Figure 4.8.

The Figure 4.9 shows that the shear stresses are high throughout the intermeshing region, where the mixing paddles interact in the twin-screw mixer and are combined with elongational flow, especially near the tip of the paddle and in the gap between the paddles.

The Figure 4.10 indicates results for particle tracking showing the distributive mixing between the upper and lower halves of twin-screw mixers. The mixing mechanism helps in better understanding the process and predicting the observations.

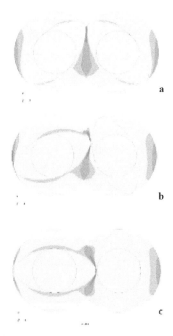

FIGURE 4.9   Mapping of the shear stress (τ_{12}) distributions where the units of stress are g/cm² in the twin-screw mixer at positions (a) 45°, (b) 67.5°, and (c) 90° from the initial position.

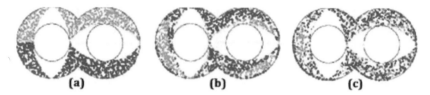

FIGURE 4.10 Particle tracking results illustrating the distributive mixing between the upper and lower halves of twin-screw mixers (a) initially, (b) after one revolution, and (c) after 10 revolutions where the red (light) colored points were arbitrarily assigned a concentration of 1 and the blue (dark) colored points a concentration of zero.

4.6 THERMAL ANALYSIS OF FOODS

Most foods are subjected to variations in their temperature during production, transport, storage, preparation and consumption, for example, pasteurization, sterilization, evaporation, cooking, freezing, chilling, etc. Temperature changes cause alterations in the physical and chemical properties of food components which influence the overall properties of the final product, for example, taste, appearance, texture and stability. Chemical reactions such as hydrolysis, oxidation or reduction may be promoted, or physical changes, such as evaporation, melting, crystallization, aggregation or gelation may occur. A better understanding of the influence of temperature on the properties of foods enables food manufacturers to optimize processing conditions and improve product quality. It is therefore important for food scientists to have analytical techniques to monitor the changes that occur in foods when their temperature varies. These techniques are often grouped under the general heading of thermal analysis. In principle, most analytical techniques can be used or easily adapted to monitor the temperature-dependent properties of foods: spectroscopic (NMR, UV-visible, IR spectroscopy, fluorescence), scattering (light, X-rays, neutrons), physical (mass, density, rheology, heat capacity) etc. Nevertheless at present, the term thermal analysis is usually reserved for a narrow range of techniques that measure changes in the physical properties of foods with temperature (TG/DTG, DTA, DSC and Transition temperature).

Carbohydrate-Protein mixtures are commonly used for the production of snack foods by extrusion. The properties and storage stability of such extrudate can be controlled if the phase behavior and the influence of intrinsic and external factor on this behavior are well understood. Most foods are mixtures of two types of biopolymer, protein and carbohydrates that in many cases are immiscible and retain their own glass transition. For such system, if a low molecular weight compound acts as plasticizer for each biopolymer, both glass transition temperatures are depressed [33]. The phase behavior of simple food systems, particularly starchy foods, has been extensively studied in the past. Yet, the majority of foods have a much more complex composition, with a protein – carbohydrate polymeric matrix that entraps low mass molecular weight components like fats, sugars and others. For such matrices, the interactions of proteins and carbohydrate with water, with the minor components, and with each other govern the structural property relationship.

In this chapter, the phase behavior of a complex shrimp protein and carbohydrate extruded product was studied and represented on a state diagram. The state diagram defines the different physical and chemical state in which the system exists as a function of moisture content and temperature [35], and can be used to predict the systems behavior during processing and storage [30].

Phase behavior of protein and carbohydrate extruded system are illustrated on a state diagram. Rice starch and protein were phase separated at macromolecular level and both retained their own phase transitions. The glass transition temperature of protein dictated the texture of the mixed system and starch contributing to the high value of mechanical properties. Water has a plasticizing effect on both biopolymers. At room temperature, extrudate with <13% moisture is glassy while at moisture >16% is rubbery.

One of the major factors that control the glass transition of food system is the water content. Phase transition behavior of rice shrimp mixture was studied at optimum temperature (150°C) at four different moisture condition (11, 13, 15, and 17%).

4.6.1 PHASE TRANSITIONS IN THE RICE – SHRIMP (CARBOHYDRATE – PROTEIN) EXTRUDATE

Glass transition temperature (T_g) was studied to evaluate the phase transition behavior of carbohydrate – protein complex food system. Glass transition temperature shows low even nil effect with the change of extrudate process temperature, whereas significant change of glass transition temperature with the change of process moisture was studied. Thus in the present study, the phase behavior at optimum temperature (150°C) and four different process moisture condition (11, 13, 15, and 17%) for the extruded system is represented. The T_g value of protein plays an important role in controlling extrudate shrinkage and cellular damage [1]. In the present study, protein glass transition was analyzed to evaluate the product quality.

The glass transition behavior of rice flour extrudate at 11% process moisture condition is represented in Figure 4.11. This figure shows inflection point at 56°C (55.649°C); onset temperature, e.g., the temperature at which inflection started was noted at 54°C (53.994°C) and C_p value at 0.218 J/g/°C. Thus glass transition temperature of rice flour was 56°C.

Baseline of DSC thermogram of rice and Chapra extrudate at 11% moisture is represented in Figure 4.12. Baseline in this figure shows two-inflection points (T_g value), which indicate two glass transition temperatures in carbohydrate-protein complex food system. Two glass transition temperatures corresponding to carbohydrate and protein system at 11% moisture conditions are clearly illustrated in Figures 4.13 and 4.14, respectively. The T_g value of extrudate at 11% moisture condition (from Figure 4.13) was 54.2°C, C_p value at 0.177 J/kg, and onset value at 53.3°C. The T_g value in Figure 4.13 resembles the T_g values of only rice sample (presented in Figure 4.11). The glass transition temperature (T_g value) of carbohydrate system in rice shrimp extrudate (i.e., Carbohydrate-protein mixture) in Figure 4.13 is marked as T_{g1}.

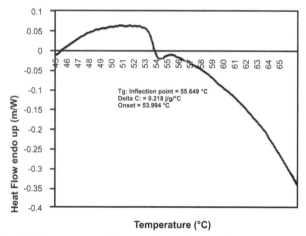

FIGURE 4.11   DSC thermogram of rice sample extruded at 150°C and 11% moisture content.

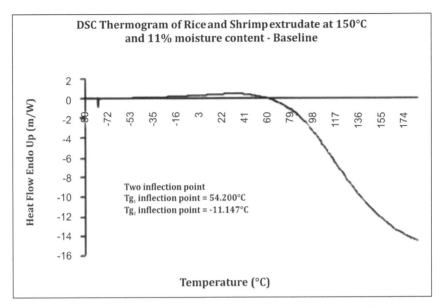

FIGURE 4.12 DSC thermogram of rice shrimp extrudate 150°C and 11% moisture content-baseline.

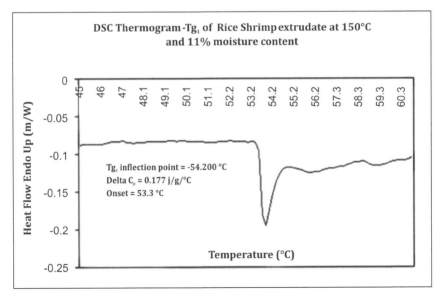

FIGURE 4.13 T_{g1} of rice shrimp extrudate at 150°C and 11% moisture content.

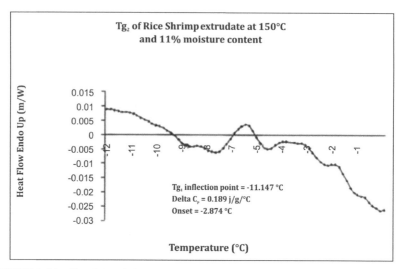

FIGURE 4.14 T_{g2} of rice shrimp extrudate at 150°C and 11% moisture content.

Thus, other T_g value of the extrudate obtained at 11% moisture content in Figure 4.14 indicates glass transition temperature of protein sample in rice-shrimp extrudate (T_{g2}). The Figure 4.14 shows that inflection point was at –11.147°C and C_p value was 0.189J/gand Onset value (inflection starting temperature) was at –2.874°C. Therefore, T_g analysis of carbohydrate and shrimp extrudate shows two separate T_g values which in turn indicate that phase behavior of carbohydrate and protein extrudate distinctly explains phase separation of two biopolymer at macromolecular level.

The glass transition temperatures for the extrudate processed at 13%, 15% and 17% moisture conditions and at optimum process temperature (150°C) are represented in Figures 4.15–4.17. The T_{g2} value at 13% process moisture condition (Figure 4.15) shows inflection point at –21°C (–21.150°C) for which C_p value was 3.531 J/g/°C, onset value was –19.269°C. The T_{g2} value at 15% moisture condition (Figure 4.16) shows inflection point at –36°C (–36.105°C) where inflection started at 19°C (–19.262°C) and C_p value was 0.204 J/g/°C. The T_{g2} value at 17% moisture condition (Figure 17) shows inflection point at –27°C (–26.700°C), where inflection started at 28°C (–27.743°C) and C_p value was 0.175 J/g/°C.

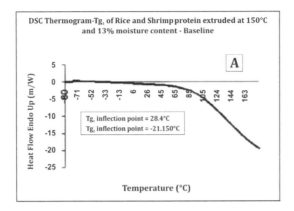

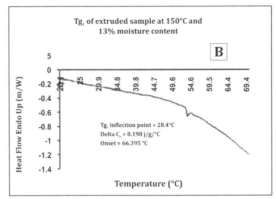

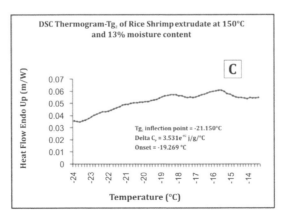

FIGURE 4.15   DSC thermogram – T_{g2} of rice shrimp extrudate at 150°C and 13% moisture content: (A) Baseline; (B) T_{g1} of rice shrimp extrudate at 150°C and 13% moisture content; and (C) T_{g2} of rice shrimp extrudate at 150°C and 13% moisture content.

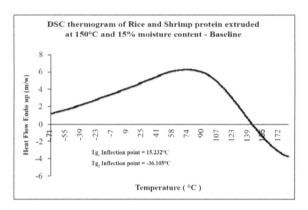

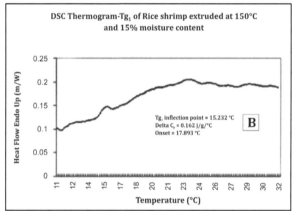

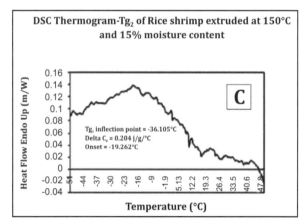

FIGURE 4.16 DSC thermogram – T_{g2} of rice shrimp extrudate at 150°C and 15% moisture content A) Baseline B) T_{g1} of rice shrimp extrudate at 130°C and 15% moisture content C) T_{g2} of rice shrimp extrudate at 130°C and 15% moisture content.

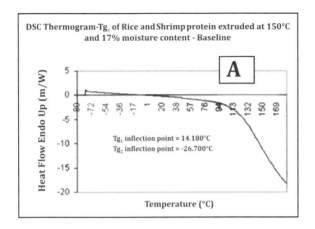

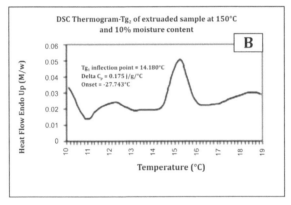

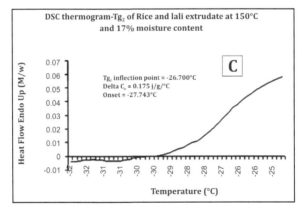

FIGURE 4.17   DSC thermogram – T_{g2} of rice shrimp extrudate at 150°C and 17% moisture content: (A) Baseline; (B) T_{g1} of rice shrimp extrudate at 150°C and 17% moisture content; (C) T_{g2} of rice shrimp extrudate at 150°C and 17% moisture content.

Very few T_g values have been reported in the literature for protein system. The T_g values reported in the literature for beef are also very contradictory ranging from −60 to −5° C [4] and for meat −23°C. In the present study, glass transition temperature varies from −11 to −36°C. Location of protein glass transition controls the physical state of product [36]. Thus, the T_{g2} values show decreasing trend with increase in moisture content from 11 to 15%, whereas slight increase in glass transition temperature was noted at 17%.

When the product temperature (T_p) is above T_g of 36°C (or the bubble temperature: highest glass transition temperature (T_g) where most expanded product can be obtained) [16], the cell wall viscosity is low enough for bubble growth or shrinkage. Differential Scanning Calorimetry (DSC) experiments indicated that T_g of Chapra (shrimp) protein at control formulation (with 15% moisture after which bubble starts to collapse), as it exited the die, was about (−36°C) while the barrel temperature was 150°C and product temperature (T_p) was maintained at 50°C. This caused the extrudates to shrink as the air inside the cells cooled down and contracted. The cells increased in size as long as T_p remained greater than T_g of 36°C. However, at a certain critical cell size, the cell walls reached their maximum extensibility (especially when barrel temperature was 170°C), beyond which they ruptured and the extrudate structure was damaged and increase in protein T_g value was also observed. When the oven temperature was 135°C, the structure was set before this critical cell size could be attained, thus no rupture of cells was observed whereas at 170°C the cell structure rupturing started. Here in this study also, all the process temperatures were so chosen that product temperature always shows higher value than that of protein glass transition temperature (Tg_2) to obtain product with less structural damage [1].

Thus, T_g value shows significant impact in the food product design. The location of the protein glass transition temperature (T_{g2}) controls the physical state of the product. The extrudates obtained at 11%, 13% and 17% moisture conditions show rubbery texture and lower expansion volume at room temperature compared to the product obtained at 15% moisture. This behavior is clearly explained from Tg_2 value at different process moisture condition. T_{g2} value at 11%, 13% and 17%

process moisture conditions is higher than that of the Tg_2 value at 15% process moisture condition and thus the difference with product temperature (50°C) is higher at 11%, 13% and 17% than that at 15% process moisture condition. As a result, product shows different characteristic than that obtained at 15% moisture (i.e., optimum process moisture condition).

These results are clearly explained in the present study, where it was estimated that $T_g + 36$°C values at 11%, 13% and 17% moisture conditions are 25°C, 15°C, and 9°C (as T_g value correspond to −11°C, −21°C, and −27°C, respectively at 11%, 13% and 17% moisture conditions), which is lower than product temperature (50°C) and higher than the $T_g + 36$°C value obtained at 15% process moisture condition (0°C) (as T_g value at 15% moisture condition was at −36°C). Thus, at 15% moisture condition, the difference of product temperature ($T_p − 50$°C) with $T_g + 36$°C value (i.e., 0°C) is highest and highest increase in cell size was observed which in turn predict highest expansion volume at that process moisture condition. Thus, extrudates obtained at 15% moisture condition are typical glassy polymers, while the extrudates obtained at lower and higher moisture condition become rubbery.

Upon direct examination by texture analyzer, the product obtained at 15% moisture condition were found brittle and glassy as also evident from lower shear force value (Table 4.1) And those obtained at 11%, 13% and 17% moisture become rubbery as also validated from high shear stress data presented in Table 4.2. Thus protein glass transition temperature has significant impact on food product design and process control.

4.7 SUMMARY

The present study represents microstructural and thermal behavior of the extrudate produced. Microstructural analysis performed by SEM in present study revealed symmetrical fashioned network like structure obtained at 150°C and 15% moisture. Phase behavior of studied complex extruded system indicated that the starch and proteins were immiscible and retained their own glass transition. The immiscibility of the system

TABLE 4.1 Effects of Different Extrusion Conditions on Shear Force

Assay	Process temperature (°C)	Feed moisture (%)	Shear force (N)
1.	120	11	11.49
2.	120	13	14.43
3.	120	15	14.74
4.	120	17	15.03
5.	120	19	15.87
6.	135	11	12.34
7.	135	13	15.52
8.	135	15	15.65
9.	135	17	18.74
10.	135	19	18.85
11.	150	11	12.76
12.	150	13	18.75
13.	150	15	19.55
14.	150	17	18.74
15.	150	19	18.67
16.	165	11	12.67
17.	165	13	18.76
18.	165	15	18.54
19.	165	17	18.06
20.	165	19	17.87

was clearly illustrated on thermogram. Immiscibility of the system in turn predicts no interaction between carbohydrate and protein part of the extrudate. This observation thus proves better stability, quality as well as digestibility of the product. Texture of matrix was dictated by the location of the protein phase glass transition. The starch phase played a major role in the extrudate texture through its contribution to the mechanical properties.

Thus the state diagram provides valuable information, which can be used as a predictive tool for food product design. It must also be concluded that the analysis and interpretation of the result is very much complicated by the multitude and interference of different transitions.

TABLE 4.2 Effects of Different Extrusion Conditions Shear Stress

Assay	Process temperature (°C)	Feed moisture (%)	Shear stress (N/m^2)
1.	120	11	3.751×10^5
2.	120	13	3.547×10^5
3.	120	15	3.437×10^5
4.	120	17	3.33×10^5
5.	120	19	3.13×10^5
6.	135	11	3.67×10^5
7.	135	13	3.45×10^5
8.	135	15	3.35×10^5
9.	135	17	3.25×10^5
10.	135	19	3.04×10^5
11.	150	11	3.33×10^5
12.	150	13	3.24×10^5
13.	150	15	3.13×10^5
14.	150	17	3.25×10^5
15.	150	19	3.26×10^5
16.	165	11	3.33×10^5
17.	165	13	3.45×10^5
18.	165	15	3.26×10^5
19.	165	17	3.28×10^5
20.	165	19	3.04×10^5

Phase transition behavior of carbohydrate-protein complex food system analysis shows immiscible characteristic of starch and proteins system and retention of their own glass transition. The immiscibility of the system was clearly illustrated on thermogram. Texture of matrix was dictated by the location of the protein phase glass transition. The starch phase played a major role in the extrudate texture through its contribution to the mechanical properties. Thus the state diagram provides valuable information, which can be used as predictive tools for food product design. Phase separation at macromolecular level shows no interaction which in turn useful information is regarding product acceptability in terms of development of undesirable flavor and odor.

ACKNOWLEDGMENTS

Author gratefully acknowledges the assistance of Dr. Suromita Bhattacharyya, Junior Research Fellow working in project G.I Activity, Patent Information Centre in formatting my work.

KEYWORDS

- bubble growth
- differential scanning calorimeter
- expansion
- extrudate
- extruder
- extrusion
- intermeshing
- nucleation
- phase transition
- preconditioning
- rheology
- scanning electron microscope
- shear force
- shear stress
- single screw extruder
- T_g
- thermogram
- twin-screw extruder
- velocity vector
- viscoelastic

REFERENCES

1. Alavi, S. H., Gogoi, B. K., Khan, M., Bowman, B. J., & Rizvi, S. S. H. (1999). Structural Properties of protein-stabilized starch-based supercritical fluid extrudates. *Food Res Intl., 32*, 107–118.

2. Alvarez-Martinez, L., Kondury K. P., Harper J. M. (1988). A general model for expansion of extruded products. *J Food Sci.*, *53*, 609–615.

3. Amon, M., & Denson, C. D. (1984). A study of the dynamics of foam growth: analysis of the growth of closely spaced spherical bubbles. *Polym. Eng. Sci.*, *24*, 1026–1034.

4. Brake, N. C. (1999). Glass transition values of muscle tissue. *J Food Sci.*, *64*(1), 10–15

5. Bressani, R., Sancher-Marroquin, A., & Morales, E. (1992). Chemical composition of grain amaranth cultivars and effects of processing on their nutritional quality. *Food Review Intl.*, *8*(1), 23–49.

6. Chang, C. N. (1992). *Study of the mechanism of starchy polymer extrudate expansion.* DPhil thesis for New Brunswick, N.J., Rutgers the State Univ. of New Jersey. 291 pp.

7. Cheng, J. J., & Manas-Zloczower, I. (1990). Flow field characterization in a banbury mixer. *Intl. Polym. Process.*, *5*(3), 178–183.

8. Cisneros, F. H., & Kokini, J. L. (2002a). A generalized theory linking barrel fill Energy and air bubble entrapment during extrusion of starch. *J Food Engr.*, *51*(2), 139–149.

9. Cisneros, F. H. (1999). Air bubble nucleation during starch extrusion. DPhil thesis for New Brunswick, N. J., Rutgers *The State Univ. of New Jersey.* 288 p.

10. Colonna, P., Doublier, J. L., Melcion, J. P., de Monredon, F., & Mercier, C. (1984). Extrusion cooking and drum drying of wheat starch, I: Physical and macromolecular modifications. *Cereal Chem.*, *61*, 538–543.

11. Colonna, P., & Mercier, C. (1983). Macromolecular modifications of manioc starch components by extrusion cooking with and without lipids. *Carbohydr. Polym.*, *3*, 87.

12. Colton, J. S., & Suh, N. P. (1987). Nucleation of microcellular foam: theory and practice. *Polymer Eng. Sci.*, *27*, 500.

13. Davidson, V. J., Paton, D., Diosady, L. L., & Laroque, G. (1984). Degradation of wheat starch in a single screw extruder: Characteristics of extruded starch polymers. *J Food Sci.*, *49*, 453.

14. Dealy, I. M., & Wissbrun, K. F. (1989). *Melt rheology and its role in plastics processing.* New York, N.Y.: *Van Nostrand Reinhold.* 680 pp.

15. Della, Valle G., Colonna, P., Patria, A., & Vergnes, B. (1996). Influence of amylose content on the viscous behavior of low hydrated molten starches. *J Rheol.*, *40*, 347–362.

16. Fan, J., Mitchell, J. R., & Blanshard, J. M. V. (1994). A computer simulation of the dynamics of bubble growth and shrinkage during extrudate expansion. *J Food Engr.*, *23*(3), 337–356

17. Fannon, J. E., Shull, J. M., & BeMiller, J. N. (1993). Interior channels of starch granules. *Cereal Chem.*, *70*, 611–613.

18. Ferket, P. (1991). Technological advances could make extrusion an economically feasible alternative to pelleting. Flavors ¾ generation, analysis, and process influence. *Proceedings of the 8th Intl Feedstuffs, 63*, 19–21

19. French, D. (1984). Organization of the starch granules. In: Whistler, R. L., BeMiller, J. N., & Paschall, E. F. (Eds.), *Starch: Chemistry and Technology*, 2nd Edition, Academic Press, New York, pp. 183–195.
20. Gallant, D. J., Bouchet, B., & Baldwin, P. M. (1997). Microscopy of starch: evidence of a new level of granule organization. *Carbohydr. Polym.*, *32*(3/4), 177–191.
21. Guy, R. C. E., & Horne, A. W. (1988) Extrusion and coextrusion of cereals. In: *Food Structure: Its Creation and Evaluation*, Blanshard, J. M. V., & Mitchel, J. R. (eds.), Butterworth Press, London, Chapter 18, pp. 331–349.
22. Hanna, M. A., Chinnaswamy, R., Gray, D. R., & Miladinov, V. D. (1997). Extrudates of starch-xanthan gum mixtures as affected by chemical agents and irradiation. *J Food Sci.*, *62*(4), 816–820.
23. Harper, J. M. (1978). Food extrusion. *Crit. Rev. Food Sci.*, *11*, 155–215.
24. Harper, J. M., & Jansen, G. R. (1981). Nutritious foods produced by low-cost technology. *LEC Report 10.*
25. Hayter, A. L., Smith, A. C., & Richmond, P. (1986). The physical properties of extruded food foams. *J Material Sci.*, *21*, 3729.
26. Hoover, R. (2001). Composition, molecular structure, and physicochemical properties of tuber and root starches: a review. *Carbohydrate Polymers*, *5*(3), 253–267.
27. Hoseney, R. C., Mason, W. R., Lai, C. S., & Guetzlaff, J. (1992). Factors affecting the viscosity and structure of extrusion-cooked wheat starch. In: Kokini, J. L., Ho, C. T., & Karwe, M. V. (eds.). *Food Extrusion Science and Technology*, New York, N.Y.: Marcel Dekker Inc., pp. 277–305.
28. Hoseney, R. C. (1985). The mixing phenomena. *Cereal Foods World*, *30*(7), 453–457.
29. Huang, H., & Kokini, J. L. (1993). Measurement of biaxial extensional viscosity of wheat flour doughs. *J Rheol.*, *37*(5), 879–891.
30. Kokini, J. L., Chang, C. N., & Lai, C. S. (1992). The role of rheological properties on extrudate expansion. In: *Food Extrusion Sci Technol.* New York, N.Y.: Marcel Dekker Inc., pp. 631–653.
31. Laxhuber, S. (1997). Drying and cooling in production of extrudates. *Kraftfutter*, *11*, 460–466.
32. Lin, S., Hseih, F., & Huff, H. E. (1997). Effects of lipids and processing conditions on degree of starch gelatinization of extruded dry pet food. *Lebens Wiss Technol.*, *30*, 754–761.
33. Matveev, Y. I. Grinberg, V. Y., & Tolstrouzov, V. B. (2000.) The plasticizing effect of water on proteins, polysaccharides and their mixtures: Glassy state of biopolymers food and seeds. *Food Hydrocolloids, 14*, 425–437.
34. Mitchell, J. R., & Areas, J. A. G. (1992). Structural changes in biopolymers during extrusion. *In: Food Extrusion Sci Technol.*, New York, N.Y.: Marcel Dekker.
35. Morales, A., & Kokini, J. L. (1999). State diagrams of soy globulins. *J Rheol.*, *43*(2), 315–325.
36. Moraru, C. I., & Kokini, J. L. (2003) Nucleation and expansion during extrusion and microwave heating of cereal foods, comprehensive reviews. In: *Food Science and Food Safety, 2*, 120–138.
37. Padmanabhan, M., & Bhattacharya, M. (1989). Extrudate expansion during extrusion cooking of foods. *Cereal Foods World*, *34*(11), 945.

38. Parker, R., & Ring, S. G. (2001). Aspects of the physical chemistry of starch. *J Cereal Sci.*, *34*(1), 1–17.
39. Phillips, T. (1994). Pet food minutes; Production update. *Petfood Indus.*, *36*, 4.
40. Riaz, M. N. (2000). Introduction to extruders and their principles. In: *Extruders in Food Applications,* by M. N. Riaz, (Ed.), CRC Press. pp. 1–23.
41. Schwartzberg, H. G., Wu, J. P., Nussinovitch, A., & Mugerwa, J. (1995). Modeling deformation and flow during vapor-induced puffing. *J Food Engr.*, *25*, 329–372.
42. Schweizer, I. I., & Reimann, S. (1986). Influence of drum drying and twin-screw extrusion cooking on wheat carbohydrate, II: Effects of lipid on physical properties degradation and complex formation of starch in wheat flour. *J Cereal Sci.*, *4*, 249–260.
43. Serrano, X. (1996). The extrusion-cooking process in animal feeding: Nutritional implications: Feed manufacturing in Southern Europe. *New Challenges Zaragoza*, *26*, 107–114.
44. Shafi, M. A., Lee, J. G., & Flumerfelt, R. W. (1996). Prediction of cellular structure in free expansion polymer foam processing. *Polym Engr Sci.*, *36*, 1950.
45. Singh, S., & Wakeling, L. (2007). Nutritional aspects of food extrusion: a review. *Intl. J Food Sci. Technol.*, *42*, 916–929.
46. Wang, N., Bhirud, P. R., & Tyler, R. T. (1999). Extrusion texturization of air-classified pea protein. *J Food Sci.*, *64*(3), 509–513.

CHAPTER 5

NON-DESTRUCTIVE TECHNIQUE OF SOFT X-RAY FOR EVALUATION OF INTERNAL QUALITY OF AGRICULTURAL PRODUCE

D. V. CHIDANAND, C. K. SUNIL, and ASHISH RAWSON

CONTENTS

5.1   INTRODUCTION

Consumers demand for quality has grown and is an important factor for safety and health. With the advent of new technologies, new methods of quality evaluation, information on nutrition, etc., there is an increase consciousness of quality. This increase in consciousness of quality demands research activities related to production of defined quality and a need for development of accurate, fast and objective quality determination of the quality characteristics. Quality is defined as degree of excellence of a product. The quality of agricultural products are classified based on the individual or combination of various properties, viz. physical, mechanical, optical, sonic, electrical, electro-magnetic, thermal, hydro and aero dynamic, etc. External quality parameters (e.g., size, shape, color, tenderness, and hardness) are evaluated based on eye judgment and hand feel. Internal qualities include texture (firmness, crispness and juiciness), nutritional parameters and defects like Borer infestation. Other internal defects like bitter pit, internal browning, water cavity, water core, frost damage, internal rotten, soft tissues are difficult to access by visual appearance, resulting in a need for technology that can determine the internal quality parameters of the produce.

To check the internal quality, destructive methods are used resulting in destruction of the food and resulting in wastage. To avoid such losses many nondestructive methods (Soft X-ray, Sonic and ultrasonic, VIS/NIR spectroscopy and MRI) have been researched for evaluation of quality of agricultural and processed products. Non-destructive quality evaluation is one of the novel techniques in the field of agriculture and food processing. This chapter discusses nondestructive technique of soft X-ray that includes importance of nondestructive methods, principle, X-ray source, applications, and recent research advances.

5.2   NON-DESTRUCTIVE METHODS OF QUALITY EVALUATION

The increase in consciousness of quality, demands research activities related to production of defined quality and a need for development of accurate, fast and objective quality determination of the quality characteristics. The quality evaluation has been done over the years, either like visual appearance, etc. The advent of new technologies and better research facilities have given rise to more accurate, fast and objective methods of quality determination. The technical growth in the twentieth century has resulted in development of different instruments and nondestructive methods for quality evaluation that include: Image analysis, Soft X-ray, Ultrasonic, Sonic, Acoustic, NMR/MRI, NIR, etc. Table 5.1 shows different nondestructive methods with quality parameters. In computer vision methods, image is captured by using modern image acquisition methods, like multispectral cameras. These images are processed by using image processing software like fuzzy logic, neural network methods, etc., for the object measurement, and classification based on size, shape, color

TABLE 5.1   Non-Destructive Techniques for Quality Evaluation

Techniques	Quality parameters	Reference
Image Analysis	Size Shape Color Texture External defects	[4, 7, 15, 38, 45, 48, 66, 78, 86]
Mechanical • Ultrasonic • Sonic • Acoustic	Firmness, tenderness, internal cavity, internal structure. Hollow heart in potato. Firmness, internal cavity density, viscoelasticity Maturity, Firmness	[30, 49, 51, 67, 68, 83]
Optics • Reflectance, transmittance and absorbance spectroscopy. • Laser spectroscopy. • Near-Infrared Spectrophotometry (NIRS)	Internal defects, color, chemical constituents, Firmness, shape Firmness, freshness, Brix value, acidity, color	[13, 31, 32, 61]
Electromagnetic • Nuclear Magnetic Resonance (NMR) • Magnetic Resonance Imaging (MRI) – MR/MRI • Impedance • Terahertz Radiation (T-rays) – 100 GHz to 30 THz of electromagnetic spectrum • X- ray Image (Soft X-ray and CT)	Moisture content, Internal defect, Internal structure, oil content, sugar content Internal cavity, moisture content, density, sugar content. Moisture content, Pesticide detection, Antibiotic detection. Internal cavity, structure, infestation, ripeness, etc.	[5, 6, 43, 24, 25, 59, 60, 63, 80]

and other parameters as shown in Table 5.1. All these nondestructive methods are applied and used for online classification in food processing operations for accurate results and objective quality determination. The nondestructive methods listed in Table 5.1 have many advantages and different applications, such as:

- Quality control of processed food and raw material.
- Removal of extraneous material during processing operation.
- Monitoring of processing and dispatch of finished goods.
- Location and determination of the extent of cracks, voids and similar defects in food products during pre and post processing.
- Determining the uniformity during processing.
- Increasing the confidence level and assessing the potential durability of the produce.

5.3   X-RAY: NON-DESTRUCTIVE QUALITY EVALUATION METHOD

The techniques [e.g., X-ray imaging, Computed Tomography (CT), Magnetic Resonance Imaging (MRI), Near Infra-red and ultrasound in conjunction with different image processing methods] have been explored for nondestructive evaluation of internal quality indicators such as internal defects, insect infestation, firmness, cavity, etc. In recent years, use of X-ray based systems is increasing, as an effective research tool for the detection of internal defects in agricultural products and is a promising technology because of nondestructive, noncontact and cheaper than NMR technology [82].

X-ray imaging is being applied for medical diagnostics, imaging, checking luggage at airport, inspecting industrial components, security, etc. X-ray application in quality evaluation and inspection is not applied commercially. X-ray imaging is also considered for one of the excellent methods to determine the internal qualities of agricultural raw materials and to inspect processed and packed foods nondestructively. The X-ray method is rapid as it requires 3–5 s to produce X-ray image [53]. The X-ray imaging method has different applications such as identifying internal quality, defects, damages, external quality, determining composition and contaminants in fruits, vegetables, seeds, nuts, and various other food products.

5.3.1   RADIATION HAZARDS

Some common effects of X-rays are: alteration in the physical properties of material (metals can be made brittle, plastics can be made stronger, and transparency of materials can be altered in some semiconductors); chemical changes (mixture of N_2 and O_2 gases gives nitrogen oxides, ethylene gas to polyethylene); biological effects (killing living organisms, preserving foods, medical oncology. One of the major problems associated with use of X-rays is that high-energy electromagnetic radiations (15 eV or more), like X-rays can ionize atoms [27] and kill biological cells. It is therefore mandatory to provide a shield between the radiation source and people working in the vicinity. Equipment designed for radiography, therefore, needs to fulfill functional as well as radiation safety requirements.

5.3.2   X-RAY DETECTION FOR QUALITY CLASSIFICATION

X-ray imaging is an established technique to detect strongly attenuating materials and research has shown the application to a number of inspection applications within the agricultural and food industries. There are two types of rays in X-ray which are known as:

- **Hard X-ray** (energies above 5–10 keV and below 0.2–0.1 nm wavelength): The shorter wavelengths closer to and overlapping the gamma rays are called *hard X-rays*. Hard X-rays are widely used to image the inside of objects, for example, in medical radiography and airport security due to their high penetrating power because of their high energy.
- **Soft X-ray** with low energy: Electromagnetic waves with wavelengths ranging from 0.1 to 10 nm with corresponding energies of about 0.12 to 12 keV are called *soft X-rays*.

The electromagnetic spectrum showing the hard and soft X-rays is shown in Fig 1. The soft X-rays are used and suitable for agricultural products, due to low penetration power and ability to reveal the internal density changes. However, due to the difference in the material properties, the penetration capacity varies through different materials. The penetration capacity and energy of the X-ray is greater when the wavelength is shorter, e.g., shorter the wavelength larger the penetration.

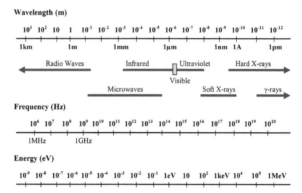

FIGURE 5.1   Electromagnetic spectrum

5.3.2.1   Attenuation Coefficient

Attenuation is a phenomenon of exponential decrease in the total energy of the X-ray beam as it traverses through the object subjected to X-ray exposure [16]. The attenuation coefficient describes how easily a material can be penetrated by a beam of light or other forms of energy or matter. The *higher* the attenuation coefficient quicker the beam gets attenuated (weakened) and *smaller* attenuation coefficient means the medium is relatively transparent to the beam. The attenuation coefficient is also called as linear attenuation coefficient, narrow beam attenuation coefficient, or absorption coefficient and measured using units of reciprocal length.

Photons in an X-ray beam, when passing through an object, are transmitted, scattered, or absorbed. The intensity of the transmitted photons is attenuated by using the following equation [16]:

$$I = I_0 \exp(-\mu z) \tag{1}$$

where, I = intensity of attenuated photons; I_0 = intensity of incident photons; μ = linear attenuation coefficient (1/cm); and z = material thickness (cm).

5.3.2.2   CT Number

The detection of foreign material using X-ray is based on the different material densities which vary depending on the absorption characteristics. The absorption characteristics of the material are expressed as

CT numbers [57]. The CT number is expressed as brightness data in an image, and is based on linear X-ray absorption coefficients [40].

$$\text{CT number} = (\mu - \mu_w)_k / \mu_w \qquad (2)$$

where, μ = object linear X-ray absorption coefficient (m^{-1}); μ_w = linear X-ray absorption coefficient of water (m^{-1}); k = constant (1000). If k = 1000, then CT number is called as Hounsfield unit.

5.4   X-RAY IMAGING SYSTEM

The first X-ray detector was used by Roentgen in 1985 which was a sheet of paper coated with barium platinocyanide. This paper fluoresced when impacted by X-rays, and led to their initial discovery, after which many different materials have been observed to react to the presence of X-rays and have led to invention and use of many different types of detectors. The modern X-ray inspection units are classified into one of three categories: *film*, *line scan machines*, and *direct detection semiconductor materials*. The film type of X-ray inspection units are the most widely used because of its high resolution and dynamic range. They are used for quality inspection of many food products other than medical and dental purposes.

Two methods are used to acquire X-ray image. In the *first method* the object to be inspected is fixed and the line scan sensor is moved with a constant speed within the exposure range of the X-ray source tube. In *second method*, the object to be inspected is moved through the inspection zone where the X-ray source tube and line scan sensor are fixed.

For X-ray imaging process, the photocathode is overlaid with a material that fluoresces in the presence of X-rays by converting the incident X-ray photons into visible light. A variation on X-ray line scans imaging that allows three-dimensional images is computed axial tomography, or CT imaging. The X-ray source rotates around the sample with detectors positioned opposite the source and multiple "slices" are progressively imaged as the sample is gradually passed through the plane of the X-rays. These slices are combined using tomographic reconstruction to form a three dimensional image. Helical or spiral CT machines incorporate faster computer systems and advanced software to process continuously changing cross sections.

As the sample moves through the X-ray circle, three-dimensional images are generated that can be viewed from multiple perspectives in real time on computer monitors. The X-ray imaging system contains four important basic elements according to Kotwaliwale et al. [40, 41]:

a. X-ray source: Produce X-rays of appropriate intensities.
b. X-ray converter: It produces a visible output proportional to the incident X-ray photons and stops X-rays from source, reaching the imaging medium.
c. Imaging medium: Photographic medium for capturing the images.
d. Casing for imaging medium: Protects the Imaging system from surrounding radiations.

5.4.1   X-RAY SOURCE

X-rays are produced, when high-energy electrons strike a target material, typically Tungsten. The two sources of the X-rays are radioactive substances and X-ray tubes. The X-ray tube is similar to a light bulb in design, except that the electrons shedding from the heated filament are subjected to a high voltage, causing them to accelerate and strike the target at high energies. With the high energy electrons decelerating in the target material, the electrons of the target atoms are excited to high energy levels and decay to their ground state with emission of X-ray photons. The bombardment in X-ray tubes must take place in a vacuum for preventing ionization of air.

The X-rays produced from the X-ray tubes are polychromatic beams, whereas the radioactive sources produce monochromatic X-rays (all photons of same energy level). Different types of X-ray tubes have been used by different researchers predominantly in agricultural produces with the advantage of producing X-rays of varying intensity. The variation in the tubes depend on the maximum voltage, current, focal spot size, window material, electrode material, tube cooling material and many other factors [41].

5.4.2   CONVERTERS/TRANSFORMERS

The Intensifying screens are used to convert X-rays into light. It produces a visible output proportional to the incident X-ray photons and stops

X-rays from source, reaching the imaging medium Different types of converters are used by different researchers like Xenon gas scintillator, phosphors, semiconductors, etc. [41]. Semiconductors act as converter as well as imaging medium, as it converts X-rays directly into electrical charge.

5.4.3 IMAGING MEDIUM

The acquisition of images is done by the imaging medium. The acquisition can be in two forms: film based or digital. Different types of imaging mediums have been for the agricultural produce. The various methods are: Photographic plate/ film, radiography film, photodiode, digital camera, line scan camera, intensifiers, fluoroscopes, intensifiers coupled with CCD camera, image digitizer, etc. [41].

5.5 X-RAY PARAMETERS

The X-rays have two characteristics which are important; *energy and current*. The energy refers to the maximum energy that an X-ray photon can possess when exiting the tube (generally between 20 and 100 KeV for food inspection) and defines the penetrating power of the X-ray beam. The current is associated with the number of X-ray photons being generated and measured in mA. The power supply balance is required between the energy and current, which has consequences on the resulting image quality as the thickness, density and absorption properties varies among the agriculture produce [29]. Based on the power supply balance, most X-ray inspection systems are limited to less than 10 mA of current. The focal spot size can be defined as the size of the target area over which X-rays are generated. The focal spot size affects the characteristics of the imaging system. The combinations of the current and voltage can be as high voltage and high current, high voltage and low current, low voltage and high current, and low voltage and low current. All the combinations have been tried for the agricultural produce. Figure 5.2 shows schematic of soft X-ray imaging equipment.

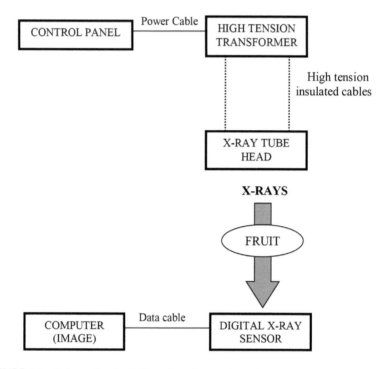

FIGURE 5.2   Schematic of soft X-ray imaging equipment.

5.6   IMAGE PROCESSING AND ANALYSIS

The image processing techniques have been widely used over the years for the machine vision of agricultural produce. These techniques in combination with computer-based algorithms have been used for decision making, task management for diagnosis of different attributes. There are many different types of algorithms which are applied and adopted for the image processing (enhancement of images) and help in decision of the quality or different attributes. These algorithms have shown an improvement over the years, with the improvement in the software and hardware. The various image-processing techniques are:

- Image representation.
- Image preprocessing.
- Image enhancement.

- Image restoration.
- Image analysis.
- Image reconstruction.
- Image data compression.

The real world image is considered to be a function of two variables. The image is expressed or presented in form of pixels and it should have headers which contain information of like bits/pixel, resolution, compression type, etc. The preprocessing is done to improve the image quality by removing noises or by enhancement of the important features of interest in the image. In image preprocessing, the image is magnified or reduced (known as scaling).

The interested part of the image is magnified or zoomed to improve the scale of display for visual interpretation or to match with the other image. The magnification is done so that the original pixel should be maintained and with same brightness value. Reduction is process of reducing the magnified or zoomed is reduced to its original size. Other processes in image pre-processing are *rotation* (used in image mosaic and image registration) and *mosaic* creation (combining two or more images to form a single large image without radiometric imbalance and is required to get the synoptic view of the entire area).

The image enhancement is applied to the digital image to correct the problems like poor contrast, edge enhancement, pseudocoloring, noise filtering, sharpening, and magnifying. It emphasizes certain image characteristics. Enhancement algorithms are interactive and application dependent. Some of the image enhancement procedures such as filters, morphological operations, and pixel-to-pixel operations are generally used to correct inconsistencies in the image which are caused by inadequate and/or nonuniform illumination. Some of the enhancement techniques are:

a. **Contrast stretching**: The contrast stretching helps in segregation of the homogeneous objects which have less difference in the original image. The narrow range is stretched to the dynamic range.

b. **Noise filtering**: The noise filtering is used for removal of unnecessary information and various noises from the image.

c. **Histogram modification**: Histogram modification helps in reflecting the characteristics of the image. By modifying the histogram, the image characteristics can be modified.

The Image analysis is a process of making quantitative measurements from an image to produce a description [65]. The examples of the image analysis can be sorting of different grades on a sorting line, reading labels in stores, etc. These image analysis techniques extract certain features from the image that helps in decision support. After the preprocessing is done, the image analysis is done which involves image segmentation, image classification and image description.

5.6.1   IMAGE SEGMENTATION

It is the process of subdividing an image into its constituent parts, which depends on the problem being solved. The segmentation techniques are used to isolate the desired object from the image and used for the measurements and decision. As an example, in insect infested mango X-ray image, the first step is to segment the contents of the mango and then to segment the insect identified based on the color difference (density difference) or infested area, which is of interest for the study. Segmentation can be achieved by following techniques: (a) thresholding; (b) edge based segmentation (using edge operators by detecting discontinuity in gray level, color and texture); and (c) region based segmentation (grouping together of similar pixels and form a region representing as single object in the image) [71]. The image thresholding techniques are the commonly used and are used for characterizing the image region based on constant reflectivity or light absorption of the surface.

5.6.2   IMAGE CLASSIFICATION

It is used for information extraction by labeling the pixel or a group of pixels on its gray scale [12, 19]. The information extraction techniques rely on the spectral reflectance properties of the image and develop algorithms for spectral analysis. The multispectral classification is performed using two methods: a. Supervised and b. Unsupervised.

5.6.3   IMAGE RESTORATION

It is a process of removal or minimization of degradations in image such as de-blurring of images, noise filtering and geometric distortion correction. The image is restored to its original quality by inverting the physical degradation methods (defocus, additive noise, linear motion and atmospheric degradation).

The processed image can be *reconstructed* to a two or higher dimensional object from the one-dimensional projections. This helps in giving the inside view of the object and helps in nondestructive testing. And further the image is *compressed* for achieving the image data, image data transfer on network, etc. For image compression methods like DCT (Discrete Cosine Transformation) and wave let based compressions are used [62]. The data or quantitative information of the object of interest is obtained from the processed image and algorithms are used to process the information and help in decision-making by the control systems.

With the help of image analysis and development of different types of algorithms, many researchers have developed automatic sorting lines for agricultural produce like pomegranate [8], tomato [18], and peach [69], etc. The primary algorithms like subtraction of images [39] taken at specific intervals used for relative movement of live insects infested kernels were developed by Gonzalez and Woods [21] in 2001. Other algorithms developed are thresholding based on pixel intensity [23], un-sharp masking and filtering [52], first order (pixel intensity) and second order (intensity change) information, filtering and averaging to remove noise [36], region of interest segmentation using edge detection filters [82], three stage algorithm comprising of image enhancement, local equalization filter and thresholding [20], adaptive unsupervised thresholding algorithm based on local gray level distribution [29], etc.

5.7   X-RAY APPLICATION

Among the machine vision techniques, X-rays have distinct advantages in nondestructive food inspection. X-rays can penetrate inside the objects and internal defects or properties can be measured and identified. A typical machine vision inspection system involves steps like acquiring an image and then segmenting the objects of interest by using various software techniques including thresholding, etc. as mentioned earlier in this chapter (Figure 5.3). The quality of the acquired digital image plays an important role in machine vision applications and it depends on imaging system components, objects, background, lighting, and noise which are explained or discussed earlier.

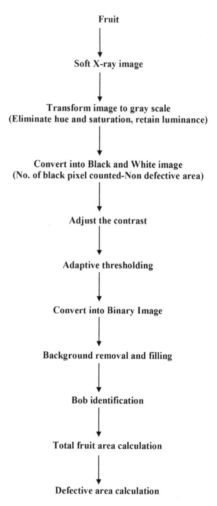

FIGURE 5.3 Image processing to identify defective area [60].

X-ray has been examined for its suitability for quality assessment of fruits by many researchers over the years. The X-rays have higher penetration, if the wavelength is shorter. The intensity of the measured light after passing the sample is dependent on the initial intensity, density and thickness of the absorber substance and its absorption coefficient (wavelength and material dependent): law of absorption. The higher the water content of vegetables, higher is the portion of absorption. The common internal disorders or defects that can be detectable by the X-ray techniques include such as: cork spot, bitter pit, water core and brown core for apple, blossom and decline, membranous stain, black rot, seed germination, freeze damage in citrus and hollow heart, bruises, black heart in potato, etc., as the selectivity of the X-ray method is limited because the method is primarily dependent on the density of the tissue and not the chemical composition [74]. Three different types of X-ray techniques are used for the quality control or evaluation: *2-dimensional radiography* (commonly used in medical diagnosis), the *line scan method* (producing images when the product passes through a vertical plane of X-rays, for example, baggage control at the airport and Roentgen CT (medical diagnosis). All the three methods produce a 3-dimensional image, from which 2-dimensional information can be computed or captured.

The X-ray application was commercially used as a routine in the detection of hollow heart in potato tubers [56]. It is also used for the measurement of water content in apples [76], change in density during ripening of tomatoes [9], pistachio nuts insect infestation [33]. X-rays units are used for the separation of the freeze-damaged citrus fruits [2]. Lim and Barigou [44] used the X-ray microCT for the visualization and analysis of the 3D cellular microstructure. The quantitative information for different parameters like spatial cell size distribution, cell wall-thickness distribution, connectivity, and voidage were obtained. Reyes *et al.* [64] used fruits with various different structures for quantitative evaluation of the X-ray potential for internal infestation caused by the weevil in the mango seeds. Soft X-ray has been used for the study of the internal infestation by fruit flies in mango [80]. Computer algorithms were developed to differentiate the sound and infested mango fruits. These algorithms can detect the defects based either on absorption of X-ray or processed binary and RGB images.

Yang et al. [84] tested the possibility of examining internal injuries of various fruit using digitized X-ray imaging analysis. The digitalized X-ray images showed that this technique can detect injuries caused by *B. dorsalis* at as early as 3 days after implantation of eggs in some fruits. They concluded that the tunnels created by larvae in infested fruits provide a good contrast on the X-ray images and are easily detected. The shape of the injury tunnels obviously differs from the internal structure of the fruit and in addition to the contrast of the image, the density contours showed remarkable uneven lines or broken areas (insect damage by pests). Matheker [50] has conducted study on the automatic insect damage detection. The kernel necrotic spots were detected in pistachio nuts by Yanniotis et al. [85] in 2011. The necrotic spots appeared as darker gray areas of almost round shape in the X-ray image.

Other applications for the nondestructive identification of defects have been reported for various commodities like apples, mango, onion, pistachio nuts, almonds, pecans, etc. [1,10, 11, 17, 22, 23, 33, 35, 39, 75]. The X–rays with an energy ranging from 15 to 80 kvp at various current levels has been reportedly used. The higher energy X-ray of 100 kvp is not suitable for radiography of foods [27]. Table 5.2 shows the applications of

TABLE 5.2   X-Ray Applications for Agricultural Produce

Commodity	Application	Parameters/References
Almonds	Pinhole detection	32 keV and 3 mA for exposure time 60 s on films
		35 keV and 30 mA for 3 ms in case of line scan [36]
Apple	Bruise detection	35 kV and 10 mA Bruise detection in apple [17].
	Browning	
	Water core	49.9 to 50 kVp and 9.9 to 13 mA Watercore damage in apples [35].
	Rot insect damage	
Apricot	Infestation	25 to 70 keV and 10 to 100 mA [87]
Citrus	Infestation	40–90 kV and 110–250 μA [29]
Gauva	Infestation	40–90 kV and 110–250 μA [29]
Mango	Seed weevil detection	40 kV and 800 mA-s [73]
	Internal quality detection	[27, 80]
Onion		70 kV, 1 mA [75]
Peach	Classification	50 kV and 300 mA [23]
	Infestation	40–90 kV and 110–250 μA [29]
	Internal quality	
Pecans	Quality evaluation	25–50 kVp and 0.25–1 mA at 460 ms [39]
Pineapple	Translucency	[22]
Pomegranate	Internal quality detection	90 mA and 100 kV [60]
Potato	Detection of blemishes	[4]
Wheat	Insect infestation	[53, 54]
	Fungal infestation	

TABLE 5.3 The Optimum Parameters for Some Familiar Imported/Exported Fruits

Type of Fruit	Voltage (kV)	Current (µA)	Integration Time (s)
Apple	75	125	0.12
Guava	75	125	0.12
Mango	75	125	0.12
Peach	65	125	0.12
Pear	75	125	0.12
Pitaya	75	125	0.12
Sunkist	70	125	0.12

X-ray in agricultural produce. Table 5.3 shows the optimum parameters for generation of soft X-ray to obtain the best quality image for identifying the defects of some familiar fruits [28]. These parameters help in obtaining better images of the fruits for further processing and decision support. Joe-Air Jiang et al. [29] have developed an adaptive, image segmentation algorithm based on the local pixel intensities and unsupervised thresholding algorithms for determining the insect infestation in several types of fruits such as citrus, peach, guava, etc. Analyzes were performed using the developed algorithm on the X-ray images obtained with different image acquisition parameters. Fruit containing high amounts of water have been deemed unsuitable for X-ray imaging.

5.8 CONCLUSIONS

The X-ray techniques are powerful tools for nondestructive internal quality evaluation. The international markets for agricultural produce are huge and growing. To conquer and to sustain these markets, there is a need for export of high quality products with no internal defects. The currently practiced methods such as sensory evaluation, color sorting, size grading, and similar other methods cannot effectively address the classification of internal defect free whole produce. Nevertheless, necessity has motivated a considerable research effort in this field spanning many decades. Many researchers have devoted considerable effort towards the development of machine vision systems for different aspects of quality evaluation and sorting of agricultural products. As a result, new algorithms and hardware architectures have been developed for high-speed extraction of features that are related to specific quality factors of fruits. The low penetration power and ability to reveal the internal density changes make soft X-rays suitable to be used for fruits. Harmful effects of X-rays are definitely a cause of concern while using these techniques, but properly designed shielding can prevent human exposure. Improvements in technology have allowed X-ray detection of internal defects that were not possible in the past. These improvements can be expected to continue into the future.

5.9   SUMMARY

The increased consumer consciousness and demand for quality has given rise to different noninvasive techniques for analysis of agricultural produce. In recent years, development of many nondestructive evaluation techniques for quality assessment have made a progress and also in development of new equipments which are safer and economical for nondestructive evaluation. Non-destructive quality evaluation is one of the novel techniques in the field of agriculture and food processing. Soft X-ray is one of the successful nondestructive technologies for quality evaluation which is gaining popularity and has a potential to be explored and applied commercially. This chapter discusses the nondestructive methods and importance; and the nondestructive technique of Soft X-ray: Principle, X-ray source, image processing, applications and recent advances.

KEYWORDS

- acoustic
- attenuation coefficient
- contrast stretching
- CT
- CT Number
- electromagnetic spectrum
- hard X- ray

- histogram modification
- image processing
- imaging medium
- MRI
- NIR
- NMR
- noise filtering
- nondestructive
- radiation
- sensory evaluation
- soft X-ray
- sonic
- terahertz radiation
- ultrasonic
- X-ray parameters
- X-ray tube
- X-ray converters/transformers
- X-ray source

REFERENCES

1. Abbott, J. (1999). Quality measurement of fruits and vegetables. *Postharvest Biology and Technology, 15*, 207–225.
2. Abbott, J. A., Lu, R., Upchurch, B. L., & Stroshine, R. L. (1997). Technologies for nondestructive quality evaluation of fruits and vegetables. *Horticultural reviews.* Chichester, U.K., Wiley, pp. 1–120.
3. Barcelon, E. G., Tojo, S., & Watanabe, K. (1999). X-Ray Computed Tomography for Internal Quality Evaluation of Peaches. *Journal of Agricultural Engineering Research, 73*, 323–330.
4. Barnes, M., Duckett, T., Cielniak, G., Stroud, G., & Harper, G. (2010). Visual detection of blemishes in potatoes using minimalist boosted classifiers. *Journal of Food Engineering, 98*, 339–346.
5. Barreiro, P., Ortiz, C., Ruiz-Altisent, M., Ruiz-Cabello, J., Fernandez-Valle, M. E., Recasens, I., & Asensio, M. (2000). Mealiness assessment in apples and peaches using MRI techniques. *Mag Reson Imag, 18*(9), 1175–1181.

6. Barreiro, P., Ruiz-Cabello, J., Fernandez-Valle, M. E., Ortiz, C., & Ruiz-Altisent, M. (1999). Mealiness assessment in apples using MRI techniques. *Mag Reson Imag, 17*(2), 275–281.

7. Blasco, J., Aleixos, N., & Moltó, E. (2003). Machine vision system for automatic quality grading of fruit. *Biosystems Engineering, 85*(4), 415–423.

8. Blasco, J., Cubero, S., Gómez-Sanchís, J., Mira, P., & Moltó, E. (2009). Development of a machine for the automatic sorting of pomegranate (*Punicagranatum*) arils based on computer vision. *Journal of Food Engineering, 90*, 27–34.

9. Brecht, J. K., Shewfelt, R. L., Garner, J. C., & Tollner, E. W. (1991). Using X-ray-computed tomography to nondestructively determine maturity of green tomatoes. *HortScience, 26*(1), 45–47.

10. Casasent, D. A., Sipe, M. A., Schatzki, T. F., Keagy, P. M., & Lee, L. C. (1998). Neural net classification of X-ray pistachio nut data. *Lebensm- Wiss u-Technol (Food Sci + Technol), 31*(2), 122–128.

11. Casasent, D. A., Talukder, A., Keagy, P., & Schatzki, T. (2001). Detection and segmentation of items in X-ray imagery. *Trans ASAE, 44*(2), 337–345.

12. Chellappa, R. (1992). *Digital Image Processing.* 2nd Edition. IEEE Computer Society Press.

13. Cho, R. W. (1996). Nondestructive quality evaluation of intact fruits and vegetables by near infrared spectroscopy. Nondestructive quality evaluation of horticultural crops. *Proceedings of the Intl. Symposium on Nondestructive Quality Evaluation of Horticultural Crops* in Kyoto, Japan; Aug 26. Tokyo, Japan: Saiwai Shobou Publishing Co., pp. 8–14.

14. Chuang, C. L., Ouyang, C. S., Lin, T. T., Yang, M. M., Yang, E. C., Huang, T. W., Kuei, C. F., Luke, A., & Jiang, J. A. (2011). Automatic X-ray quarantine scanner and pest infestation detector for agricultural products. *Computers and Electronics in Agriculture, 77*, 41–59.

15. Costa, C., Menesatti, P., Paglia, G., Pallottino, F., Aguzzi, J., & Rimatori, V. (2009). Quantitative evaluation of Tarocco sweet orange fruit shape using optoelectronic elliptic Fourier based analysis. *Postharvest Biology and Technology, 54*(1), 38–47.

16. Curry, T. S., Dowdey, J. E., & Murry, R. C. (1990). *Christensen's Physics of Diagnostic Radiology.* Lea and Febiger, Malvern, PA.

17. Dienar, R. G., Mitchell, J. P., Rhoten, M. L. (1970). Using an X-ray image scan to sort bruised apple. *Agric Eng, 51*, 356–361.

18. Edan, Y., Pasternak, H., Shmulevich, I., Rachmani, D., & Guedalia, D. (1997). Color and firmness classification of fresh market tomatoes. *Journal of Food Science, 62*, 793–796.

19. Ernest, L. (1979). *Computer Image Processing and Recognition.* Academic Press.

20. Fornal, J., Jelinski, T., Sadowska, J., Grundas, S., Nawrot, J., Niewiada, A., Warchalewski, J., & Blaszczak, W. (2007). Detection of granary weevil Sitophilus granarius (L.) eggs and internal stages in wheat grain using soft X-ray and image analysis. *J Stored Prod Res, 43*, 142–148.

21. Gonzalez, R. C., & Woods, R. E. (2001). *Digital Image Processing.* 2nd Edition. Prentice Hall, Upper Saddle River.

22. Haff, R. P., Slaughter, D. C., Sarig, Y., & Kader, A. (2006). X-ray assessment of translucency in pineapple. *Journal of food processing and preservation, 30*, 527–533.

23. Han, Y. J., Bowers, S. V., & Dodd, R. B. (1992). Nondestructive detection of split-pit peaches. *Trans ASAE, 35*(6), 2063–2067.

24. Hills, B. P., Wright, K. M., & Gillies, D. G. (2005). A low-field, low-cost Halbach magnet array for open-access NMR. *J Magn Reson, 175*(2), 336–339.

25. Hua, Y. F., & Zhang, H. J. (2010). Qualitative and quantitative detection of pesticides with terahertz time-domain spectroscopy. *IEEE Transactions on Microwave Theory and Techniques, 58*(7), 2064–2070.

26. ISO 9000, 2005. *Quality management systems–Fundamentals and vocabulary.*

27. Jha, S. N. (2010). Nondestructive evaluation of food quality: Theory and Practices. e-ISBN 978-3-642-15-796-7, pp. 101–148.

28. Jiang, J. A., Ouyang, C. S., Lin, T. T., Chang, H. Y., Yang, M. M., Yang, E. C., Kuel, J. F., & Chen, T. W. (2006). Application of LabView platform to X-ray automatic quarantine system for fruits. *ISMAB, 3*, 747–754.

29. Jianga, J. A., Changa, H. Y., Wua, K. H., Ouyanga, C. S., Yang, M. M., Yang, E. C., Chen, T. W., & Lin, T. T. (2008). An adaptive image segmentation algorithm for X-ray quarantine inspection of selected fruits. *Computers and Electronics in Agriculture, 60*, 190–200.

30. Jivanuwong, S. (1998). *Nondestructive Detection of Hollow Heart in Potatoes Using Ultrasonics.* M.S. Thesis, Virginia Polytechnic Institute and State University.

31. Kawano, S. (1999). Non-destructive methods for quality analysis-especially for fruits and vegetables. XXXIV Vortragstagung Zerstörungsfreie Qualitätsanalyze; Freizing-Weihenstephan; Mar 22–23. Deutsche Gesellschaftfür Qualitätsforschung (Pflanzliche Nahrungsmittel) e.V. pp. 5–12.

32. Kawano, S. (2002). Sample presentations for near infrared analysis of intact fruits, single grains, vegetable juice, milk and other agricultural products. In: *Near Infrared Spectroscopy.* Davies, A. M. C., & Cho, R. K. (eds.). Proceedings of the 10th Intl Conference. Chichester, U.K., NIR Publications, p. 519.

33. Keagy, P. M., Parvin, B., & Schatzki, T. F. (1996). Machine recognition of navel orange worm damage in X-ray images of pistachio nuts. *Lebenson Wiss Technol, 29*(1–2), 140–145.

34. Khoshroo, A., Keyhani, A., Rafiee, S. H., Zoroofi, R. A., & Zamani, Z. (2009). Pomegranate quality evaluation using machine vision. *Acta Hort., 818*, 347–352.

35. Kim, S., & Schatzki, T. (2000). Apple water core sorting using X-ray imagery: I. Algorithm development. *Transaction of the ASAE, 44*(4), 997–1003.

36. Kim, S., & Schatzki, T. (2001). Detection of pinholes in almonds through X-ray imaging. *Transaction of the ASAE, 43*(6), 1695–1702.

37. Kleynen, O., Leemans, V., & Destain, M. F. (2005). Development of a multispectral vision system for the detection of defects on apples. *Journal of Food Engineering, 69*, 41–49.

38. Koc, A. B. (2007). Determination of watermelon volume using ellipsoid approximation and image processing. *Postharvest Biology and Technology, 45*, 366–371.

39. Kotwaliwale, K., Paul, R., Weckler, Gerald, H., Brusewitz, Glenn, A., & Maness, N. O. (2007). Non-destructive quality determination of pecans using soft X-rays. *Postharvest Biology and Technology, 45*, 372–380.

40. Kotwaliwale, N., Subbiah, J., Weckler, P. R., Brusewitz, G. H., & Kranzler, G. A. (2007). Calibration of a soft X-ray digital imaging system for biological materials. *Trans ASABE, 50*(2), 661–666.

41. Kotwaliwale, N., Singh, K., Kalne, A., Jha, S. N., Seth, N., & Kar, A. (2014). X-ray imaging methods for internal quality evaluation of agricultural produce, *J Food Sci Technol*, *51*(1), 1–15.

42. Krutz, G. W., Gibson, H. G., Cassens, D. L., & Zhang, M. (2000). Color vision in forest and wood engineering. *Landwards, 55*, 2–9.

43. Lammertyn, J., Dresselaers, T., Van Hecke, P., Jancsok. P., Wevers, M., & Nicolai, B. M. (2003). MRI and X-ray CT study of spatial distribution of core breakdown in 'Conference' pears. *Magn Reson Imag, 21*(7), 805–815.

44. Lim, K. S., & Barigou, M. (2004). X-ray microcomputed tomography of cellular food products. *Food Res Int, 37*(10), 1001–1012.

45. Liming, X., & Yanchao, Z. (2010). Automated strawberry grading system based on image processing. *Computers and Electronics in Agriculture, 71*(S1), S32–S39.

46. Lin, T. T., Jiang, J. A., Ouyang, C. S., & Chang, H. Y. (2005). *Integration of An Automatic* X-ray *Scanning System for Fruit Quarantine*. Department of Bio-Industrial Mechatronics Engineering, National Taiwan University.

47. Lluís, P., Marcilla, A., Rojas-Argudo, C., Alonso, M., Jacas, J. A., & Ángel del Río, M. (2007). Effects of X-ray irradiation and sodium carbonate treatments on postharvest Penicillium decay and quality attributes of clementine mandarins. *Postharvest Biology and Technology, 46*(3), 252–261.

48. López-García, F., Prats, J. M., Ferrer, A., & Valiente, J. M. (2006). Defect Detection in Random Color Textures using the MIA T2 Defect Maps. *Lecture Notes in Computer Science, 4142*, 752–763.

49. Lu, R., & Abbott, J. A. (1996). A transient method for determining dynamic viscoelastic properties of solid foods. *Trans ASAE, 39*(4), 1461–1467.

50. Matheker, S. K. (2010). *Development of a New Local Adaptive Thresholding Method and Classification Algorithms for X-ray Machine Vision Inspection of Pecans*. PhD Thesis, Graduate College of the Oklahoma State University.

51. Mizrach. A., Galili. N., & Rosenhouse, G. (1989). Determination of fruit and vegetable properties by ultrasonic excitation. *Transactions of ASAE, 32*(6), 2053–2058.

52. Morita, K., Tanaka, S., Thai, C. N., & Ogawa, Y. (1997). Development of soft X-ray imaging for detecting internal defects in food and agricultural products. *Proc. Sensors for Nondestructive Testing: Measuring Quality of Fresh Fruits and Vegetables-Florida* – Feb 18–21, pp. 305–315.

53. Neethirajan, S., Jayas, D. S., & White, N. D. G. (2007a). Detection of sprouted wheat kernels using soft X-ray image analysis. *Journal of Food Engineering, 81*, 509–513.

54. Neethirajan, S., Jayas, D. S., & White, N. D. G. (2007b). Dual energy X-ray image analysis for classifying vitreousness in durum wheat. *Postharvest Biology and Technology, 45*, 381–384.

55. Njorje, J. B., Ninomiya, K., Kondo, N., & Totita, H. (2002). Automated fruit grading system using image processing. *Proceedings of the 41st SICE Annual Conference*, pp. 1346–1351.

56. Nylund, R. E., & Lutz, J. M. (1950). Separation of hollow heart potato tubers by means of size grading, specific gravity, and X-ray examination. *American Potato Journal, 27*(6), 214–222.

57. Ogawa, Y., Morita, K., Tanaka, S., Setoguchi, M., & Thai, C. N. (1998) Application of X-ray CT for detection of physical foreign materials in foods. *Trans ASAE, 41*(1), 157–162.

58. Oliveria, C. Q., Ferreira, R. S., & Estrada, J. (2005). Digital image processing applied to inspection of internal disorders in mangoes cv Tommy Atkins. *International Nuclear Atlantic Conference – INAC* Santos, SP, Brazil, August 28 to September 2, ISBN: 85-99141-01-5.

59. Parasoglou, P., Parrott, E. P. J., Zeitler, J. A., Rasburn, J., Powell, H., & Gladden, L. F. (2009). Quantitative moisture content detection in food wafers. *Infrared, Millimeter, and Terahertz Waves. IRMMW-THz. 34th International Conference.*

60. Payal, G., & Sunil, C. K. (2014). Quality analysis of pomegranate by nondestructive soft X-Ray method. *J Food Process Technol*, 5(6).

61. Quilitzsch, R., & Hobert, E. (2003). Fast determination of apple quality by spectroscopy in the near infrared. *J Appl Bot*, 77(5/6), 172–176.

62. Rao, K. M. M. (2006). *Overview of Image Processing.* Readings in image processing, NRSA, Hyderabad.

63. Redo-Sanchez, A., & Zhang, X. C. (2011). Assessment of terahertz spectroscopy to detect antibiotic residues in food and feed matrices. *Analyst, 136*(8), 1733–1738.

64. Reyes, M. U., Paull, R. E., Armstrong, J. W., Follett, P. A., & Gautz, L. D. (2000). Non-destructive inspection of mango fruit (*Mangifera indica* L.) with soft X-ray imaging. In: *Proceedings of the Sixth International Mango Symposium*, Subhardrabandhu, S. & Pichakum, A. (eds.). pp. 787–792.

65. Richards, J. A., & Jia, X. (1999). *Remote Sensing Digital Analysis.* Enlarged edition. Springer Verlag.

66. Rupali, S. J., & Patil, S. S. (2013). A Fruit Quality Management System Based On Image Processing. *IOSR Journal of Electronics and Communication Engineering, 8*(6), 1–5.

67. Sarkker, N., & Wolfe, R. R. (1993). Potential of ultrasonic measurements in food quality evaluation. *Transactions of ASAE, 26*(2), 624–629.

68. Shmulevich, I., Galili, N., & Benichou, N. (1995). Development of a nondestructive method for measuring the shelf-life of mango fruit. *Proceedings Food Processing Automation IV Conference*, Chicago, IL, 3–5 November. American Society for Agricultural Engineering, St. Joseph, MI, pp. 275–287.

69. Singh, N., Delwiche, M. J., Johnson, R. S., & Thompson, J. (1992). Peach maturity grading with color computer vision. *ASAE Paper* 92-3029.

70. Singh, N., Michael, J., Delwiche, & Scott, J. R. (1993). Image analysis methods for realtime color grading of stone fruit. *Computers and Electronics in Agriculture*, pp. 71–84.

71. Sonka, M., Hlavac, V., & Boyle, R. (1999). *Image Processing, Analysis, and Machine Vision.* California, USA: PWS Publishing.

72. Sun, D. W. (2000). Inspecting pizza topping percentage and distribution by a computer vision method. *Journal of Food Engineering, 44*, 245–249.

73. Thomas, A. Z., & Rosemary, L. (1986). Detection of potato tuber diseases and defects vegetable crops. *Vegetable MD online. The Cornell Plant Pathology Vegetable Disease Web Page.* Information Bulletin 205.

74. Tollner, E. W., Affeldt, H. A., Brown, G. K., Chen, P., Galili, N., Haugh, C. G., Notea, A., Sarig, Y., Schatzki, T. F., Shmulevich, I., & Zion, B. (1994). Nondestructive detection of interior voids, foreign inclusions and pests. In: *Sarig Y, Brown G, editors. Nondestructive technologies for quality evaluation of fruits and vegetables. Proceedings of the Intl. Workshop funded by the United States-Israel Binational Agricultural Research and Development Fund (BARD)*; Spokane, Wash., Jun 15. ASAE: St. Joseph, Mich. pp. 86–96.

75. Tollner, E. W., Gitaitis, R. D., Seebold, K. W., & Maw, B. W. (2005). Experiences with a food product X-ray inspection system for classifying onions. *Trans ASAE, 21*(5), 907–912.

76. Tollner, E. W., Hung, Y. C., Upchurch, B. L., & Prussia, S. E. (1992). Relating X-ray absorption to density and water content in apples. *Trans ASAE, 35*(6), 1921–1928.

77. Trater, A. M., Alavi, S., & Rizvi, S. S. H. (2005). Use of noninvasive X-ray micro tomography for characterizing microstructure of extruded biopolymer foams. *Food Research International, 38,* 709–719.

78. Unay, D., & Gosselin, B. (2007). Stem and calyx recognition on 'Jonagold' apples by pattern recognition. *J. Food Eng., 78,* 597–605.

79. USDA. (2007). *United States Standards for Grades of Mangos.* U.S. Dept. Agric., Agric. Mktg.

80. Veena, T., Chidanand, D. V., Alagusundaram, K., & Kedarnath, V. (2015). Quality analysis of mango fruit with fruit fly insect by nondestructive soft X-ray method. *International Journal of Agricultural Science and Research, 5*(3), 37–46.

81. Wang, H. H., & Sun, D. W. (2002). Correlation between cheese meltability determined with a computer vision method and with Arnott and Schreiber. *Journal of Food Science, 67*(2), 745–749.

82. Yacob, Y., Ahmad, H., Saad, P., Aliana, R., Roaf, A., & Ismail, S. (2005). A comparison between X-ray and MRI in postharvest Non-Destructive detection method. *Proceedings of ICIMU 05,* Malaysia.

83. Yamamoto, H., Lwamoto. M., & Haginuma, S. (1980). Acoustic impulse response method for measuring natural frequency of intact fruits and preliminary applications to internal quality evaluation of apples and watermelons. *Journal of Texture Studies, 11,* 117–136.

84. Yang, E.-C., Yang, M.-M., Liao, L.-H., & Wu, W.-Y. (2006). Non-Destructive Quarantine Technique- Potential Application of Using X-ray Images to Detect Early Infestations Caused by Oriental Fruit Fly (*Bactrocera dorsalis*) (Diptera: Tephritidae) in Fruit. *Formosan Entomol., 26,* 171–186.

85. Yanniotis, S., Proshlyakov, A., Revithi, A., Georgiadou, M., & Blahovec, J. (2011). X-ray imaging for fungal necrotic spot detection in pistachio nuts. *Procedia* 11th International Congress on Engineering and Food (ICEF11). *Procedia Food Science, 1,* 379–384.

86. Zhao, X., Burks, T. F., Qin, J., & Ritenour, M. A. (2009). Digital microscopic imaging for citrus peel disease classification using color texture features. *Applied Engineering in Agriculture, 25*(5), 769–776.

87. Zwiggelaar, R., Bull, C. R., Mooney, M. J., & Czarnes, S. (1997). Detection of soft materials by selective energy X-ray transmission imaging and computer tomography. *J Agric Eng Res, 66*(3), 203–212.

PART III

ROLE OF ANTIOXIDANTS IN FOODS

CHAPTER 6

IN VITRO ANTIOXIDANT EFFICACY: SELECTED MEDICINAL PLANTS OF GUJARAT

POOJA MOTERIYA, JALPA RAM, MITAL J. KANERIA, and
SUMITRA CHANDA

CONTENTS

6.1   INTRODUCTION

Cellular oxidation results in production of reactive oxygen species (ROS), in animal and human biological systems which are responsible for oxidative damage in cell constituents thus producing various degenerative diseases and aging. This oxidative modification promotes age related disorders, including cardiovascular diseases. Such damage in human biological system can be prevented by cellular antioxidants [40]. Various macromolecules like protein, DNA and lipids are damaged by free radicals and actions of radicals are opposed by antioxidants by suppressing their formation or by scavenging them [31]. When the human immune system is suppressed by an overproduction of oxidizing agents, then the damage of nucleic acids, lipids and protein can result in tissue injury. For this reason, the search for natural antioxidants has become necessary in order to protect human body from free radicals and to retard the development of chronic diseases [39].

Free radicals are molecules with unpaired electrons and in their quest to find another electron they are very reactive and cause damage to surrounding molecules. They are various types of free radicals such as: hydroxyl radical, superoxide anion radical, singlet oxygen, nitric oxide radical, hypo chloride radical, lipid peroxide, etc. All these are capable of reacting with membrane lipids, nucleic acids, proteins, enzymes and other small molecules resulting cellular damage. There are two types of free radicals: endogenous (originates within organisms) and exogenous (coming from outsides).

Endogenous free radicals include: normal aerobic respiration, stimulates polymorpho nuclear leukocytes, macrophages and peroxisomes. Exogenous free radicals include tobacco smoke, ionizing radiation, certain pollutants, organic solvents poor diet, ultra violet rays and pesticides. They are capable of attacking the healthy cells of the body, causing them to loose their structure and function. These may cause many diseases like cancer [1], liver diseases [29], inflammation [15], diabetes [19] etc. Free radicals also cause food spoilage, rancidity, deteriorate shelf life, quality of food, etc.

Antioxidants are the molecules that scavenge free radicals present in the cell. They donate their electrons to stabilize the harmful free radicals and offer protection. Natural antioxidants are present in the plants that are able to scavenge free radicals from the body. In recent years, there has been growing interest in finding natural antioxidants in plants because they show strong protection against the damages of the cellular organelles caused by free radical induced oxidative stress, inhibit oxidative damage and may prevent many diseases [13, 33].

Synthetic antioxidants, such as butylated hydroxytoluene (BHT), butylated hydroxyanisole (BHA) and propyl gallate (PG), have been widely used around the world for decades [3]. But their use is becoming unpopular because of their toxicity, adverse side effects, carcinogenic nature, etc. [30]. Therefore, the search for natural antioxidants is gaining popularity and much work in this direction is going on. Antioxidants are found in all parts of the plants like roots, stem, bark, leaves, peels, seeds, flowers, etc. With increasing recognition of herbal medicines as an alternative form of health care, screening of medicinal plants for biologically active compounds has become an important aspect of research today.

Considering the above, in this chapter presents seven plants that were screened for their antioxidant property. The plants were selected on the basis of their use in traditional medicine. The plants selected for this study were: *Aerva lanata* Linn., *Terminalia bellirica* (Gaertn.) Roxb., *Terminalia chebula* Retz., *Zea mays* L., *Terminalia catappa* L., *Tribulus terrestris* L., and *Boerhaavia diffusa* Linn.

6.2   MATERIAL AND METHODS

6.2.1   PLANT MATERIAL

A. lanata was collected from sea-shore region of Veraval; *Terminalia* species were collected from Jamnagar; *Z. mays* corn hair was collected from the local market of Rajkot. *T. terrestris* and *B. diffusa* were purchased from Rajkot. The plants were thoroughly washed, separated, and dried under shade. The dried plant parts were homogenized to fine powder and stored in airtight bottles which were later used for extraction.

6.2.2   DECOCTION EXTRACTION METHOD

Five grams of dried powder was extracted with 100 ml of ultra pure distilled water at 100°C for 30 min in water bath [9, 22]. It was filtered with 8 layers of muslin cloth and centrifuged at 5000 rpm in centrifuge (Remi centrifuge, India) for 10 min. The supernatant was collected and solvent was evaporated to dryness. The residue was weighed to obtain extractive yield and it was stored in airtight bottle at 4°C.

6.2.3 QUANTITATIVE PHYTOCHEMICAL ANALYSIS

6.2.3.1 Determination of Total Phenol Content (TPC)

The amount of total phenol content was determined by Folin-ciocalteu's reagent method [26]. The extract of 0.5 ml and 0.1 ml of Folin-ciocalteu's reagent (0.5 N) were mixed and the mixture was incubated at room temperature for 15 min. Then, 2.5 ml of sodium carbonate (2 M) solution was added and further incubated for 30 min at room temperature and the absorbance was measured at 760 nm (Systronics, India), against a blank sample. The calibration curve was made by preparing Gallic acid solution in distilled water. Total phenol content is expressed in terms of Gallic acid equivalent (mg/g of extracted compound). The assay was carried out in triplicate and the mean values with ± SEM are presented in this chapter.

6.2.3.2 Determination of Total Flavonoid Content (TFC)

The amount of flavonoid content was determined by aluminum choloride colormetric method [10]. The reaction mixture of 3.0 ml consisted of 1.0 ml of sample (1 mg/ml), 1.0 ml methanol, 1.0 ml of aluminum chloride (1.2%) and absorbance was measured at 415 nm using a UV-VIS spectrophotometer (Systronics, India), against a blank sample. The calibration curve was made by preparing quercetin solution in methanol. The flavonoid content is expressed in terms of quercetin equivalent (mg/g of extracted compound). The assay was carried out in triplicate and the mean values with ± SEM are presented in this chapter.

6.2.4 ANTIOXIDANTS ASSAYS

6.2.4.1 Determination of 2,2-diphenyl-1-picrylhydrazyl (DPPH) Free Radical Scavenging Assay

The free radical scavenging activity was measured by using DPPH by the modified method of Mc Cune and Jonh [25]. The reaction mixture 3.0 ml consisted of 1.0 ml methanol, 1.0 ml of different concentrations of extracts diluted by methanol and 1.0 ml DPPH (0.3 mM), and was incubated for 10 min in dark, after which the absorbance was measured at 517 nm UV-VIS Spectrophotometer (Systronics, India), against a blank sample. Ascorbic acid was used as positive control [23]. The percentage inhibition was calculated by comparing the results of the test and the control. Percentage of inhibition was calculated using the formula:

$$\% \text{ Inhibition} = [1 - (A/B)] \times 100 \tag{1}$$

where, B is the absorbance of the blank (DPPH plus methanol) and A is absorbance of the sample (DPPH, methanol, plus sample).

6.2.4.2 Determination of 2,2'-Azino-bis-(3-ethyl) benzothiazoline-6-sulfonic acid (ABTS) Radical Cation Scavenging Assay

The ABTS radical cation scavenging activity was determined by the method described by Re et al. [35]. ABTS radical cation were produced by reaction of ABTS (7 mM) with potassium sulfate (245 mM) and incubating the mixture at room temperature in the dark for 16 h. The working solution obtained was further diluted with methanol to give an absorbance of 0.85 ± 0.20. The 1.0 ml of different concentrations of extracts diluted by methanol was added to 3.0 ml of ABTS working solution. The reaction mixture was incubated at room temperature for 4 min and then the absorbance was measured at 734 nm using a UV-VIS Spectrophotometer (Systronics, India), against a blank sample. Ascorbic acid was used as a positive control [44]. Percentage of inhibition was calculated using the formula described in Eq. (1).

6.2.4.3 Ferric Reducing Antioxidant Power Assay (FRAP)

The reducing ability was determined by FRAP assay [7]. FRAP assay is based on the ability of antioxidant to reduce Fe^{3+} to Fe^{2+} in the presence of TPTZ forming an intense blue Fe^{2+}-TPTZ complex with an absorption maximum at 593 nm. This reaction is pH-dependent (optimum pH 3.6, 1 part 10 mM TPTZ and the reaction mixture is incubated at 37°C for 10 min. Then the absorbance was measured at 593 nm using a UV-VIS spectrophotometer (Systronics, India), against a blank sample FeSO$_4$ was used as a positive control [41]. The antioxidant capacity based on the ability to reduce ferric ions of sample was calculated from the linear calibration curve and expressed as M/g FeSO$_4$ equivalents per gram of extracted compound.

6.3   RESULTS AND DISCUSSION

6.3.1   EXTRACTIVE YIELD

The extractive yield of different plants is given in Figure 6.1. The extractive yield varied among the different parts of different plants. The maximum extractive yield was in *A. lanata* leaf (Figure 6.1A) and minimum was in *T. bellirica* stem (Figure 6.1B).

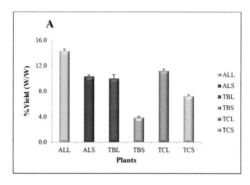

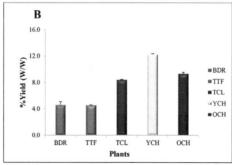

FIGURE 6.1   The extractive yield of leaf (top: A) and stem (bottom: B) extracts by decoction extraction method.

6.3.2 QUANTITATIVE PHYTOCHEMICAL ANALYSIS (QPA)

In this study, decoction extracts of seven different plants were evaluated for their total phenol content (TPC) and total flavonoid content (TFC). The amount of TPC and TFC varied among different plants and their parts. TPC and TFC of different plant extracts are given in Figure 6.2. In all plants, TPC was significantly more than that of TFC. In *A. lanata* leaf extract, TPC was slightly more than the stem extract and TFC was almost similar but less in both leaf and stem extracts as compared to TPC (Figure 6.2A). In *T. bellirica* leaf extract, TPC was more than the stem extract and TFC was almost similar but comparatively very less than TPC in both leaf and stem extracts (Figure 6.2B). In both leaf and stem extracts of *T. bellirica*, TPC was considerably more than that of leaf and stem extracts of *A. lanata*. In *T. chebula*, TPC and TFC of leaf and stem extracts almost had similar content but TPC was considerably more than TFC in both leaf and stem (Figure 6.2C). Both leaf and stem extracts of *T. chebula* had considerably more TPC than both leaf and stem extracts of *T. bellirica*. In *Z. mays* corn hair extracts, TPC was more in old corn hair as compared to young corn hair and TFC was almost similar but comparatively less in both young and old corn hair extracts

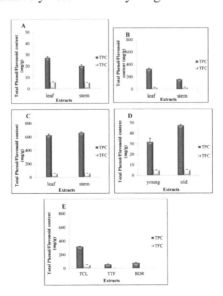

FIGURE 6.2 Total phenol and flavonoid content of *A. lanata* leaf and stem (A); *T. bellirica* leaf and stem (B); *T. chebula* leaf and stem (C); *Z. mays* hair young and old (D); and *T. catappa* leaf, *T. terrestris* fruit and *B. diffusa* root (E).

(Figure 6.2D). This plant had TPC and TFC similar to that of *A. lanata* but slightly more content than A. lanata plant parts. In *T. catappa* leaf extract, TPC was more as compared to *T. terrestris* fruit and *B. diffusa* root extracts. TFC was almost negligible in *T. terrestris* fruit extract and *B. diffusa* root extract. *T. catappa* leaf extract some TFC was found (Figure 6.2E).

6.3.3   ANTIOXIDANT ACTIVITY

Different antioxidant compounds may act through different distinct mechanism against different oxidizing agents and hence any one method cannot correctly evaluate the antioxidant efficacy of plant extracts [42]. Hence, in this study, three different antioxidant assays (DPPH, ABTS and FRAP) with different mechanisms of action were done to compare the antioxidant capacity of decoction extracts of seven plants and their different parts.

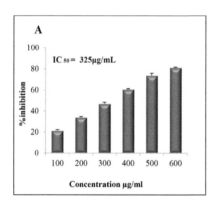
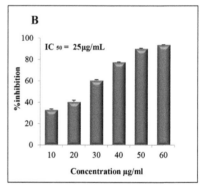
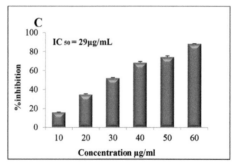

FIGURE 6.3 DPPH free radical scavenging activity of *A. lanata* stem (A); *T. bellirica* leaf (B); and stem (C).

In this study, seven different plants and their parts were evaluated for their DPPH free radical scavenging activity. Out of 11 extracts, six extracts showed DPPH free radical scavenging activity while IC_{50} value of other extracts was more than 1000 µg/ml. The DPPH activity of *A. lanata* stem extract is given in Figure 6.3A. The concentration range of *A. lanata* stem extract ranged from 100–600 µg/ml and IC_{50} value was 325 µg/ml (Figure 6.3A). The DPPH activity of *T. bellirica* leaf and stem extracts is given in Figures 6.3B and 6.3C, respectively. The concentration range of *T. bellirica* leaf and stem extracts ranged from 10–60 µg/ml and IC_{50} value were 25 µg/ml and 29 µg/ml, respectively, (Figures 6.3B and 6.3C). The DPPH activity of *T. chebula* leaf extract, *T. chebula* stem extract and *T. catappa* leaf extract is given in Figure 6.4. In all the three plant extracts,

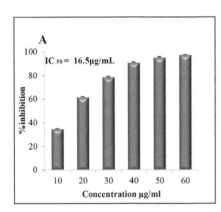
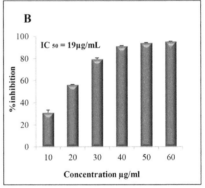
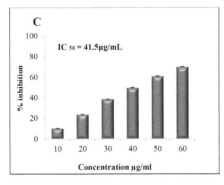

FIGURE 6.4   DPPH free radical scavenging activity of *T. chebula* leaf (A) and stem (B) and *T. catappa* leaf (C).

the concentration ranged from 10–60 µg/ml and their IC_{50} value was 16.5 µg/ml, 19 µg/ml and 41.5 µg/ml, respectively (Figure 6.4). The IC_{50} value of *T. chebula* leaf extract is very much near to that of IC_{50} value of standard ascorbic acid (11.5 µg/ml).

ABTS cation radical scavenging activity of seven plant extracts and standard ascorbic acid is given in Figure 6.5. IC_{50} value of standard ascorbic acid was 6.4 µg/ml. ABTS cation radical scavenging activity of *A. lanata* leaf and stem extracts and *T. bellirica* leaf and stem extracts is given in Figure 6.5. The concentration range of *A. lanata* leaf and stem extracts ranged from 100–600 µg/ml and 20–120 µg/ml and IC_{50} value was 410 µg/ml and 86 µg/ml, respectively (Figures 6.5A and 6.5B). *A. lanata* stem extract showed better ABTS cation radical scavenging

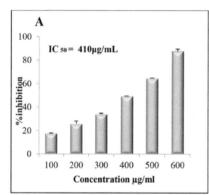
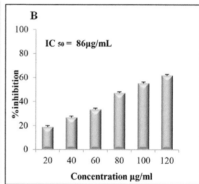
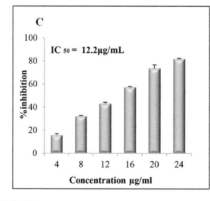
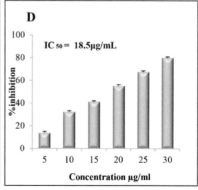

FIGURE 6.5 ABTS radical cation scavenging activity of *A. lanata* leaf (A) and stem (B); *T. bellirica* leaf (C) and stem (D).

activity than *A. lanata* leaf extract. The concentration range of *T. bellirica* leaf and stem extracts ranged from 4–24 µg/ml and 5–30 µg/ml and IC_{50} value was 12.2 µg/ml and 18.5 µg/ml, respectively (Figures 6.5C and 6.5D). *T. bellirica* leaf extract showed better ABTS cation radical scavenging activity than stem extract. In both the plants, the part that showed better activity was different.

ABTS cation radical scavenging activity of *T. chebula* leaf and stem extracts and *Z. mays* corn hair young and old extracts is given in Figure 6.6. The concentration range of *T. chebula* leaf and stem extracts ranged from 1–6 µg/ml and 2–12 µg/ml and IC_{50} value was 5.7 µg/ml and 7.1 µg/mL, respectively (Figures 6.6A and 6.6B). The concentration range of *Z. mays* corn hair young and old extracts ranged from 80–480 µg/ml and 25–150 µg/ml and IC_{50} value was 220 µg/ml and 105 µg/ml, respectively

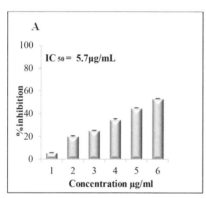
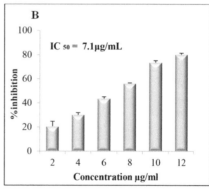
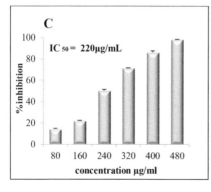
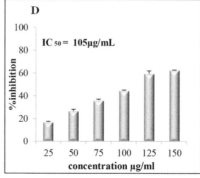

FIGURE 6.6   ABTS radical cation scavenging activity of *T. chebula* leaf (A) and stem (B); and *Z. mays* hair young (C) and old (D).

(Figures 6.6C and 6.6D). *T. chebula* extracts showed best ABTS cation radical scavenging activity which was comparable to that of standard ascorbic acid (IC$_{50}$ value = 6.4 μg/ml). ABTS cation radical scavenging activity of *T. chebula* leaf was better than that of standard ascorbic acid. *Z. mays* corn hair old extract showed better activity than young corn hair extract. ABTS cation radical scavenging activity of *T. catappa* leaf, *T. terrestris* fruit and *B. diffusa* root extracts is given in Figure 6.7. The concentration of *T. catappa* leaf, *T. terrestris* fruit and *B. diffusa* root extracts ranged from 4–24 μg/ml, 100–600 μg/ml and 70–420 μg/ml and IC$_{50}$ value was 13.8 μg/mL, 425 μg/mL and 235 μg/mL, respectively (Figure 6.7). When all the 3 plant extracts are compared, *T. chebula* leaf extract showed best ABTS cation radical scavenging activity. Its IC$_{50}$ value was near to that of standard ascorbic acid.

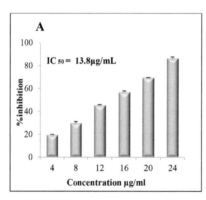
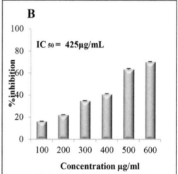
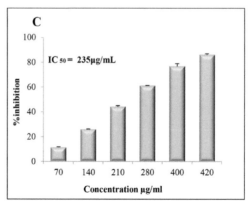

FIGURE 6.7 ABTS radical cation scavenging activity of *T. catappa* leaf (Top left: A), *T. terrestris* fruit (Top right: B), and *B. diffusa* root (Bottom: C).

FRAP of different plant extracts is given in Figure 6.8. The FRAP of seven plant extracts showed good FRAP activity except *A. lanata* leaf and stem extracts and *Z. mays* young corn hair extract. *A. lanata* leaf and stem extracts had almost similar FRAP activity but, comparatively very poor activity (Figure 6.8A). FRAP activity of *T. bellirica* leaf extract was slightly more than stem extract (Figure 6.8B). FRAP activity of *T. chebula* leaf and stem extracts was almost similar which was considerably more than the FRAP activity of leaf and stem of *T. bellerica* (Figure 6.8C). The FRAP activity of *Z. mays* young corn hair was negligible as compared to *Z. mays* old corn hair (Figure 6.8D). The FRAP activity of *Z. mays* old corn hair was more than that of *A. lanata* extracts but very much less than that of *T. bellerica* and *T. chebula* extracts. The FRAP activity of *T. catappa* leaf extract was comparable to other two *Terminalia* species and it showed considerably more activity than *T. terrestris* fruit and *B. diffusa* root extracts (Figure 6.8E). *T. catappa* leaf extract showed FRAP activity as compared to *T. Terrestris* fruit and *B. diffusa* root extracts; while *T. terrestris* fruit and *B. diffusa* root extracts showed moderate FRAP activity (Figure 6.8E).

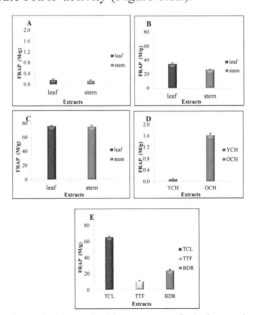

FIGURE 6.8 Ferric reducing antioxidant power of *A. lanata* leaf and stem (A), *T. bellirica* leaf and stem (B), *T. chebula* leaf and stem (C), *Z. mays* hair young and old (D) and *T. catappa* leaf, *T. terrestris* fruit and *B. diffusa* root (E).

6.4   DISCUSSIONS

In the present study, the extraction was done by decoction extraction method [16, 22]. Different plants showed different extractive yield and maximum was in *A. lanata* leaf. The extractive yield depends on solvent, time and temperature of extraction as well as chemical nature of sample. Under the same time and temperature conditions, the solvent used and the chemical property of the sample are the two most important factors [37].

Phenolic compounds which are secondary metabolites in plants are one of the most widely occurring group of phytochemicals that exhibit antiallergenic, antimicrobial, antiartherogenic, antithrombotic, antiinflammatory, vasodilatory and cardio protective effects [4, 27]. Due to the presence of conjugated ring structures and hydroxile groups; many phenolic compounds have the potential to function as antioxidants by scavenging or stabilizing free radicals involved in oxidative processes through hydrogenation or complexing with oxidizing species that are much stronger than those of vitamins C and E [5]. Flavonoids are the most common group of phenolic compounds widely distributed in all parts of plants, particularly in photosynthetic cells. It contains different structures, degrees of hydroxylation, polymerization, substitutions and conjugations. Flavonoids also exhibit different chemical properties due to their varying characteristics. All of the flavonoids contain a simple C6-C3-C6 carbon skeleton. Moreover, flavonoids exhibit a wide range of pharmacological activities, such as antioxidative, antidiabetic, antiinflammatory, antiallergic, antiviral, gastoprotective, and hepatoprotective activities [43].

Irrespective of part or plant, TPC was more than TFC; maximum TPC and TFC was found in *T. chebula* stem. When all the seven plant extracts are compared, TPC was maximum in *T. chebula* extracts followed by *T. bellirica* and *T. catappa* extracts as compared to other plant extracts (Figure 6.2). In *Z. mays* extracts also, old corn hair extract had more TPC as compared to young corn hair. Comparison of TPC and TFC content of plant extracts given an indication of its antioxidant efficacy since it is reported that there is a positive correlation between phenolic content and antioxidant activities [8]. Therefore, the results suggests or indicate that the best antioxidant activity will be shown by *Terminalia* species especially *T. chebula*.

The antioxidant capacity of different plants is different chiefly because they contain different phytoconstituents in different concentration.

They also vary in different parts of the same plant. Further, the mechanism of action of different antioxidant assays is different and hence it is always advisable and becomes necessary to evaluate more than one antioxidant assay when comparing the antioxidant efficacy of different plants. Hence in the present work, three different antioxidant assays were done.

DPPH radical is generally used as a stable free radical to determine antioxidant activity of natural compounds. DPPH is a stable nitrogen centered free radical which has been used to evaluate the radical quenching capacities of natural antioxidants, in a relatively short time, compared with other methods [2]. Radical scavenging activity is reported in terms of IC_{50} values, which indicates the concentration of antioxidant required to scavenge 50% of radicals in the reaction mixture. All the three Terminalia species showed good DPPH free radical scavenging activity with IC_{50} values ranging from 16.5 µg/ml to 41.5 µg/ml. The lowest IC_{50} values were of *T. chebula* leaf and stem indicating it to be the best to scavenge DPPH free radical. Additionally, the DPPH free radical scavenging activity of all the seven plant extracts showed a direct correlation with TPC. Floegel et al. [12], Khodja et al. [17] and Laith et al. [20] also showed a positive correlation between DPPH and phenolic content.

ABTS assay is an excellent tool for determining the antioxidant activity of hydrogen-donating antioxidants and of chain breaking antioxidants [21]. In the present work seven different plants and their parts were evaluated for their ABTS cation radical scavenging activity. Amongst all the plants and their parts, the best ABTS cation radical scavenging activity was shown by *T. chebula* leaf extract followed by *T. chebula* stem extract. This antioxidant activity also showed a direct correlation with TPC. *Rebaya* et al. [36] showed that ethanolic extract of *H. halimifolium* flowers rich in phenolic content showed good ABTS cation radical scavenging activity. Kim et al. [18] and El Hadjali et al. [11] suggested that plant extracts with more phenolic content showed higher ABTS radical scavenging activity and are more effective in termination of free radical reactions.

Different studies have indicated that the electron donation capacity, reflecting the reducing power, of bioactive compounds is associated with antioxidant activity [6, 38]. The presence of reductants such as antioxidant substances in the antioxidant sample causes the reduction of the Fe^{3+}/ferricyanide complex to the ferrous form. Therefore, Fe^{2+} can be monitored by measuring the formation of Perl's Prussian blue at 700 nm FRAP assay takes advantage of an electron transfer reaction in which a ferric salt is used as an oxidant [14]. Maximum FRAP activity was is *T. chebula*. Despite the fact that different antioxidant assays (DPPH, ABTS and FRAP) done in this work, have different reaction mechanisms and do not necessarily measure the same activity [32], the three methods clearly indicated that the studied plants possess variable but considerable antioxidant and antiradical activities [34].

All the seven plants showed a direct correlation between TPC and FRAP activity. The plant extracts which had more TPC showed more FRAP activity and vice versa. Similar results are reported by other researchers. Maizura et al. [24] showed a positive correlation between TPC and FRAP in some plant extracts. Moteriya et al. [28] showed that methanolic extract of *G. superba* rich in phenolic content showed good FRAP activity. Padalia and Chanda [30] reported positive correlation between TPC and FRAP in *Tagates erecta* flowers. It is noteworthy that the extracts showed good antioxidant activity in spite of being crude extracts. They can be further purified with different solvents and extraction methods to extract their fullest efficacy. Work in this direction is in progress under supervision of authors.

6.5 CONCLUSIONS

Eleven plant extracts belonging to seven plants were evaluated for their total phenol and flavonoid content and 3 antioxidant assays. All of them showed a significant positive correlation between antioxidant activity and total phenol content. The best activity was shown by *Terminalia* species may be because of its high phenol and flavonoid content. The antioxidant activity was as good as the standards used and in some cases even better than the standard. They can definitely be considered as a very promising natural source of antioxidants to develop drugs to treat various stress related diseases and disorders.

6.6 SUMMARY

The objective of this study was to evaluate the antioxidant abilities of seven different plant extracts (*Aerva lanata, Terminalia bellirica,*

Terminalia chebula, *Terminalia catappa*, *Zea mays*, *Tribulus terrestris* and *Boerhaavia diffusa*). Extraction was done by decoction extraction method. Three *in vitro* assays were employed, for example, 2,2-diphenyl-1-picryl hydrazole (DPPH), 2,2 Azino-bis-(3-ethyl) benzothiazoline-6-sulphonic acid (ABTS) and Ferric reducing antioxidant power (FRAP) to evaluate the antioxidant efficacy of different parts of seven plants. Total phenol and flavonoid content was also measured by Folin-ciocalteu's reagent method and aluminum chloride colorimetric method, respectively. All the three *Terminalia* species showed highest total phenol and flavonoid content and antioxidant activities. There was a direct correlation between total phenol content and antioxidant activity. The antioxidant efficacy of these plants can be contributed to its high phenol and flavonoid content. They can be used as a natural source of antioxidants to cure various diseases and disorders caused especially by stress.

KEYWORDS

- ABTS
- *aerva lanata*
- antioxidant activity
- *boerhaavia diffusa*
- decoction extracts
- DPPH
- extractive yield
- FRAP
- fruit
- IC_{50} values
- *in vitro* activity
- leaf
- medicinal plants
- oxidative stress
- quantitative phytochemical analysis
- radical

- root
- ROS
- stem
- *terminalia bellirica*
- *terminalia catappa*
- *terminalia chebula*
- TFC
- TPC
- *tribulus terrestris*
- *zea mays*

REFERENCES

1. Afolayan, A. J., & Jimoh, F. O. (2008). Nutritional quality of some wild leafy vegetables in South Africa. *International Journal of Food Sciences and Nutrition, 60*, 424–431.
2. Ahn, G. N., Kim, K. N., Cha, S. H., Song, C. B., Lee, J., & Heo, M. S. (2007). Antioxidant activities of phlorotannins purified from *Ecklonia cava* on free radical scavenging using ESR and H_2O_2-mediated DNA damage. *European Food Research and Technology, 226*, 71–79.
3. Almey, A., Khan, A. J., Zahir, S., Mustapha, S. K., Aisyah, M. R., & Kamarul, R. K. (2010). Total phenolic content and primary antioxidant activity of methanolic and ethanolic extracts of aromatic plants leaves. *International Food Research Journal, 17*, 1077–1084.
4. Alpinar, K., Ozyurek, M., Kolak, U., Guclu, K., Aras, C., Altun, M., Selic, S. E., Berker, K. L., Bektasoglu, B., & Apak, R. (2009). Antioxidant capacity of some food plants widely grown in Ayvalic of Turkey. *Food Science and Technology Research, 15*, 59–64.
5. Amic, D., Davidivic-Amic, D., Beslo, D., & Trinajstic, N. (2003). Structural-radical scavenging activity relationship of flavonoids. *Croatica Chemica Acta., 76*, 55–61.
6. Arabshahi-Delouee, S., & Urooj, A. (2007). Antioxidant properties of various solvent extract of mulberry (*Morus indica* L.) leaves. *Food Chemistry, 102*, 1233–1240.
7. Benzie, I. F., & Strain, J. J. (1999). The ferric reducing ability of plasma (FRAP) as a measure of "antioxidant power": the FRAP assay. *Analytica Biochemistry, 239*, 70–76.
8. Chanda, S., & Dave, R. (2009). *In vitro* models for antioxidant activity evaluation and some medicinal plants possessing antioxidant properties – An overview. *African Journal of Microbiology Research, 3*, 981–996.
9. Chanda, S., Amrutiya, N., & Rakholiya, K. (2013). Evaluation of antioxidant properties of some Indian vegetable and fruit peels by decoction extraction method. *American Journal of Food Technology, 8*, 173–182.

10. Chang, C. C., Yang, M. H., Wen, H. M., & Chern, J. C. (2002). Estimation of total flavonoid content in *Propolis* by two complementary colorimetric methods. *Journal of Food and Drug Analysis*, 10, 178–182.

11. El Hadj ali, I. B., Bahri, R., Chaouachi, M., Boussaid, M., & Skhiri, F. H. (2014). Phenolic content, antioxidant and alleopathic activities of various extracts of *Thymus Numidicus* Poir. Organs. *Industrial Crops and Products*, 62, 188–195.

12. Floegel, A., Kim, D. O., Chung, S. J., Koo, S. I., & Chun, O. K. (2011). Comparison of ABTS/DPPH assays to measure antioxidant capacity in popular antioxidant-rich US foods. *Journal of Food Composition and Analysis*, 24, 1043–1048.

13. Fusco, D., Colloca, G., Lo Monaco, M. R., & Cesari, M. (2007). Effects of antioxidant supplementation on the aging process. *Clinical Interventions in Aging*, 2, 377–387.

14. Gulcin, I., Elmastas, M., & Aboul-Enein, H. Y. (2012). Antioxidant activity of clove oil- A powerful antioxidant source. *Arabian Journal of Chemistry*, 5, 489–499.

15. Huang, D. H., Chen, C., Lin, C., & Lin, Y. (2005). Antioxidant and antiproliferative activities of water spinch (*Ipomoea aquatic* Forsk.) constituents. *Botanical Bulletin-Academia Sinica Taipei*, 46, 99–106.

16. Kaneria, M., Bapodra, M., & Chanda, S. (2012). Effect of extraction technique and solvent on antioxidant activity of pomegranate (*Punica granatum* L.) leaf and stem. *Food Analytical Methods*, 5, 396–404.

17. Khodja, N. K., Makhlouf, L. B., & Madani, K. (2014). Phytochemical screening of antioxidant and antibacterial activities of methanolic extracts of some Lamiaceae. *Industrial Crops and Products*, 61, 41–48.

18. Kim, J., Choi, J. N., Kang, D., Son, G. H., Kim, Y. S., Choi, H. K., Kwon, D. Y., & Lee, C. H. (2011). Correlation between antioxidative activities and metabolite changes during Cheonggukjang fermentation. *Bioscience, Biotechnology and Biochemistry*, 75, 732–739.

19. Kwon, Y., Apostolidis, E., & Shetty, K. (2008). *In vitro* studies of egg plant (*Solanum melongena*) phenolics as inhibitors of key enzymes relevant for type 2 diabetes and hypertension. *Bioresource Technology*, 99, 2981–2988.

20. Laith, A. A., Alkhuzai, J., & Freije, A. (2015). Assessment of antioxidant activities of three wild medicinal plants from Bahrain. *Arabian Journal of Chemistry*, (in press) *doi: 10.1016/j.arabjc.2015.03.004.*

21. Leong, L. P., & Shui, G. (2002). An investigation of antioxidant capacity of fruits in Singapore markets. *Food Chemistry*, 76, 69–75.

22. Li, H. B., Jiang, Y., Wongm, C. C., Cheng, K. W., & Chen, F. (2007). Evaluation of two methods for the extraction of antioxidants from medicinal plants. *Analytical and Bioanalytical Chemistry*, 388, 483.

23. Liu, J., Wang. C., Wang. Z., Zhang, C., Lu, S., & Liu, J. (2011). The antioxidant and free radical scavenging activities of extract and fractions from corn silk (*Zea mays* L.) and related flavones glycosides. *Food Chemistry*, 126, 261–269.

24. Maizura, M., Aminah, A., & Wan Aida, W. M. (2011). Total phenolic content and antioxidant activity of kesum (*Polygonum minus*), ginger (*Zingiber officinale*) and turmeric (*Curcuma longa*) extract. *International Food Research Journal*, 18, 529–534.

25. Mc Cune, L. M., & Johns, T. (2002). Antioxidant activity in medicinal plants associated with the symptoms of *diabetes mellitus* used by the indigenous peoples of the North American boreal forest. *Journal of Ethnopharmacology*, 82, 197–205.

26. Mc Donald, S., Prenzler, P. D., Autolovich, M., & Robards, K. (2001). Phenolic content and antioxidant activity of olive extracts. *Food Chemistry, 73*, 73–84.

27. Middleton, E., Kandaswami, C., & Theoharides, T. C. (2000). The effect of plant flavonoids on mammalian cells: implications for inflammation, heart diseases and cancer. *Pharmacological Reviews, 52*, 673–751.

28. Moteriya, P., Ram, J., Rathod, T., & Chanda, S. (2014). *In vitro* antioxidant and antibacterial potential of leaf and stem of *Gloriosa superba* L. *American Journal of Phytomedicine and Clinical Therapeutics, 2*, 703–718.

29. National Pharmacopoeia Committee (2010). Pharmacopoeia of the people's republic of China, Beijing. *Medical Science and Technology,* Press of China.

30. Padalia, H., & Chanda, S. (2014). Evaluation of antioxidant efficacy of different fraction of *Tagetes erecta* L. flower. *IOSR Journal of Pharmacy and Biological Sciences, 9*, 28–37.

31. Pratap, S., & Pandey, S. (2014). A review on herbal antioxidants. *Journal of Pharmacognosy Phytochemistry, 1*, 26–37.

32. Prior, R. L., Wu, X. L., & Schaich, K. (2005). Standardized methods for the determination of antioxidant capacity and phenolics in foods and dietary supplements. *Journal of Agricultural and Food Chemistry, 53*, 4290–4302.

33. Quinming, Y., Xianhui, P., Weibao, K., Hong, Y., Yidan, S., & Zhang, L. (2010). Antioxidant activities of malt extract from barley (*Hordeum vulgare* L.) towards various oxidative stress *in vitro* and *in vivo*. *Food Chemistry, 118*, 84–89.

34. Rakholiya, K., Kaneria, M., Nagani, K., Patel, A., & Chanda, S. (2015). Comparative analysis and simultaneous quantification of antioxidant capacity of four *Terminalia* species using various photometric assays. *World Journal of Pharmaceutical Research, 4*, 1280–1296.

35. Re, R., Pellegrini, N., Proteggentle, A., Pannala, A., Yang, M., & Rice-Evans, C. (1999). Antioxidant activity applying an improved ABTS radical cation decolorization assay. *Free Radical Biology and Medicine, 26*, 1231–1237.

36. Rebaya, A., Belghit, S. I., Baghdikian, B., Leddet, V. M., Mabrouki, F., Olivier, E., Cherif, J. K., & Ayadi, M. T. (2014). Total phenolic, total flavonoid, tannin content, and antioxidant capacity of *Halimium halimifolium* (Cistaceae). *Journal of Applied Pharmaceutical Science, 5*, 52–57.

37. Shimada, K., Fujikawa, K., Yahara, K., & Nakamaru, T. (1992). Antioxidant properties of Xanthan on the autoxidation of soybean oil in cyclodextrin emulsion. *Journal of Agricultural and Food Chemistry, 40*, 945–948.

38. Siddhuraju, P., Mohan, P. S., & Becker, K. (2002). Studies on antioxidant activity of Indian laburnum (*Cassia fistula* L.): a preliminary assessment of crude extract from stem bark, leaves, flowers and fruit pulp. *Food Chemistry, 79*, 61–67.

39. Sun, L., Zhang, J., Lu, X., Zhang, L., & Zhang, Y. (2011). Evaluation to the activity of total flavonoid extract from persimmon (*Diospyros kaki* L.) leaves. *Food and Chemical Toxicology, 49*, 2689–2696.

40. Tewari, I., Sharma, I., & Gupta G. L. (2014). Synergistic antioxidant activity of three medicinal plants *Hypericum perforatum, Bacopa monnieri, Camellia sinensis. Indo American Journal of Pharmaceutical Research, 4*, 2563–2568.

41. Thondre, P. S., Ryan, L., & Henri, C. J. K. (2011). Barley β-glucan extracts as rich sources of polyphenols and antioxidants. *Food Chemistry, 126*, 72–77.

42. Xiao, Y., Wang, L., Rui, X., Li, W., Chen, X., Jiang, M., & Dong, M. (2015). Enhancement of the antioxidant capacity of soy whey by fermentation with *Lactobacillus plantarum* B1–6. *Journal of Functional Foods, 12*, 33–44.

43. Yao, L. H., Jiang, Y. M., Shi, J., Tomas-Barberan, F. A., Datta, N., Singanusong, R., & Chen, S. S. (2004). Flavonoids in food and their health benefits. *Plant Foods for Human Nutrition, 59*, 113–122.

44. Zhou, H. C., Lin, Y. M., Li, Y. Y., Li, M., Wei, S. D., Chai, W. M., & Tam, N. F. (2011). Antioxidant properties of polymeric proanthocyanidins from fruit stones and pericarps of *Litchi chinensis* Sonn. *Food Research International, 44*, 558–564.

CHAPTER 7

ANTIOXIDANT ACTIVITIES OF SOME MARINE ALGAE: CASE STUDY FROM INDIA

K. D. RAKHOLIYA, J. T. PATEL, V. D. VORA, G. S. SUTARIA,
R. M. PATEL, R. A. DAVE, and M. J. KANERIA

CONTENTS

7.1 INTRODUCTION

Oxidative stress depicts the existence of products called free radicals and reactive oxygen species (ROS), which are formed under normal physiological conditions but become deleterious when not being eliminated by the endogenous systems. In fact, oxidative stress results from an imbalance between the generation of ROS and endogenous antioxidant systems. ROS such as superoxide anions, hydroxyl radicals and hydrogen peroxides are cytotoxic and give rise to tissue injuries [18, 41, 47]. Excessive amount of ROS is harmful because they initiate bimolecular oxidation, which leads to cell death and creates oxidative stress. In addition, oxidative stress causes inadvertent enzyme activation and oxidative damage to cellular system [10, 68].

Marine algae are widely spread throughout the coastal land around many continents focused on the different aspects concerned with their nature and growth. Several species are reported to play a role in prevention of bio-fouling phenomena; others are involved in the chemical, pharmaceutical and food industries. Marine algae have been consumed since ancient times, which are rich in vitamins, minerals, dietary fibers, proteins, polysaccharides, and various functional polyphenols. Moreover, marine algae are considered to be a rich source of antioxidant substances [7, 14, 24].

Marine algae, like all photosynthesizing plants are exposed to a combination of light and high oxygen concentrations, which lead to the formation of free radicals and other strong oxidizing agents. The element of the photosynthetic apparatus is vulnerable to photodynamic damage, because polyunsaturated fatty acids are important structural components of the thylakoid membrane [39, 61].

Seaweed contains several bioactive compounds, such as pigments (carotenoids, chlorophylls and tocopherols), sulfated polysaccharides (fucoidan), amino acids, and mono- and polyphenols [17, 23, 34, 55, 56]. Marine algae produce high amounts of polyphenolic secondary metabolites, phlorotannins [13, 65, 66], which are the main contributors to the overall antioxidant activity of extracts from algae [13, 17, 22, 25, 33, 35, 44, 57, 65, 66].

Hence, many types of seaweeds have been examined to identify new and effective antioxidant compounds, as well as to elucidate the mechanisms of cell proliferation, antiinflammation and apoptosis. Recently, many countries are more closely looking at algae as a potent target for bioactive substances because they have showed some possibility for valuable uses and applications [6, 36, 58].

In the present chapter, the antioxidant activity of acetone extract of some marine algae collected from the Gujarat coast in India, have been assessed.

7.2   MATERIALS AND METHODS

7.2.1   COLLECTION AND EXTRACTION

The algae were collected from the coastal region of Gujarat, India (Table 7.1). The marine algae were thoroughly washed and shade dried,

TABLE 7.1   List of the Marine Algae and Corresponding Class Collected to Evaluate Their Antioxidant Potential

Algae	Class
Caulerpa recimosa	Chlorophyceae
Caulerpa scalpelliformis	Chlorophyceae
Caulerpa taxifolia	Chlorophyceae
Champia indica	Rhodophyceae
Codium decorticatum	Chlorophyceae
Cystoseira indica	Phaeophyceae
Dictiyota dichotoma	Phaeophyceae
Dictyopteris delicatula	Phaeophyceae
Gracilaria corticata	Rhodophyceae
Halymenia venusta	Rhodophyceae
Helimeda macroloba	Chlorophyceae
Hypnea valentiae	Rhodophyceae
Padina boergesenii	Phaeophyceae
Sargassum cinereum	Phaeophyceae
Sargassum tenerimum	Phaeophyceae
Scinaia hatei	Rhodophyceae
Ulva lactuca	Chlorophyceae
Ulva reticulate	Chlorophyceae

homogenized to fine powder and stored in airtight bottles. The dried powder of marine algae was extracted by cold percolation method [45, 50]. Ten gram of dried powder was taken in 100 ml of hexane in a conical flask, plugged with cotton wool and then kept on a rotary shaker at 120 rpm for 24 h. After 24 h, the extract was filtered with eight layers of muslin cloth, centrifuged at 5000 rpm for 10 min. The supernatant was collected and the solvent was evaporated. The residue was then taken in 100 ml of acetone in a conical flask. Then the procedure followed was same as above, and the dry extract was stored at 4°C in airtight bottles.

7.2.2   DETERMINATION OF TOTAL PHENOL CONTENT

The amount of total phenol content (TPC), in acetone extract was determined by Folin-Ciocalteu's reagent method [40]. The extract (0.5 ml) and 0.1 ml Folin-Ciocalteu's reagent (0.5 N) were mixed, and the mixture was incubated at room temperature for 15 min. Then, 2.5 ml saturated sodium carbonate solution was added and further incubated for 30 min at room temperature, and the absorbance was measured at 760 nm using a UV-VIS Spectrophotometer (Shimadzu, Japan) against a blank sample. Total phenol content is expressed in terms of gallic acid equivalent (mg/g of extracted compound).

7.2.3 DETERMINATION OF DPPH FREE RADICAL SCAVENGING CAPACITY

The free radical scavenging activity of acetone extract was measured by using DPPH free radical – modified method as described by McCune and Johns [42]. The reaction mixture (3.0 ml) consisted of 1.0 ml DPPH in methanol (0.3 mM), 1.0 ml methanol and 1.0 ml of different concentrations (2 to 1000 µg/ml) of extract diluted by methanol; and it was incubated for 10 min. in dark, after which the absorbance was measured at 517 nm using a UV-VIS Spectrophotometer (Shimadzu, Japan), against a blank sample. Ascorbic acid (2 to 16 µg/ml) was used as positive control [37, 52]. Percentage of inhibition was calculated using the following formula:

$$\% \text{ Inhibition} = [1 - (A/B)] \times 100 \tag{1}$$

where, B = absorbance of blank (DPPH plus methanol), A = absorbance of sample (DPPH, methanol plus sample).

7.2.4 DETERMINATION OF SUPEROXIDE ANION RADICAL SCAVENGING CAPACITY

The superoxide (SO) anion radical scavenging activity of acetone extract was measured by the modified method as described by Robak and Gryglewski [54]. Superoxide radicals are generated by oxidation of NADH and assayed by the reduction of NBT. The reaction mixture (3.0 ml) consisted of 1.0 ml of different concentrations (20 to 1000 µg/ml) of extract diluted by distilled water, 0.5 ml Tris-HCl buffer (16 mM, pH 8), 0.5 ml NBT (0.3 mM), 0.5 ml NADH (0.936 mM) and 0.5 ml PMS (0.12 mM).

The superoxide radical generating reaction was started by the addition of PMS solution to the reaction mixture. The reaction mixture was incubated at 25°C for 5 min and then the absorbance was measured at 560 nm using a UV-VIS Spectrophotometer (Shimadzu, Japan), against a blank sample. Gallic acid (50 to 225 µg/ml) was used as a positive control [52, 54]. Percentage of inhibition was calculated using Eq. (1).

7.2.5   DETERMINATION OF ABTS RADICAL CATION SCAVENGING CAPACITY

The ABTS cation radical scavenging activity of acetone extract was measured by the modified method as described by Re et al. [53]. ABTS radical cations are produced by reacting ABTS (7 mM) and potassium persulfate (2.45 mM) and incubating the mixture at room temperature in the dark for 16 h. The ABTS working solution obtained was further diluted with methanol to give an absorbance of 0.85 ± 0.20 at 734 nm. 1.0 ml of different concentrations (1 to 1000 µg/ml) of acetone extract diluted by methanol was added to 3.0 ml of ABTS working solution. The reaction mixture was incubated at room temperature for 4 min and then the absorbance was measured at 734 nm using a UV-VIS Spectrophotometer (Shimadzu, Japan) against a blank sample. Ascorbic acid (1 to 10 µg/ml) was used as a positive control [52, 71]. Percentage of inhibition was calculated using Eq. (1).

7.2.6   FERRIC REDUCING ANTIOXIDANT POWER (FRAP)

The reducing ability of acetone extract was determined by *ferric reducing antioxidant power* (FRAP) assay of Benzie and Strain [5]. FRAP assay is based on the ability of antioxidants to reduce Fe^{3+} to Fe^{2+} in the presence of TPTZ, forming an intense blue Fe^{2+}-TPTZ complex with an absorption maximum at 593 nm. This reaction is pH-dependent (optimum pH 3.6). 100 µl extract was added to 3.0 ml FRAP reagent (10 parts 300 mM sodium acetate buffer at pH 3.6, 1 part 10 mM TPTZ in 40 mM HCl, and 1 part 20 mM $FeCl_3$), and the reaction mixture was incubated at 37°C for 10 min. Then the absorbance was measured at 593 nm using a UV-VIS Spectrophotometer (Shimadzu, Japan), against a blank sample. The calibration curve was made by preparing a $FeSO_4$ (100 to 1,000 μM/ml) solution in distilled water [52, 63]. The antioxidant capacity based on the ability to reduce ferric ions of sample was calculated from the linear calibration curve and expressed as mM $FeSO_4$ equivalents per gram of extracted compounds.

7.3   RESULTS AND DISCUSSION

Phenols are very important constituents because of their scavenging ability due to their hydroxyl group and known as powerful chain breaking antioxidants [12, 29, 31, 52]. The action of polyphenols is believed to be mainly due to their redox properties, which play an important role in adsorbing and neutralizing free radicals, quenching single and triplet oxygen, or decomposing peroxidase. Many phytochemical posses significant antioxidant capacity that are associated with lower occurrence and lower mortality rates of several human diseases [3, 11, 59].

In the present work, acetone extract of 18 different marine algae were evaluated for to estimate their total phenol content. The total phenol content of acetone extracts of screened marine algae is shown in Table 7.2. The amount of total phenol content acetone extract of 18 different marine algae was varied greatly. Amongst all the algae, phenol content ranged from 9.61 to 99.63 mg/g (Table 7.2). Amongst all the algae, the highest total phenol content was in *Champia indica* (99.63 mg/g, Table 7.2) followed by *Halymenia venusta* (92.85 mg/g, Table 2). The lowest amount of total phenol content was in *Padina boergesenii* (9.61 mg/g, Table 7.2) followed by *Caulerpa taxifolia* (10.52 mg/g, Table 7.2).

Antioxidant compounds act by several mechanisms such as, inhibition of generation and scavenging activity against reactive oxygen species (ROS), inhibition of oxidative enzymes and reducing ability. Free radicals and other reactive species cause the oxidation of biomolecules and loss of function of enzymes, which leads to cell injury and eventually necrotic cell death or apoptosis [8, 19, 32]. Therefore, antioxidant activities of different marine algae were analyzed by the different assays, such as DPPH free radical, superoxide anion radical, ABTS radical cation and ferric reducing antioxidant power.

TABLE 7.2 Total Phenol Content and Antioxidant Activity of Acetone Extract of 18 Different Algae

Algae name	TPC (mg/g)	DPPH	SO	ABTS	FRAP (mM/g)
		IC_{50} values (µg/ml)			
Caulerpa recimosa	62.27	685	>1000	540	843
Caulerpa scalpelliformis	21.54	>1000	>1000	630	351
Caulerpa taxifolia	10.52	>1000	>1000	>1000	38
Champia indica	99.63	174	520	380	1051
Codium decorticatum	18.25	>1000	>1000	775	105
Cystoseira indica	45.38	545	>1000	860	426
Dictiyota dichotoma	51.27	830	>1000	650	485
Dictyopteris delicatula	64.52	510	>1000	435	762
Gracilaria corticata	85.54	765	790	685	957
Halymenia venusta	92.85	335	580	530	913
Helimeda macroloba	75.59	845	>1000	870	359
Hypnea valentiae	81.24	670	910	865	578
Padina boergesenii	9.61	>1000	>1000	>1000	31
Sargassum cinereum	42.25	580	875	740	843
Sargassum tenerimum	17.36	>1000	>1000	825	234
Scinaia hatei	36.59	670	855	630	931
Ulva lactuca	26.53	>1000	>1000	975	128
Ulva reticulate	22.15	930	>1000	780	653
Standard	—	**11.4**	**185**	**6.5**	—

According to the mode of action, the antioxidants can be classified as free radical terminators, chelators of metal ions capable of catalyzing lipid oxidations or as oxygen scavengers that react with oxygen in closed systems. Among the various methods to evaluate the radical scavenging activity of natural compounds, DPPH method received more attention due to its fast, reliable results, relatively simple, stable and the DPPH is available commercially in high purity [27, 30, 38]. Concentration of sample at which the inhibition percentage reaches 50% is its IC_{50} value. IC_{50} value is negatively related to the antioxidant activity, as it expresses the amount of antioxidant needed to decrease its radical concentration by 50%. The lower the IC_{50} value, the higher is the antioxidant activity of the tested sample. In the present work, acetone extract of 18 different marine algae were evaluated for their DPPH free radical scavenging activity. Out of 18 algal extracts investigated, 6 algal extracts showed IC_{50} values more than 1000 µg/ml (Table 7.2), while the remaining 12 extracts showed varied levels of DPPH free radical scavenging activity (Table 7.2). IC_{50} values ranged from 174 to 930 µg/ml (Table 7.2). Amongst all algal extracts, the lowest IC_{50} value was of *Champia indica* (IC_{50} value = 174 µg/ml, Table 7.2) and the highest IC_{50} value was of *Ulva reticulate* (IC_{50} value = 930 µg/ml, Table 7.2). Ascorbic acid was used as standard and its IC_{50} value was 11.4 µg/ml (Table 7.2).

Superoxide is the first reduction product of molecular oxygen, a highly toxic radical. These radicals are produced abundant amount in all aerobic cells by several enzymatic and nonenzymatic pathways and attack a number of biomolecules including DNA [2, 49, 67]. It also forms an important source of other deterious radicals such as hydroxyl and hydro peroxide, which initiate free radical chain reactions [20, 26, 69]. In the PMS/NADH-NBT system, the superoxide anion derived from dissolved oxygen from PMS/NADH coupling reaction reduces NBT. The decrease of absorbance at 560 nm with antioxidants thus indicates the consumption of superoxide anion in the reaction mixture [27, 29, 52, 64]. In the present work, acetone extract of 18 different marine algae were evaluated for their superoxide anion radical scavenging activity. Out of 18 algal extracts investigated, 12 algal extracts showed IC_{50} values more than 1000 µg/ml (Table 7.2), while the remaining 6 extracts showed varied levels of superoxide anion radical scavenging activity (Table 7.2). IC_{50} values ranged from 520 to 910 µg/ml (Table 7.2). Amongst all algal extracts, the lowest IC_{50} value was of *Champia indica* (IC_{50} value = 520 µg/ml, Table 7.2) followed by *Halymenia venusta* (IC_{50} value = 580 µg/ml, Table 7.2) and the highest IC_{50} value was of *Hypnea valentiae* (IC_{50} value = 910 µg/ml, Table 7.2). Gallic acid was used as standard and its IC_{50} value was 185 µg/ml (Table 7.2).

ABTS assay is better to assess the antiradical capacity of both hydrophilic and lipophilic antioxidants because it can be used in both organic and aqueous solvent systems as compared to other antioxidant assays [16, 51]. $ABTS^{·+}$ is a blue chromophore generated from the oxidation of ABTS by potassium persulfate, in the presence of the plant extract, preformed cation radical gets reduced and employs a specific absorbance at 734 nm, a wavelength remote from the visible region and it requires a short reaction time. This assay is an excellent tools for determining the antioxidant activity of hydrogen donating antioxidants (scavengers of aqueous phase radicals) and of chain-breaking antioxidants (scavengers of lipid peroxyl radicals) [1, 43, 48]. In the present work, acetone extract of 18 different marine algae were evaluated for their ABTS radical cation scavenging activity. Out of 18 algal extracts investigated, only 2 algal extracts showed IC_{50} values more than 1000 µg/ml (Table 7.2), while the remaining 16 extracts showed varied levels of ABTS radical cation scavenging activity (Table 7.2). IC_{50} values ranged from 380 to 975 µg/ml (Table 7.2). Amongst all algal extracts, the lowest IC_{50} value was of *Champia indica* (IC_{50} value = 380 µg/ml, Table 7.2) and the highest IC_{50} value was of *Ulva lactuca* (IC_{50} value = 975 µg/ml, Table 7.2). Ascorbic acid was used as standard and its IC_{50} value was 6.5 µg/ml (Table 7.2).

Ferric reducing antioxidant power (FRAP) assay is commonly used for the routine analysis of single antioxidant and total antioxidant activity of plant extracts [28, 70]. Antioxidative activity has been proposed to be related to its reducing power. Therefore, the antioxidant potential were estimated for their ability to reduce TPTZ–Fe (III) complex to TPTZ–Fe (II). FRAP assay as used by several authors for the assessment of antioxidant activity of various samples [9, 52, 60, 62]. The FRAP assay treats the antioxidants contained in the samples as reductants in a redox linked colorimetric reaction, and the value reflects the reducing power of antioxidants. The procedure is relatively simple and easy to standardize. The antioxidant potentials of different samples were estimated by their ability to reduce the TPTZ–Fe(III) complex to the TPTZ–Fe(II) complex. At low pH (optimum pH 3.6) a ferric salt, $Fe(III)(TPTZ)_2Cl_3$ (TPTZ) (as an oxidant), is reduced by antioxidants to its intense blue colored form Fe^{2+}-TPTZ complex (Fe^{2+} tripyridyltriazine) with maximum absorbance at 593 nm [5, 15, 21]. The FRAP of acetone extracts of screened algae is shown in Table 7.2. The FRAP of algal extracts revealed that all of them showed varied level of activity. The activity ranged from 31 to 1051 mM/g. Among all the algal extracts, highest activity was in *Champia indica* (1051 mM/g, Table 7.2), while the lowest activity was in *Padina boergesenii* (31 mM/g, Table 7.2) and *Caulerpa taxifolia* (38 mM/g, Table 7.2).

There are many reports that there is a direct correlation between phenol content and antioxidant activity [4, 26, 46, 48, 52]. In the present study also, the acetone extract of marine algae screened had more phenol content and correspondingly more antioxidant activity, as well as *C. indica* had highest phenol content and showed good antioxidant activity as compared to other algae screened.

7.4 CONCLUSIONS

Marine algae have several active chemicals such as antioxidant compounds. In the present investigation, the acetone extract of *Champia indica* showed good antioxidant activity. Thus, it can be a good source of antioxidant compounds. Marine organisms are currently undergoing detailed investigations with the objective of isolating biologically active molecules along with the search for new compounds. Moreover, the Gujarat coast is a potential source of a variety of biologically active marine organisms and it is hoped that the present results will provide a starting point for investigations aimed at exploiting new natural antioxidant substances.

7.5 SUMMARY

In the present investigation, antioxidant potential of 18 marine algae from Gujarat coast – India were evaluated. Algal samples were collected, dried and crushed to make it fine powder. Algae powder was defatted with hexane and then extracted using acetone by cold percolation extraction method. Antioxidant activity of acetone extract of 18 algae was evaluated by measuring the quenching ability on DPPH free radical, superoxide anion radical, ABTS radical cation, ferric reducing antioxidant power and total phenolic content. As a result of the study, *Champia indica* acetone extract was found to have a good antioxidant potential as well as high phenolic content. Based on the results, tested algae appear to be good natural antioxidant agent. Identification of the active compounds of these algal species will lead to their evaluation of considerable commercial potential in food production, medicine and cosmetic industry.

KEYWORDS

- **ABTS**
- algae
- **antioxidant activity**
- **bioactivity**
- *champia indica*
- chlorophyceae
- **cold percolation extraction**
- **DPPH**
- **FRAP**
- *gracilaria corticata*
- **gujarat**
- *halymenia venusta*
- **hexane, acetone**
- *hypnea valentiae*
- india
- **oxidative stress**
- **phaeophyceae**
- **radicals**
- rhodophyceae
- **ROS**
- *sargassum cinereum*
- *scinaia hatei*
- seaweed
- **SO**
- **TPC**

REFERENCES

1. Adedapo, A. A., Jimoh, F. O., Koduru, S., Masika, P. J., & Afolayan, A. J. (2008). Evaluation of the medicinal potentials of the methanol extracts of the leaves and stems of *Halleria lucida*. *Bioresource Technology*, *99*, 4158–4163.

2. Alrahmany, R., & Tsopmo, A. (2012). Role of carbohydrases on the release of reducing sugar, total phenolics and on antioxidant properties of oat bran. *Food Chemistry, 132,* 413–418.

3. Amarowicz, R., & Pegg, R. B. (2001). Assessment of the antioxidant and prooxidant activities of tree nut extracts with a pork model system. *Animal Reproduction and Food Research, 10,* 745–747.

4. Babbar, N., Oberoi, H. S., Sandhu, S. K., & Bhargav, V. K. (2012). Influence of different solvents in extraction of phenolic compounds from vegetable residues and their evaluation as natural sources of antioxidants. *Journal of Food Science and Technology, 51,* 2568–2575.

5. Benzie, I. F., & Strain, J. J. (1996). The ferric reducing ability of plasma (FRAP) as a measure of "antioxidant power": the FRAP assay. *Analytical Biochemistry, 239,* 70–76.

6. Blunt, J. W., Copp, B. R., Munro, M. H. G., Northcote, P. T., & Prinsep, M. R. (2005). Marine natural products. *Natural Product Reports, 22,* 15–61.

7. Cardozo, K. H. M., Guaratini, T., Barros, M. P., Falcao, V. R., Tonon, A. P., Lopes, N. P., Campos, S., Torres, M. A., Souza, A. O., Colepicolo, P., & Pinto, E. (2007). Metabolites from algae with economical impact. *Comparative Biochemistry and Physiology, 46,* 60–78.

8. Chanda, S., & Dave, R. (2009). *In vitro* models for antioxidant activity evaluation and some medicinal plant possessing antioxidant properties: An overview. *African Journal of Microbiological Research, 3,* 981–996.

9. Chanda, S., Amrutiya N., & Rakholiya, K. (2013). Evaluation of antioxidant properties of some Indian vegetable and fruit peel by decoction extraction method. *American Journal of Food Technology, 8,* 173–182.

10. Chanda, S., Dave, R., & Kaneria, M. (2011). *In vitro* antioxidant property of some Indian medicinal plants. *Research Journal of Medicinal Plant, 5,* 169–179.

11. Chanda, S., Dudhatra, S., & Kaneria, M. (2010). Antioxidative and antibacterial effects of seeds and fruit rind of nutraceutical plants belonging to the Fabaceae family. *Food and Function, 1,* 308–315.

12. Chandini, S. K., Ganesan, P., & Bhaskar, N. (2008). *In vitro* antioxidant activities of three selected brown seaweeds of India. *Food Chemistry, 107,* 707–713.

13. Chkhikvishvili, I. D., & Ramazanov, Z. M. (2000). Phenolic substances of brown algae and their antioxidant activity. *Applied Biochemistry and Microbiology, 36,* 289–291.

14. Darcy-Vrillon, B. (1993). Nutritional aspects of the developing use of marine macroalgae for the human food industry. *International Journal of Food Sciences and Nutrition, 44,* S23–S35.

15. Daur, I. (2015). Chemical composition of selected Saudi medicinal plants. *Arabian Journal of Chemistry, 8,* 329–332.

16. Durmaz, G. (2012). Freeze-dried ABTS$^{\cdot+}$ method: A ready-to-use radical powder to assess antioxidant capacity of vegetable oils. *Food Chemistry, 133,* 1658–1663.

17. Farvin, K. H. S., & Jacobsen, C. (2013). Phenolic compounds and *in vitro* antioxidant activities of selected species of seaweed from Danish coast. *Food Chemistry, 138,* 1670–1681.

18. Feillet-Coudray, C., Sutra, T., Fouret, G., Ramos, J., Wrutniak-Cabello, C., Cabello, G., Cristol, J. P., & Coudray, C. (2009). Oxidative stress in rats fed a high-fat high sucrose diet and preventive effect of polyphenols: involvement of mitochondrial and NAD(P)H oxidase systems. *Free Radical Biology and Medicine, 46,* 624–632.

19. Fernandez-Agullo, A., Pereira, E., Freire, M. S., Valentao, P., Andrade, P. B., Gonzalez-Alvarez, J., & Pereira, J. A. (2013). Influence of solvent on the antioxidant and antimicrobial properties of walnut (*Juglans regia* L.) green husk extracts. *Industrial Crops and Products, 42,* 126–132.

20. Gulcin, I., Huyut, Z., Elmastas, M., & Aboul-Enein, H. Y. (2010). Radical scavenging and antioxidant activity of tannic acid. *Arabian Journal of Chemistry, 3,* 43–53.

21. Hajimahmoodi, M., Sadeghi, N., Jannat, B., Oveisi, M. R., Madani, S., Kiayi, M., Akrami, M., Akrami, M. R., & Ranjbar, A. M. (2008). Antioxidant activity, reducing power and total phenolic content of Iranian olive cultivar. *Journal of Biological Sciences, 8,* 779–783.

22. Hermund, D. B., Yesiltas, B., Honold, P., Jonsdottir, R., Kristinsson, H. G., & Jacobsen, C. (2015).Characterization and antioxidant evaluation of Icelandic *F. vesiculosus* extracts *in vitro* and in fish-oil-enriched milk and mayonnaise. *Journal of Functional Foods,* doi: 10.1016/j.jff.2015.02.020.

23. Holdt, S. L., & Kraan, S. (2011). Bioactive compounds in seaweed: Functional food applications and legislation. *Journal of Applied Phycology, 23,* 543–597.

24. Ito, K., & Hori, K. (1989). Seaweed: chemical composition and potential food uses. *Food Reviews International, 5,* 101–144.

25. Jimenez-Escrig, A., Jimenez-Jimenez, I., Pulido, R., & Saura-Calixto, F. (2001). Antioxidant activity of fresh and processed edible seaweed. *Journal of Agricultural and Food Chemistry, 81,* 530–534.

26. Kaneria, M., & Chanda, S. (2013b). Evaluation of antioxidant and antimicrobial capacity of *Syzygium cumini* L. leaves extracted sequentially in different solvents. *Journal of Food Biochemistry, 37,* 168–176.

27. Kaneria, M. J., & Chanda, S. V. (2013a). The effect of sequential fractionation technique on the various efficacies of pomegranate (*Punica granatum* L.). *Food analytical Techniques, 6,* 164–175.

28. Kaneria, M. J., Bapodara, M. B., & Chanda, S. V. (2012b). Effect of extraction techniques and solvents on antioxidant activity of Pomegranate (*Punica granatum* L.) leaf and stem. *Food Analytical Methods, 5,* 396–404.

29. Kaneria, M. J., Rakholiya, K. D., & Chanda, S. V. (2014). *In vitro* antimicrobial and antioxidant potency of two nutraceutical plants of Cucurbitaceae. In: Gupta, V. K. (Ed.) *Traditional and Folk Herbal Medicine: Recent Researches–Vol.2,* Daya Publication House, India, pp. 221–247.

30. Kaneria, M., Baravalia, Y., Vaghasiya, Y., & Chanda, S. (2009). Determination of antibacterial and antioxidant potential of some medicinal plants from Saurashtra region, India. *Indian Journal of Pharmaceutical Sciences, 71,* 406–412.

31. Kaneria, M., Kanani, B., & Chanda, S. (2012c). Assessment of effect of hydroalcoholic and decoction methods on extraction of antioxidants from selected Indian medicinal plants. *Asian Pacific Journal of Tropical Biomedicine, 2,* 195–202.

32. Kaneria, M., Rakholiya, K., & Chanda, S. (2012a). *Search for natural antimicrobial and antioxidant agents.* Lambert Academic Publishing GmbH & Co. KG, Germany.

33. Kim, A. R., Shin, T. S., Lee, M. S., Park, J. Y., Park, K. E., Yoon, N.Y., Kim, J. S., Choi, J. S., Jang, B. C., Byun, D. S., Park, N. K., & Kim, H. R. (2009). Isolation and identification of phlorotannins from *Ecklonia stolonifera* with antioxidant and antiinflammatory properties. *Journal of Agricultural and Food Chemistry, 57,* 3483–3489.

34. Kim, E. Y., Kim, Y. R., Nam, T. J., & Kong, I. S. (2012). Antioxidant and DNA protection activities of a glycoprotein isolated from a seaweed, *Saccharina japonica*. *International Journal of Food Science and Technology, 47*, 1020–1027.

35. Kosanic, M., Rankovic, B., & Stanojkovic, T. (2015). Biological activities of two macroalgae from Adriatic coast of Montenegro. *Saudi Journal of Biological Sciences, 22*, 390–397.

36. Kuda, T., Taniguchi, E., Nishizawa, M., & Araki, Y (2002). Fate of water-soluble polysaccharides in dried *Chorda filum* a brown alga during water washing. *Journal of Food Composition and Analysis, 15*, 3–9.

37. Liu, J., Wang, C., Wang, Z., Zhang, C., Lu, S., & Liu, J. (2011). The antioxidant and free radical scavenging activities of extract and fractions from corn silk (*Zea mays* L.) and related flavone glycosides. *Food Chemistry, 126*, 261–269.

38. Loizzo, M. R., Tundis, R., Bonesi, M., Menichini, F., Mastellone, V., Avallone, L., & Menichini, F. (2012). Radical scavenging, antioxidant and metal chelating activities of *Annona cherimola* Mill. (cherimoya) peel and pulp in relation to their total phenolic and total flavonoid contents. *Journal of Food Composition and Analysis, 25*, 179–184.

39. Matsukawa, R., Dubinsky, Z., Kishimoto, E., Masaki, K., Masuda, Y., Takeuchi, T., Chihara, M., Yamamoto, Y., Niki, E., & Karube, I. (1997). A comparison of screening methods for antioxidant activity in seaweeds. *Journal of Applied Phycology, 9*, 29–35.

40. Mc Donald, S., Prenzler, P. D., Autolovich, M., & Robards, K. (2001). Phenolic content and antioxidant activity of olive extracts. *Food Chemistry, 73*, 73–84.

41. McCord, J. M., & Fridovich, I. (1998). Superoxide dismutase: the first 20 years (1968–1988). *Free Radical Biology and Medicine, 5*, 363–369.

42. McCune, L. M., & Johns, T. (2002). Antioxidant activity in medicinal plants associated with the symptoms of diabetes mellitus used by the indigenous peoples of the North American boreal forest. *Journal of Ethnopharmacology, 82*, 197–205.

43. Nithiyanantham, S., Siddhuraju, P., & Francis, G. (2013). A promising approach to enhance the total phenolic content and antioxidant activity of raw and processed *Jatropha curcas* L. kernel meal extracts. *Industrial Crops and Products, 43*, 261–269.

44. O'Sullivan, A. M., O'Callaghan, Y. C., O'Grady, M. N., Queguineur, B., Hanniffy, D., Troy, D. J., Kerry, J. P., & O'Brien, N. M. (2011). *In vitro* and cellular antioxidant activities of seaweed extracts prepared from five brown seaweeds harvested in spring from the west coast of Ireland. *Food Chemistry, 126*, 1064–1070.

45. Parekh, J., & Chanda, S. (2007). *In vitro* antibacterial activity of the crude methanol extract of *Woodfordia fructicosa* Kurz. Flower (Lythraceae). *Brazilian Journal of Microbiology, 38*, 204–207.

46. Perez, M. B., Banek, S. A., & Croci, C. A. (2011). Retention of antioxidant activity in gamma irradiated Argentinian sage and oregano. *Food Chemistry, 126*, 121–126.

47. Radhakrishnan, S., Bhavan, P. S., Seenivasan, C., Shanthi, R., & Muralisankar, T. (2014). Replacement of fishmeal with *Spirulina platensis*, *Chlorella vulgaris* and *Azolla pinnata* on nonenzymatic and enzymatic antioxidant activities of *Macrobrachium rosenbergii*. *The Journal of Basic and Applied Zoology, 67*, 25–33.

48. Rakholiya, K. D., Kaneria, M. J., & Chanda, S. V. (2014b). Mango pulp: a potential source of natural antioxidant and antimicrobial agent. In: Gupta, V. K. (Ed.) *Medicinal Plants: Phytochemistry, Pharmacology and Therapeutics–Vol. 3*, Daya Publication House, India, pp. 253–284.

49. Rakholiya, K., Kaneria, M., & Chanda, S. (2011). Vegetable and fruit peels as a novel source of antioxidants. *Journal of Medicinal Plants Research, 5,* 63–71.

50. Rakholiya, K., Kaneria, M., & Chanda, S. (2014a). Inhibition of microbial pathogens using fruit and vegetable peel extracts. *International Journal of Food Science and Nutrition, 65,* 733–739.

51. Rakholiya, K., Kaneria, M., & Chanda, S. (2014c). *Antimicrobial and antioxidant potency of Mangifera indica L. leaf.* Lambert Academic Publishing GmbH & Co. KG, Germany.

52. Rakholiya, K., Kaneria, M., Nagani, K., Patel, A., & Chanda, S. (2015). Comparative analysis and simultaneous quantification of antioxidant capacity of four *Terminalia* species using various photometric assays. *World Journal of Pharmaceutical Science, 4,* 1280–1296.

53. Re, R., Pellegrini, N., Proteggentle, A., Pannala, A., Yang, M., & Rice-Evans, C. (1999). Antioxidant activity applying an improved ABTS radical cation decolorization assay. *Free Radical Biology and Medicine, 26,* 1231–1237.

54. Robak, J., & Gryglewski, R. J. (1988). Flavonoids are scavengers of superoxide anions. *Biochemical Pharmacology, 37,* 837–841.

55. Rodriguez-Bernaldo de Quiros, A., Frecha-Ferreiro, S., Vidal-Pérez, A. M., & Lopez-Hernandez, J. (2010). Antioxidant compounds in edible brown seaweeds. *European Food Research and Technology, 231,* 495–498.

56. Rodriguez-Jasso, R. M., Mussatto, S. I., Pastrana, L., Aguilar, C. N., & Teixeira, J. A. (2014). Chemical composition and antioxidant activity of sulfated polysaccharides extracted from *Fucus vesiculosus* using different hydrothermal processes. *Chemical Papers, 68,* 203–209.

57. Ruperez, P., Ahrazem, O., & Leal, J. A. (2002). Potential antioxidant capacity of sulfated polysaccharides from edible marine brown seaweed *Fucus vesiculosus*. *Journal of Agricultural and Food Chemistry, 50,* 840–845.

58. Singh, S., Kate, B. N., & Banerjee, U. C. (2005). Bioactive compounds from cyanobacteria and microalgae: an overview. *Critical Reviews in Biotechnology, 25,* 73–95.

59. Souza, J. N. S., Silva, E. M., Loir, A., Rees, J.-F., Rogez, H., & Larondelle, Y. (2008). Antioxidant capacity of four polyphenol-rich Amazonian plant extracts: a correlation study using chemical and biological *in vitro* assays. *Food Chemistry, 106,* 331–339.

60. Sowndhararajan, K., Siddhuraju, P., & Manian, S. (2011). Antioxidant and free radical scavenging capacity of the underused legume, *Vigna vexillata* (L.) A. Rich. *Journal of Food Composition and Analysis, 24,* 160–165.

61. Sukenik, A., Zmora, O., & Carmeli, Y. (1993). Biochemical quality of marine unicellular algae with special emphasis lipid composition: II. *Nannochloropsis* sp. *Aquaculture, 117,* 313–326.

62. Tai, Z., Cai, L., Dai, L., Dong, L., Wang, M., Yang, Y., Cao, Q., & Ding, Z. (2011). Antioxidant activity and chemical constituents of edible flower of *Sophora viciifolia*. *Food Chemistry, 126,* 1648–1654.

63. Thondre, P. S., Ryan, L., & Henry, C. J. K. (2011). Barley β-glucan extracts as rich sources of polyphenols and antioxidants. *Food Chemistry, 126,* 72–77.

64. Wang, J., Wang, Y., Liu, X., Yuan, Y., & Yue, T. (2013). Free radical scavenging and immunomodulatory activities of *Ganoderma lucidum* polysaccharides derivatives. *Carbohydrate Polymers, 9,* 33–38.

65. Wang, T., Jonsdottir, R., & Olafsdottir, G. (2009). Total phenolic compounds, radical scavenging and metal chelating of extracts from Icelandic seaweed. *Food Chemistry, 116*, 240–248.

66. Wang, T., Jonsdottir, R., Liu, H., Gu, L., Kristinsson, H. G., Raghavan, S., & Olafsdottir, G. (2012). Antioxidant capacities of phlorotannins extracted from the brown algae *Fucus vesiculosus. Journal of Agricultural and Food Chemistry, 60*, 5874–5883.

67. Waris, G., & Alam, K. (2004). Immunogenicity of superoxide radical modified-DNA: studies on induced antibodies and SLE anti-DNA autoantibodies. *Life Sciences, 75*, 2633–2642.

68. Wright, E., Scism-Bacon, J. L., & Glass, L. C. (2006). Oxidative stress in type 2 diabetes: the role of fasting and postprandial glycaemia. *International Journal of Clinical Practice, 60*, 308–314.

69. Wu, J. H., Tung, Y. T., Chyu, C. F., Chien, S. C., Wang, S. Y., Chang, S. T., & Kuo, Y. H. (2008). Antioxidant activity and constituents of extracts from the root of *Garcinia multiflora. Journal of Wood Science, 54*, 383–389.

70. Xu, W., Zhang, F., Luo, Y., Ma, L., Kou, X., & Huang, K. (2009). Antioxidant activity of a water-soluble polysaccharide purified from *Pteridium aquilinum. Carbohydrate Research, 344*, 217–222.

71. Zhou, H. C., Lin, Y. M., Li, Y. Y., Li, M., Wei, S. D., Chai, W. M., & Tam, N. F. (2011). Antioxidant properties of polymeric proanthocyanidins from fruit stones and pericarps of *Litchi chinensis* Sonn. *Food Research International, 44*, 613–620.

CHAPTER 8

OMEGA-3 PUFA FROM FISH OIL: SILVER BASED SOLVENT EXTRACTION

KIRUBANANDAN SHANMUGAM

CONTENTS

8.1   INTRODUCTION

Omega-3 *Poly Unsaturated Fatty Acids* (PUFA) are a kind of nutritional lipids and a family of unsaturated fatty acids, which contains a final carbon-carbon double bond at the n–3 position, e.g., the third bind from the methyl end of the fatty acid. Additionally, it is used as nutritional additives in various human foods. Because the human body is not capable of producing n−3 fatty acids, de novo but it can form 20-carbon unsaturated n−3 fatty acids (like EPA) and 22-carbon unsaturated n−3 fatty acids (like DHA) from the 18-carbon n−3 fatty acid α-linolenic acid. These conversions would take place competitively with Omega n−6 fatty acids that are essential and similar to chemical analogs that are derived from linoleic acid. The both n−3 α-linolenic acid and n−6 linoleic acid are essential nutritional lipids, which must be obtained from food. There are three essential omega-3 fatty acids important nutritionally to humans such as α-linolenic acid (18:3, n−3; ALA), eicosapentaenoic acid (20:5, n−3; EPA), and docosahexaenoic acid (22:6, n−3; DHA).

As an alternative to the commonly used dietary foods such as flaxseeds, walnuts, strawberries and raspberries as a source of omega-3 fatty acids, supplementation has been recommended. Due to plentiful availability, fish oil is mostly considered as the feed stock for separation of Omega-3 PUFA using a variety of conventional processes. The concentrated form of Omega-3 PUFA from fish oil is an alternate for nutritional lipids and developed food additives. In addition to that, the advantage of concentrated forms of ω-3 PUFA is that they are devoid of saturated and mono saturated fatty acids. Now days, Omega-3 fatty acids have been important part of human diet throughout evolution. Production of ω-3 PUFA may be achieved using a number of techniques. These products may be in the form of free fatty acids, methyl and ethyl esters or acylglycerols.

Both omega-3 and omega-6 PUFA are precursors of hormone-like compounds known as eicosanoids, which are involved in numerous biological processes in the human body. It is suggested that the typical 'Western' diet, which is relatively high in Omega 6 PUFA and low in Omega-3 PUFA, may not supply the appropriate balance of PUFA for the proper biological function of the body. Consequently, this imbalance would cause a variety of diseases such as cardiovascular, hypertension, inflammatory and autoimmune disorders, depression and certain disrupted neurological functions. Long chain PUFA are now considered 'conditionally essential' for infant growth and development. The long-chain Omega 3-PUFA, namely eicosapentaenoic acid (EPA), docosapentaenoic acid (DPA) and docosahexaenoic acid (DHA) may be acquired mainly from sea foods or derived from the α-linolenic acid by a series of chain elongation and desaturation. EPA and DHA are synthesizing mainly by both uni- and multicellular marine plants such as phytoplankton and algae. Due to bioaccumulation, they are eventually transferred through the food web and are incorporated into lipids of aquatic species such as fish and marine mammals particularly those living in the cold waters of the Atlantic region at low temperatures, in such environments [1, 3, 13–15].

In this chapter, marine lipids/oils are suggested as best-feed stock to produce the concentrate form of Omega-3-PUFA.

8.1.1 SEPARATION OF OMEGA-3 PUFA FROM FISH OIL

Marine oils are the abundant source for Omega-3 PUFA and were used as the raw material for the production of Omega 3-PUFA concentrate and EPA/DHA-ethyl esters. Since marine oils are complex mixtures of fatty acids with varying chain lengths and degrees of different unsaturation, so separation of individual fatty acids is challenging for the production of highly concentrated Omega-3 components. Therefore, commercial production of marine oil concentrates with high percentages of EPA and DHA is now a major challenge for food scientists and biotechnologists engaged in Omega-3 extraction and production. There are numerous techniques for the production of Omega-3 PUFA but only few can be used for cost effective and large-scale production to meet the growing demand. However, the advantages and disadvantages of this technology will be discussed later in this chapter. The existing methods to recover Omega-3 PUFA from fish oil are:

- Chromatographic methods
- Distillation methods
- Low temperature crystallization process
- Enzymatic separation

- Supercritical extraction
- Urea complexation

The concentration of Omega-3 PUFA by urea complexation is more common as the final product is in the form free acid or simple ester. However, Urea complexation has numerous disadvantages such as the formation of carcinogenic alkyl carbamates compounds [26, 28] due to reactivity between fatty acids with urea and lower yield of Omega-3 PUFA [3, 4].

8.1.2 *SOLVENT EXTRACTION PROCESS FOR ISOLATION OF OMEGA-3 PUFA*

Solvent extraction is a method of separation of various biological compounds from natural resources. It is a mass transfer operation in which a solution is brought into contact with a second solvent, essentially immiscible with the first in order to bring the transfer of one or more solutes into the second solvent. The separations that can be achieved by this method are simple, convenient and rapid to perform and they are clean as much as the small interfacial area certainly precludes any phenomena analogous to the conventional separations. In this process, One liquid phase is the feed consisting of numerous solute dissolved and a carrier. The other phase is a solvent to be used for extracting the desired solute from the feedstock by contacting each other in a suitable contacting system. The solvent extraction is understood to be a transfer of the solute from the feed to a solvent. During and at the end of extraction processes, the feed deprived of solute becomes a raffinate and the solvent containing solute is called extract.

8.1.3 *IMPLEMENTATION OF LIQUID–LIQUID EXTRACTION WITH CHEMICAL REACTION*

In case of solvent extraction without chemical reaction, mass transfer rate depends on the effective contacting between feed and solvent and mixing between these two phases; where as in the solvent extraction with chemical reaction, the mass transfer rate depends on the reaction rate and mixing in the contacting system [2]. This concept has been implemented in the recovery of Omega-3 PUFA from fish oil by extracting with silver nitrate solution. One particular application where this was highlighted was for extractions involving high-value solvents, such as the removal of DHA-et from an organic solvent using silver ion solutions with reversible chemical reaction. A number of LLE examples appears in the literature using silver-salt solutions to complex with the unsaturated esters and selectively extract Omega-3 PUFA such as DHA and EPA. The challenge, often reported with this approach, is the reduced solubility of the target compounds in the aqueous phase, commonly employed as a carrying medium of the silver salts, resulting in a surface-active complexing mechanism that requires significant mass transfer areas between phases to be effective. Mini-fluidic platforms offer a potential solution to this constraint, with interfacial areas ranging from 100 to 10000 m^2/m^3 reactor volume.

8.1.4   OBJECTIVES OF THIS RESEARCH STUDY

In this book chapter, a mini-fluidic/microfluidic platform was used to explore the feasibility of liquid–liquid extraction of Omega-3 PUFA from fish oil ethyl esters using a silver nitrate solution as a potential solvent. Moreover, the hydrodynamics of mini-fluidics contact system was explained during this process and compared to what was observed in the previous literature, as much of the previous reports focused on idealized systems containing a purified Omega-3 compound dissolved in a high-purity organic solvent. Solvent based extraction methods for separation of Omega-3 PUFA was performed both in a batch-wise stirred tank reactor and in an innovatively constructed Slug flow based mini-fluidic reactor experimental system constructed with food grade 1/16th inch ID Tygon tubing [5, 6].

The scientific purpose of this study was to explore the feasibility of mini-fluidic reactor technology for extraction of Omega-3 PUFA using concentrated aqueous silver nitrate solution as an alternative to either primary extraction technology (i.e., the urea precipitation process) or secondary purification technology such as chromatography or molecular distillation. In addition to that, the extraction at micro scale for recovery of DHA-et and EPA-et dissolved in organic solvent with silver ion is also discussed and compared microfluidic extraction with mini-fluidic extraction. Additionally, the performance of silver-based LLE for recovering Omega-3 PUFA from fish oil ethyl esters is explored and the scaling strategies for silver based solvent extraction process, specifically as it relates to the flow pattern and the ancillary processing steps, which would be needed for full-scale operation, is commented.

8.2   LIMITATION OF CONVENTIONAL PROCESS OVER THE MINI MICROFLUIDIC EXTRACTION

An overview of the more predominant concentration processes encountered in literature is provided here in an effort to illustrate the relative processing parameters, advantages and disadvantages (Figure 8.1). The processes reviewed include urea precipitation, low-temperature crystallization, molecular distillation, supercritical extraction, chromatographic methods, enzymatic enrichment, and solvent extraction. Much of the content of this review

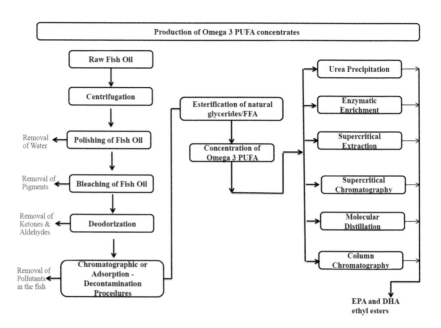

FIGURE 8.1   Production of Omega-3 PUFA by various conventional processes.

is based on the recent study by Lembke [4], and is discussed for the purpose of identifying disadvantages of these applications for each approach.

8.2.1 UREA PRECIPITATION

From the point view of economic feasibility, the urea complexation process is conventional and the yield of Omega-3 PUFA is around 65% in the industrial process. Furthermore, urea crystallization is performed on fatty acids or ethyl esters, when the saturated and monounsaturated components are removed through formation of adducts with urea, leaving behind a fraction enriched in PUFA, particularly the highly unsaturated components EPA and DHA. Although it is efficient and cheap, yet urea crystallization is not recommended for the production of food grade ingredients because of risk of producing carcinogenic alkyl carbamates.

8.2.2 LOW TEMPERATURE CRYSTALLIZATION

Low temperature crystallization of lipids, especially triglycerides and fatty acids, is based on the separation of triglycerides and fatty acids or their methyl/ethyl esters according to their melting points in different organic solvents at very low temperatures (−50°C to −70°C). The low-temperature crystallization has to use large amounts of flammable solvents while also operating under a heavy refrigeration load, presenting serious shortcomings for large scale industrial applications [4].

8.2.3 MOLECULAR DISTILLATION

Fractional distillation is another feasible process for isolation of Omega-3 PUFA from fish oil under ultra-low pressure and low temperature. In this process, the separation is based on the difference in the boiling point and molecular weight of fatty acids under the conditions of high temperature (180–250°C) and reduced pressure (0.1 to 1 mm of Hg). Although molecular Distillation is an efficient process to produce the highly concentrated form of Omega-3 PUFA from fish oil ethyl esters, yet it is cost expensive and then the starting natural material triglyceride is lost in the process.

8.2.4 SUPERCRITICAL EXTRACTION

Even though supercritical extraction is very efficient than other conventional processes, yet research into this process is ongoing due to the significant advantages and extraction performance observed. There remains, however, the overriding cost of pressurizing both the oil feedstock and the recycled CO_2 to achieve supercritical conditions.

8.2.5 CHROMATOGRAPHY

Liquid chromatography was employed to improve the yield of EPA and DHA from fish oils and has the highest selectivity for separating specific chain lengths and degrees of unsaturation. Liquid chromatography is, however, a dilution process requiring the feedstock to be diluted in significant quantities of organic solvent (mobile phase) prior to injection into a separation column. Column effluents with specific residence times are then removed and reconcentrated, potentially introducing oxidative stress on PUFAs [4, 33].

8.2.6 ENZYMATIC ENRICHMENT

The lipase enzyme is used for esterification, hydrolysis and exchange of free fatty acids in the esters. The direction and the choice of the reaction can be influenced by the choice of experimental conditions. The reaction is reversible and under low water activity conditions, the enzyme functions 'in reverse' that is the synthesis of an ester bond rather than its hydrolysis. Due to the cost of enzymes, the process is expensive.

8.3 SOLVENT EXTRACTION OF EPA AND DHA OMEGA-3 PUFA FROM FISH OILS

Silver nitrate salt-solutions have previously been used as an aqueous solvent for the purification of the ethyl ester of PUFA (PUFA-Et) developed by Yazawa et al. [7]. In this process, silver ions interact with the double bonds in unsaturated fatty acids to produce complexes, which are soluble in ionic/polar carrying phases. Since the separation is performed exclusively via a reversible chemical reaction, this technique is very simple and would be a promising method for the purification of PUFAs. Similar principles have been applied in chromatography-based separations, where Fagan et al. [8] reported that the ethyl esters of EPA and DHA from tuna oils can be isolated using a polymeric silver cation exchange as the stationary phase. In this case, greater than 88% of separation occurred and suggested the potential for reasonable yields. Limited reports are available of extraction performance using silver salts to isolate polyunsaturated fatty acids from a mixture of FFA, with previously known methods not providing for a sufficiently selective and/or efficient process for concentrating omega-3 fatty acids. US Patent [9] reported that silver salt solutions could be used in the extraction of PUFA from marine oils, describing a potential process that is summarized as follows (Figure 8.2):

a. Combining the fatty acid oil mixture and an aqueous silver salt (such as $AgNO_3$ or $AgBF_4$) solution to form an aqueous phase and an organic phase, wherein in the aqueous phase, the aqueous silver salt solution forms reversibly a complex with the at least one omega-3 fatty acid;

b. Separating the aqueous phase from the organic phase from the contacting systems;

c. Extracting the aqueous phase with a displacement liquid, or increasing the temperature of the aqueous phase to at least 30°C, or a combination of extracting with a displacement liquid and increasing the temperature, resulting in the formation of at least one extract;

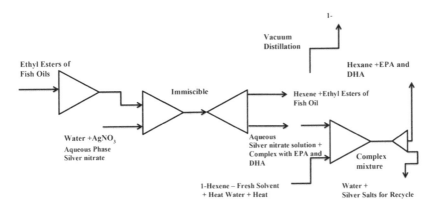

FIGURE 8.2   The simplified process for liquid–liquid extraction of EPA and DHA.

    d. Combining the aqueous phase with water, or extracting the aqueous phase with supercritical CO_2, or a combination of combining the aqueous phase with water and extracting the aqueous phase with supercritical CO_2, to dissociate the complex, wherein an aqueous phase comprising the silver salt and at least one solution in which a fatty acid concentrate forms; and

    e. Separating the at least one solution comprising the fatty acid concentrate from the aqueous phase comprising the silver salt.

Within available literature, it was observed that PUFA–ME (Poly Unsaturated Fatty acids methyl esters) with the high degree of saturation such as DHA-ME and EPA-ME selectively bind to the aqueous phase of silver nitrate solution. Li et al. [10] reported that the extraction of polyunsaturated fatty acids from oils with hydrophobic ionic liquids (imidazolium-based) containing silver salts (silver tetra fluoroborate) as the extraction medium yielded greater extraction performance than silver nitrate based solutions. In their work, the percentage of PUFA of methyl esters in the extract phase approached 93.1%. Moreover, Li et al. [10, 11] demonstrated improved solvent recovery and salt recirculation when an ionic liquid was employed with a hexene-based secondary extraction, although the method employed may also be suitable for aqueous-based salt systems. Temperature has also been found to significantly affect solvent extraction with silver nitrate solutions. In the study by Teramoto et al. [42], the distribution ratios of polyunsaturated fatty acid ethyl esters between aqueous $AgNO_3$ extraction phase and an organic phase of esters dissolved in heptane were found to increase drastically as the temperature was lowered. Qualitative evidence in literature would suggest that hydrophobic ionic liquids may potentially offer advantages over water when used as the dissolving phase for the silver salts.

Li et al. [11] reported that $AgBF_4$ exhibited high extraction potential in the hydrophobic ionic-liquids but little or no extraction capacity in the hydrophilic ionic liquids. It was observed that the increase in alkyl chain in the ionic liquid (imidazolium ionic liquid) caused better hydrophobicity with $AgBF_4$ and leads to better extraction potential and also minimum extraction time. The preparation of ionic liquid, however, is a tedious process and requires inert gas atmospheres such as argon blanketing and heating at 70°C. Additionally, there is a chance of decomposition of ionic liquid conjugated silver salts under harsh environments. The ionic liquids with silver salts are also light-sensitive, and must be stored under dark conditions. A significant limitation of the solvent extraction approach is the need for large amounts of expensive silver-salts, both for high recovery and freezing point depression when an aqueous carrier phase is used at low temperatures. While $AgNO_3$ was historically employed for this method, Li et al. [10] recently reported a number of other novel π-complexing sorbents for extraction of Omega-3 PUFA from oils. Covalently immobilizing ionic liquids (ILs) onto silica and then coating this silica- supported ILs with silver salts, with applications primarily in chromatography, prepared novel π-complexing sorbents. The salts explored (Table 8.1) do, however, offer potential alternatives for a solvent-solvent extraction system.

8.3.1 SILVER-BASED SOLVENT BASED METHODS EXTRACTION – REACTION KINETICS

One of the advantages of an Omega-3 PUFA separating from fish oil that consists of fatty acid ethyl esters/ an aqueous concentrated silver nitrate solution/organic solvent is the permeation facilitation of PUFAs due to Π bond complex formation between silver ion and the Carbon –Carbon

TABLE 8.1 Silver-Salts for the Selective Extraction of Unsaturated Fatty Acids

Salt compound	In water	In Ionic Liquids	Remarks
$AgBF_4$	US Patent 2012/038833 [9]	M. Li [11]	$AgBF_4$ in water less extraction potential than in IL
$AgNO_3$	M. Teramoto [42]	M. Li [11]	Cheap silver salt compared to other options and Least performance in Ionic Liquids and Lowest extraction potential among silver salts
$AgClO_4$	—	M. Li [11]	Poor performance than $AgBF_4$
$(CH_3)_3COO\,Ag$	—	M. Li [11]	Good Extraction performance in IL
CF_3SO_3Ag	—	M. Li [11]	Good Extraction performance in IL
$CF_3COO\,Ag$	—	M. Li [11]	Good Extraction performance in IL

double bonds available in the PUFA (Figure 8.3). On the other hand, saturated fatty acids in the fish oil do not react with silver ion and remain
in the oil phase. The reason being that saturated fatty acid esters do not
form Π bond complex and not dissolving in the aqueous phase. According
to Table 8.1, Li et al. [11] reported that $(CH3)_3COOAg$, CF_3SO_3Ag and
CF_3COOAg showed moderate extraction potential and the separation per-
formances were in the order of $AgBF_4 > AgCF_3SO_3 >> AgNO_3$.

8.3.2 MOLECULAR MECHANISM OF REACTION BETWEEN OMEGA-3 PUFA WITH SILVER ION

In silver-based solvent extraction, silver ions in the aqueous phase play
a major role in the extraction processes and bind with the PUFAs double
bond form reversible complexes in the aqueous phase. In the study with
silver, the complexes are of the charge-transfer type in which the unsaturated compound acts as an electron donor and the silver ion as an electron
acceptor [12, 44].

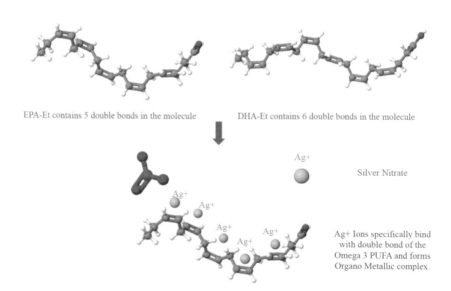

EPA-Et contains 5 double bonds in the molecule   DHA-Et contains 6 double bonds in the molecule

Ag+

Silver Nitrate

Ag+ Ions specifically bind
with double bond of the
Omega 3 PUFA and forms
Organo Metallic complex.

FIGURE 8.3   Reaction mechanism of silver ion interaction with PUFA.

$$DHA/EPA\text{-}Et + AgNO_3 \rightarrow DHA/EPA:Ag^{2+} \text{ complex} + \text{Fish oil-Et}$$
$$(\text{Aqueous phase})$$

Additionally, the formation mechanism of the complex between dou-ble bonds and the silver ion is still a topic of interest receiving significant attention within scientific research intensively, since these complexes play an important role in process development based on solvent based Liquid–Liquid Extraction and silver based chromatography methods. However, quantitative data (e.g., equilibrium constants) exist only for a number of short chains mono-and di-olefins. The stability of complex with organic compound decreases with the increasing chain-length. Generally, the rate of complexation is very rapid but the complexes are unstable and exist in equi-librium with the free form of the olefin. The coordination forces between Omega-3 PUFA and silver ion could seem to be very weak. These particular properties of complexation between a double bond and a silver ion are favor-able for use in chromatography/extraction. A significant unknown within this bonding process is the relative ratio of Ag ions required to cause one mole of DHA and EPA to solubilize within the extraction process [12, 44].

1 mole of DHA (356 g) + 6 Moles Silver ions = DHA-et –Ag complex

1 mole of EPA (330 g) + 5 moles Silver Ions = EPA-et – Ag complex

The nature of the complexation was based on a charge transfer complex is formed between the metal ion and the Π electrons of the double bond(s) in the unsaturated organic molecule. To be obscure, the preferred model could assume formation of a sigma type bond between the occupied 2p orbitals of an olefinic double bond and the free 5s and 5p orbitals of the transition metal ion (Ag+), and a (probably weaker) Π acceptor back bond between the occupied 4d orbitals of the metal ion and the free antibonding 2p pi* orbitals of the olefinic bond. The major advantage of using silver ion (Ag (I)) as the complexing agent with Olefins is that compared to other transition metals is relatively inexpensive and stable. The salt silver nitrate is a crystalline com-pound and stable for a very long period when kept in the dark and also easily dissolvable in water, methanol or acetonitrile for various analytical purposes. Most importantly, the formed complexes with fatty acids have weak bonds, therefore, the complexation reaction is reversible (Figure 8.4) and is therefore highly suitable for use in solvent extraction processes [12, 44].

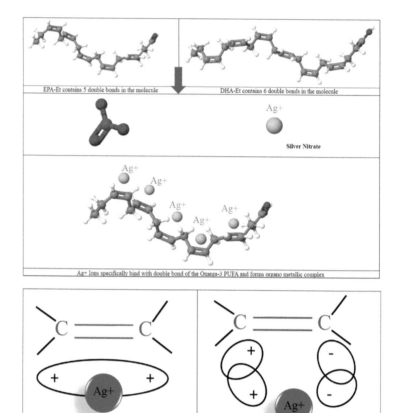

EPA-Et contains 5 double bonds in the molecule

DHA-Et contains 6 double bonds in the molecule

Ag+

Silver Nitrate

Ag+

Ag+

Ag+

Ag+

Ag+

Ag+ Ions specifically bind with double bond of the Omega-3 PUFA and forms organo metallic complex

FIGURE 8.4 The Dewar model of interaction between Ag$^+$ and an olefinic double bond.

8.3.3 PROCESS CHEMISTRY OF SILVER BASED SOLVENT EXTRACTION

Fish oils have attracted attention as a source of Omega-3 polyunsaturated fatty acids (PUFAs) such as eicosapentaenoic acid (EPA, C20:5w3) and docosahexaenoic acid (DHA, C22:6w3) since they have favorable physiological activity and therapeutic advantages. However, it is very difficult to separate the desired components from fish oils consisting of various fatty acids with different numbers of double bonds and carbon atoms. On the other hand, solvent extraction can be operated under mild conditions and is suitable for mass production. Recently, it was reported that ethyl esters of PUFAs such as DHA and EPA could be successfully extracted into silver nitrate solutions since silver ion can complex with carbon-carbon double bonds of PUFAs [42, 44].

8.3.3.1   Process Variables that Affect Solvent Extraction Processes

Many variables do affect the reaction between the silver ion and Omega-3 PUFA and explained here for optimizing the process for maximum recovery of Omega-3 PUFA from fish oil ethyl esters [42].

a. Effect of silver ion concentration on solvent extraction
In this solvent extraction process, the concentration of silver ions in the aqueous phases is directly proportional to the number of Carbon–Carbon double bond present in the fatty acids. On the basis of these concepts, the following extraction mechanism is postulated. Firstly, PUFA-E dissolves physically in the aqueous phase. Then, the dissolved PUFA-E forms a complex with Ag+ in the aqueous phase.

$$(PUFA)_{Organic\ Phase\ -\ Fish\ Oil\ Phase} \leftrightarrow (PUFA)_{Aqueous\ Phase\ -\ Silver\ nitrate\ solution\ Phase}$$

The partition coefficient of this reactive extraction is defined as follows:

$$K_D = \frac{(PUFA - E)_{AqueousPhase(SilverNitrateSolution)}}{(PUFA - E)_{OrganicPhase(FishOilPhase)}}$$

$$(PUFA)_{Aqueous\ Phase} + n\,(Ag^+)_{Aqueous\ Phase} \leftrightarrow (PUFA - nAg^+)^{n+}_{Aqueous\ Phase}$$

$$\beta n = \frac{(PUFA - nAg+)n +_{AqueousPhase}}{(PUFA)_{AqueousPhase}\,n(Ag^+)_{Aqueous}}$$

The PUFA-Es with a higher degree of unsaturation can be selectively extracted into silvernitrate solutions. However, the ability of PUFA-ET to form a complex with silver ion has a crucial effect on distribution ratio.

b. Effect of solvent in organic phase
The addition of organic solvents into the oil phase (Fish Oil Phase) affects the extraction process and its equilibrium. When the saturated

hydrocarbons as the solvent are added into the organic phase of the extraction process, the distribution ratio increases with an increase in carbon number of the solvent. The distribution ratio of EPA-Et is considerably dependent on the solvents in the oil phase [42].

c. Effect of addition of polar solvent into the aqueous phase
The addition of methanol, ethanol and propanol to the aqueous phase $AgNO_3$ solution remarkably increase distribution ratio of extraction processes [42].

d. Effect of temperature on extraction processes
It should be noted that D increases remarkably with a decrease in temperature. It is deduced that PUFA-E is extracted into silver nitrate solutions as a complex (PUFA-E-nAg)" + where n is the number of carbon-carbon double bonds in PUFA-E. The number of double bonds in PUFA-Es has a decisive effect on the distribution ratio, and PUFA-Es with a high degree of unsaturation such as DHA-Et and EPA-Et can be selectively extracted into silver nitrate solutions. The distribution ratio of EPA-Et is considerably dependent on the solvents in the oil phase. The distribution ratio can be remarkably increased by an addition of water-soluble alcohols such as methanol, ethanol, and n-propanol. Temperature dependence of D is very large, and lower temperature is favorable for the extraction [42].

8.3.4 MASS TRANSFER APPROACH IN REACTION KINETICS

Mass transfer and mixing play a crucial role in the solvent extraction processes, where DHA-Et and silver ion reacts (Figures 8.5 and 8.6). Teramoto et al. [42] considered that DHA-Et reacts with the silver ion in an aqueous bulk. However, DHA-Et would adsorb on oil/water interface. Kamio et al. [17] confirmed that the interaction between interfacial tension and DHA-Et concentration in the organic phase and proved that the interfacial tension decreased with increasing the DHA-Et concentration. This fact indicates that DHA-Et adsorbs on the liquid–liquid interface. Because DHA-Et adsorbs on the interface, there is every possibility that DHA-Et reacts with the silver ion at the liquid–liquid interface. In this case, Interfacial tension plays a crucial role in the mass transfer. The following stepwise processes are considered for the case of extraction of Omega-3 PUFA with silver nitrate solution:

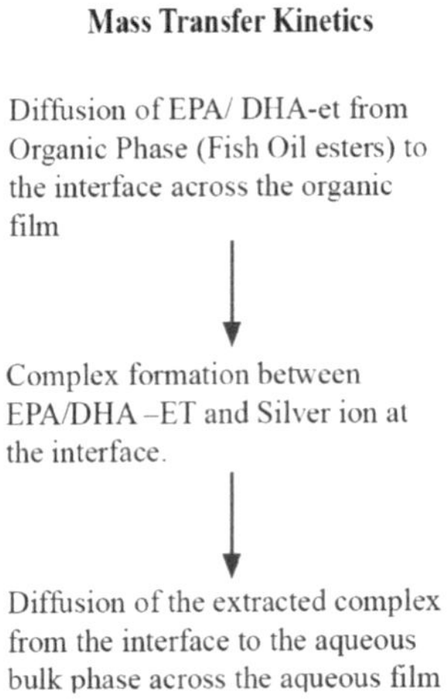

Mass Transfer Kinetics

Diffusion of EPA/ DHA-et from
Organic Phase (Fish Oil esters) to
the interface across the organic
film

↓

Complex formation between
EPA/DHA –ET and Silver ion at
the interface.

↓

Diffusion of the extracted complex
from the interface to the aqueous
bulk phase across the aqueous film

FIGURE 8.5   Stepwise processes in mass transfer in solvent extraction processes.

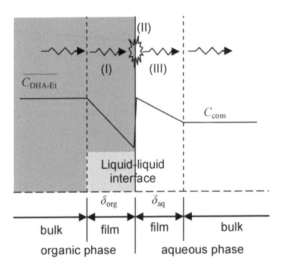

FIGURE 8.6   Schematic representation of DHA-Et concentration around Oil/Water interface of a single compartment. Mass transfer rates for Liquid–liquid extraction are often characterized conveniently in a two-film model, involving a complex formation at the Liquid- Liquid interface.

- Diffusion of DHA-Et from organic bulk to the interface across the organic laminar film;
- Diffusion of DHA-Et across the aqueous laminar film to the aqueous bulk; and

- Complex-formation reaction between DHA-Et and silver ion in the aqueous bulk.

8.4 CONCEPT OF EXTRACTION AT MICRO/MINI SCALE

Extraction is a major unit operation in food processing. In most cases in any extraction process, it includes two main steps namely the formation of dispersion and followed by phase separation. A variety of conventional extraction equipment including mixer settler cascades, counter current, sieve plate, packed, pulsed and rotating disc extraction columns are used in the production of numerous chemical and fine chemicals. The main parameters governing extraction processes are mass transport across the phase boundary, interfacial kinetics, and mass transport phenomena within the bulk phase. Since mass transport by diffusion from one phase into the other is greatly facilitated by a large specific surface area, efficient dispersion is an indispensable prerequisite to achieving total phase saturation, characterized by thermodynamic equilibrium. The specific surface area itself is inversely related to the droplet size of the disperse phase. Hence, mass transport by diffusion should be completed within milliseconds due to the large surface to volume ratio and the short diffusion length. The performing extraction process at micro/mini scale has many advantages such as extreme cost effective since it requires fewer solvents for extraction process than the conventional process. It can be used to thoroughly explore the entire process design space of a process and easy to alter the optimum operating conditions to achieve the optimum yield of desired components in the extraction process. Furthermore, it is very efficient since it could be assembled a parallel architecture approach for scaling up the extraction process [18–20].

8.4.1 MINIATURIZATION – PROCESS INTENSIFICATION FOR EXTRACTION PROCESS

Process intensification, resulting in the larger profit and smaller plant, is becoming a highly promising direction in chemical engineering. While doing so safety, health and environment aspects and energy conservation themes are integrated. Process Intensification is described as the miniaturization of numerous unit operations in the chemical processing and biochemical processing which helps to achieve the following aims [18]:

- A reduction in energy use;
- A reduction in capital expenditures (through the building lower plant cost);
- A reduction in plant profile (Height) and plant foot print (area);
- Environmental benefits;
- Safety benefits (low volume of hazardous or toxic compounds in process).

Liquid–liquid extraction-solvent extraction, is a method to separate compounds based on their distribution in two different immiscible liquids (usually water and an organic solvent). In the extraction process, the phase ratio between the organic phase and the aqueous phase is depended on the recovery and the distribution coefficient in the two phases. As well-known when the phase ratio is far away from 1:1, the extraction process becomes more difficult for its low mass transfer interfacial area and low hold-up of the dispersed phase. Therefore, external energy input and high extraction column is generally needed for high phase ratio extraction processes in the case of caprolactam extraction, phosphoric acid purification, hydrogen peroxide production and oil refining processes. It is highly required to develop an effective method to intensify the extraction processes with high phase ratio. These considerations could be implemented in miniaturization of extraction process at laboratory scale for the recovery of Omega-3 PUFA from fish oil. Mini-fluidic flow reactor are an approved tool for process intensification, as they offer numerous advantages, such as enhanced heat and mass transfer, an increased surface to volume ratio, reduced material consumption, and stable laminar flow, even at high shear rates and further easy to tune flow patterns. However, conventional separation techniques cannot be easily transferred to the micro/mini scale, as surface forces dominate over body forces on these length scales, so that separation based on gravity cannot be performed within the environment at micro/mini scale.

Bond number, which correlates body forces to surface tension/ Interfacial tension forces, describes this kind of behavior and is much smaller than 1 for micro/mini systems. Several researches have been exploiting novel separation principles, such as separation by capillary forces and by different wetting properties of the outlet micro/mini channels. Among the diverse separation processes, extraction has proven to work particularly well within the micro/mini environment. Liquid–liquid extraction processes at mini/micro scale exploit differences in solubility of one or more components within two immiscible liquids. They consist of two subsequent steps namely contacting and separation of the two liquid phases, which both strongly depend on the flow pattern.

8.4.2 ADVANTAGES OF MINI/MICRO SCALE PROCESS IN LIQUID–LIQUID EXTRACTION PROCESS

In general, a micro reactor/mini-fluidic reactor has numerous advantages of continuous processing over the batch processing. Continuous processing allows steady state operation, so that large-scale production is given by parallization of micro/mini flow rectors so called numbering up instead of scaling up the micro scale/mini scale processes. It could provide several advantages such as high surface-to-volume area; enhanced mass and heat transfer, laminar flow conditions, uniform residence time, back mixing minimized (increased precision and accuracy), high-throughput and use of very small amounts of materials, low manufacturing, operating, and maintenance costs (if mass produced), and low power consumption, minimal environmental hazards and increased safety and "scaling-out" or "numbering-up" instead of scaling-up. Compared to the conventional extraction process, Mini/micro scale extraction process has less power requirement (Table 8.2).

TABLE 8.2   Power Input Requirement for Various Liquid–Liquid Contactors

Contactor	Power input, KJ/m³
Agitation Extraction Column	0.5–150
Mixer Settler	150–250
Rotating disk impinging streams contactor	175–250
Impinging stream extractor	35–1500
Centrifugal extractor	850–2600
Micro reactor*	0.2–20

*Kashid [19].

8.4.2.1 Overcoming of Hydrodynamics Problem

In conventional extraction equipment, there is an inability to condition the drop size precisely and the non-uniformities that result because of the complexities of the underlying hydrodynamics. As consequence, it affects optimal performance.

8.4.2.2 Solvent Inventory

Solvent Inventory is the main problem in conventional extractors. In large size conventional industrial extractors, a large amount of solvent is required. Less solvent is required in mini-channel. Reduction of characteristic plant dimensions in micro/mini reactors offers a powerful for overcoming bottlenecks in heat and mass transfer and well-defined flow patterns with better temperature conditions.

8.4.3 LIMITATIONS OF CONVENTIONAL EXTRACTION PROCESS

Liquid–liquid extraction is a widely used separation process in various laboratory and chemical, petroleum, pharmaceutical, hydrometallurgical, and food industries. This is mainly because of its cost-effectiveness, in comparison with alternative separations processes. Extraction involves three basic steps as follows:

- Contacting and mixing of the two immiscible liquid phases- organic and aqueous phases,
- Formation and maintaining the droplets or films of the dispersed phase, and
- Subsequently separating the two phases from each other after extraction processes.

Further solvent recovery and raffinate cleanup operations may also be required. The performance of an extraction process is dependent on many factors, such as selection of solvent, operating conditions, mode of operation, extractor type, and an assortment of design criteria. A wide variety of extraction equipment is available: mixer-settlers, centrifugal extractors, and columns. A mixer-settler consists of a mixer (agitated tank) in which the aqueous and organic liquids are contacted, followed by gravity separation in a shallow basin called a settler, where the liquids disengage into individual layers and are discharged separately. In centrifugal extractors, the two immiscible liquids of different densities are rapidly mixed in the annular space between a rotor and the stationary housing. The separation efficiency is very much higher in the centrifugal extractor than in a mixer-settler.

The third class of extraction equipment, columns, is usually used industrially in its countercurrent mode. Columns may be divided into two types: static columns (e.g., spray column, sieve plate column, and packed column) and agitated columns (e.g., rotating disk contactor, Schreiber column, Kuhni column, Karr column, and pulsed column). A common drawback of conventional equipment is the inability to condition the drop size precisely and the nonuniformities that result because of the complexities of the underlying hydrodynamics. This leads to uncertainties in extractor design and often imposes severe limitations on the optimal performance that can be achieved [33, 36].

8.4.4 MINI-FLUIDICS APPROACH ON SILVER BASED SOLVENT EXTRACTION OF OMEGA-3 PUFA

Liquid–liquid extractions (LLE) with reversible chemical reactions are encountered with in a broad range of industrial applications where various design configurations such as stirred tank, mixer-settler and rotating disc contactors are adopted based on mixing requirements and mass transfer characteristics. The most commonly used extraction equipment in LLE is the stirred tank reactor which has certain pros and cons of its design and operating conditions such as inadequate mixing and high energy requirements. Other stage-wise LLE operations, such as spray, packed and sieve tray columns, rely heavily on the type of column internals and flow rate of the liquid phases involved in the extractions processes, which can often result in poor mass transfer, flow mal-distribution and inefficient extraction performance [19].

An evolving technology in the area of LLE is the recent development of mini-fluidics/microfluidics-based contacting systems which are potentially used for mass transfer limited and/or reversible exothermic reactions.

Multiphase liquid–liquid flow in mini/microfluidics and the effect of hydro-dynamics on reaction conversion and mixing in the process continues to be an area of active research. While the fundamental parameters governing an extraction processes are mass transfer across the phase boundary, interfacial kinetics and mixing between light and bulk phases, these parameters may be strongly affected by geometry, orientation, temperature, contaminants, and other elements of a process which can be controlled to varying extents during scale-up.

In LLE with chemical reactions, the reaction rate depends on the interfacial area and mixing characteristics. Conventional stirred tank reactors can achieve specific interfacial areas from 100 to 1000 m^2/m^3, depending on the reactor volume and the type of mechanical agitator/impeller. In conventional CSTRs, the interfacial area of the dispersion generated by stirring is not sufficient for applications which are either highly mass-transfer limited or strongly exothermic. As a consequence, diluents and/or longer residence times may be required to achieve high conversions.

In conventional systems, gravity is the dominant force acting on a multiphase system, where buoyancy force has a significant effect on the flow pattern and hydrodynamics. In mini/microfluidic systems, the surface forces increase until gravitational forces no longer have a significant impact on the hydrodynamics of the system. Interfacial tension forces begin to play a more predominant role in determining flow properties due to the reduction of channel diameter. Therefore, miniaturization of the systems causes a shift in dominant forces which results in the flow stabilities. Based on hydraulic diameter size, the channel are classified into conventional ($D_h > 3$ mm), Mini-channels (200 μm $< D_h < 3$ mm) and micro channels (10 μm $< D_h < 20$) [24].

Single-phase flow in mini-fluidics is typically laminar or transitional, where viscous effects dominate over inertial forces. Therefore, active mixing devices are difficult to incorporate into a given design, prompting the development of numerous passive mixing configurations and operating strategies. Laminar flow at mini scales offers effective diffusion for mixing two immiscible phases due to the reduced path length, provided sufficient interfacial areas can be maintained [40]. The most common flow profile for multiphase applications in these systems is slug/Taylor flow or emulsified flow, with stratified flow rarely observed due to the increased dominance of surface forces relative to gravity [21–24].

These flow patterns have both pros and cons in their application. In the case of stratified flow, the flow is laminar and the mass transfer between two phases is controlled by diffusion. In the case of slug/Taylor flow pattern, the mass transfer between two fluids is improved by internal circulation which takes place in each plug of liquid. This creates good mass transfer while minimizing mass transfer between adjacent slugs, creating a plug flow profile in the reactors. For emulsion-based flow, mass transfer is excellent but separation of the finished products can be challenging. According to literature review, slug flow patterns have offered better performance than other flow patterns from a practical processing perspective [24].

To produce concentrated pharmaceutical and food grade Omega-3 PUFA, efficient extraction and concentration methods are required and also continually researched for improvement on performance of separation. Fish oils are abundant in Omega 3–PUFA and have traditionally been used as the feedstock for preparation of Omega-3 PUFA concentrate. Since fish oils contain a complex mixture of fatty acids with various chain lengths and degrees of unsaturation, so that separation of individual fatty acids is challenging for production of highly concentrated Omega-3 components. With the successful implementation of a number of technologies at the industrial scale, the challenge now faced is now to develop more cost–effective procedures to produce Omega-3 PUFA concentrates to meet the growing demand [35, 37].

Recently, micro/mini-fluidic technology has been developed to enhance the performance of liquid–liquid extraction and other applications such as micro total analysis, nuclide separation systems and mini and micro chemical plants. The flow patterns generated in mini-fluidic technology have been shown to offer high mass transfer areas and consistent extraction performance, while minimizing solvent inventory requirements compared to conventional extraction systems (Figure 8.7). This has led to its adoption for a number of temperature sensitive reactions such as nitration and halogenation for transitioning from batch to continuous processing technology [24, 29]. Given the reversible and exothermic nature of the silver-PUFA complexation reaction and the desire to operate at lower temperatures if possible, the compact framework of the mini-fluidic system should enable more efficient temperature control while reducing both solvent inventory and ambient temperature effects on operating efficiency. Kamio et al. [16, 17] also described liquid–liquid extractions of DHA-Et from an

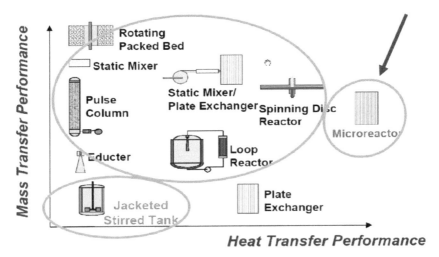

FIGURE 8.7 Increased heat and mass transfer in micro reactor (Source: J.C. Schouten, Symposium on Micro Process Engineering for Catalysis and Multi-phases, Eindhoven University of Technology, February 2006).

organic carrier using silver salt solutions in a micro reactor framework. In their work, contact times on the order of 10–20 seconds were sufficient to reach equilibrium at the conditions tested, leading to the current interest in this approach. This section explores a number of approaches on mini-fluidic technologies which have been considered for processing of marine oils, with the intent of providing a brief overview of the state of the field.

Miniaturization is an alternate approach to process intensification which aids to achieve a reduction in energy used for operations, capital expenditures, plant profile in terms of height and area and further environmental benefits [18]. This method has been attempted for silver-based solvent extraction of Omega-3 PUFA from fish oil ethyl esters. Execution of silver-based solvent extraction in mini-channels results in significant interfacial surface area and a shift in dominant forces governing fluid flow. In fluid–fluid chemical reaction based extraction systems, the organic and aqueous phases are generally immiscible in nature. Since the two immiscible phases must contact each other before reacting, both mass transfer and the chemical kinetics impact the overall rate expression. Increased mass transfer can be achieved through either an increase in local shear (mixing) or interfacial area (mixing or geometric scale reduction), or a decrease in diffusive path length (geometric scale reduction).

The net impact of these adjustments within a variety of technologies is summarized in Table 8.3 as overall volumetric mass transfer coefficient $K_L a$ values. A wide variety of conventional extraction equipment is accessible in solvent extraction, such as mixer settlers, centrifugal extractors, spray columns and agitated columns. However, a significant limitation in conventional systems is the inability to control the drop size precisely and the lack of plug-flow characteristics as the process is scaled to larger capacities. Mixing and solvent inventory are key issues in the separation performance in extraction, with large reactor volumes requiring significant secondary storage capacity [35].

TABLE 8.3 Specific Interfacial Areas for Conventional and Microreactors

Type of conventional reactor	Specific interface area, $[m^2/m^3]$	Type of microreactor	Specific interface area, $[m^2/m^3]$
Packed column	10–350	Micro bubble column	5,100
Countercurrent flow	10–1700	(1100 μm × 170 μm)	
Co-current flow			
Bubble columns	50–600	Micro bubble column (300 μm × 100 μm)	9,800
Spray columns	10–100	Micro bubble column (50 μm × 50 μm)	14,800
Mechanically stirred bubble columns	100–2000	Falling film microreactor (300 μm × 100 μm)	27,000
Impinging jets	90–2050		

Within the past decade, mini-fluidic contacting systems have been applied to various applications of process engineering. Based on a review of extraction methods of potential benefits of operating mini-channel scales, limited analysis has been performed on solvent extraction of Omega-3 PUFA. Of particular note was a lack of hydrodynamic studies evaluating the impact on yield of Omega-3 PUFA. Table 8.4 summarized the application of mini-fluidics in silver-based solvent extraction and approximate yields to which this work will be compared. Table 8.5 presents the overview of silver ions based extraction of Omega-3 PUFA in different processes.

TABLE 8.4 $K_L a$ Values for Different Types of Contactors

Conventional extractors	Extraction system	$K_L a$ ($\times 10^{-4}$ s^{-1})
Agitated vessel	Water (c)-iodine-CCl$_4$ (d)	0.16–16.6
	Sulfate ore (c)-uranium-kerosene (d)	2.8–17
Rotating disc contactor	Water (c)-succinic acid-n-butanol (d)	57
	Water (c)-acetic acid-methyl Isobutyl ketone (d)	20–120
	Water (c)-acetone-DCDE (d)	63–266
Rotating agitated column	n-hexane (c)-acetone-water (d)	0.15
	Toluene (c)-acetone-water (d)	0.2–1.0
	Water (c)-furfural-toluene (d)	105
Spray column	Water (c)-acetone-benzene (d)	8–60
	Water (c)-adipic acid-ether (d)	20–70
	Water (c)-acetic acid-benzene (d)	17.5–63
	Water (c)-acetic acid-nitrobenzene (d)	7–32
Packed column	CCl$_4$ (c)-acetone-water (d)	7.4–24
	Kerosene (c)-acetone-water (d)	5.8–61
	Methyl isobutyl ketone (c)-uranyl Nitrate-water (d)	14.7–111
	Toluene (c)-diethyl amine-water (d)	5–14.7
	Vinyl acetate (c)-acetone-water (d)	7.5–32
Perforated plate column	Water (c)-acetaldehyde-vinyl acetate (d)	28.5
Impinging streams	Water (c)-iodine-kerosene (d)	15–2100
	Kerosene (d)-acetic acid-water (c)	500–3000
	Water (c)-iodine-kerosene (d)	560–2000
Rotating discs impinging streams contactor	Water (c)-iodine-kerosene (d)	1187–3975
	Kerosene-acetic acid-water (c)	1364–4456
	Water (c)-succinic acid-n-butanol (d)	1311–9815
Mini-channel (0.5, 0.75, 1 mm)	Water-iodine-kerosene	1311–9815
	Kerosene-acetic acid-water	4058–14730
	Water-succinic acid-n-butanol	200–3200

*ID varies from 0.5–1 mm, c – continuous phase, d – Dispersed phase.

TABLE 8.5 Overview of Silver Ions Based Extraction of Omega-3 PUFA in Different Processes

System	Flow pattern	Residence Time	Yield	Reference
∏ Complexing Sorbents	Solid Phase extraction in Mistral Multi-Mixer	30 mins at 20°C	89–91% Stripping in solvents	Li [10]
Imidazolium-based ionic liquids Containing silver tetra fluoroborate	Mixing	25 mins at 20°C	Distribution ratio-320	Li [10]
Ionic Liquids based AgBF₄ in Mistral Multi-Mixer	Mixing	30 mins at 20°C	80% in stripping Solvents	Li [11]
Silver Ion Loaded Porous Hollow fiber membrane	Flow through packed column	N/A	100% recovery	Shibasaki [39]
Silver nitrate Solution	Hydrophobic Hollow fiber membrane	N/A, 303 K	N/A	Kubota et al. [46]
Silver Nitrate Solution	Micro reactor Slug flow	20 sec		Kamio [18]
Silver nitrate solution	Slug flow* based Mini-fluidic reactor	36 sec at 10°C	80%	Donaldson [5]

Donaldson [5] reported that the mini-fluidic experimental set up designed for slug flow pattern. However, in experimental conditions, a Stratified flow pattern was observed due to physical properties of the liquids.

Solvent extraction of Omega-3 PUFA can be operated on mild condition such lower temperature and is suitable of large-scale production. It is reported [42] that ethyl esters of PUFAs such as DHA and EPA could be successfully extracted into silver nitrate solutions since silver ion can complex with carbon-carbon double bonds of PUFAs. It is reported that the extractions of ethyl and methyl esters of PUFA such as EPA and DHA from various organic solvents into the aqueous silver nitrate solutions were performed. It was observed that EPA and DHA specifically bound to the silver ion as it has highest degree of unsaturation. In addition to that, the distribution ratio (D) increased drastically as temperature decreased and also by addition of water-soluble alcohols such as methanol, ethanol, and n-propanol to aqueous solutions. The distribution ratio depends on the solvent used for dissolution of organic phases.

The closest work to that presented here is an extraction of pure components of ethyl ester of Docosahexaenoic acid (DHA-Et) and Eicosapentanoic acid (EPA-Et) from an organic carrier into an aqueous silver nitrate solution in a 0.5 mm T-micro reactor [17]. In the work of Kamio et al. [17], extraction equilibrium was reached after 20 seconds, with large mass transfer rates observed due to the high interfacial areas present and internal recirculation within the liquid slugs. The DHA-Et reaction against silver ion follows first order kinetics with the respect to oil phase and is 2.5 orders with respect to silver ion concentration.

The creation of slug flow in the micro reactor especially in the case of Omega-3 PUFA extractions offers better interfacial area and enhanced mixing between the organic and aqueous phase. The enhancement was such that at lower temperatures (268 K), slug flow pattern had faster extraction rates than emulsion-based methods without the associated phase separation difficulties. While promising in nature, the work by Kamio et al. [17] focused specifically on highly idealized mixtures, and did not explore the impact of commercial fish oils which contain a broad spectrum of organic compounds.

8.5   DESIGN CRITERIA FOR EXTRACTION AT MINI/MICRO SCALE EXTRACTION PROCESS

Micro/Mini–fluidic reactor (MR) technology does enable continuous processes based on plug flow reactors with the minimal volume of reactants, rapid dynamic responses and robustness, good temperature control, efficient mixing, etc. There are many design parameters, which does control mini/micro fluidic extraction process such as residence time, channel diameter and hydrodynamics.

8.5.1   CHANNEL DIAMETER

The effect of inner diameter of mini/micro channel affects mass transfer in solvent extraction at mini/micro scale. In addition to that, liquid–liquid mass transfer processes at mini/micro scale is complicated. Mass transfer equation for micro channel with "T" junction as contacting system between organic and aqueous phase is given below:

$$K_{La} = 2.12 \times 10^{-6} \left[q^{-0.34} u^{0.53} ID^{-1.99} \right] \qquad (1)$$

The above equation is valid for ID between 1.2 mm and 0.8 mm the experiments were conducted at the volume flow rate ratio q of 3–24 and the total flow velocity u of 1.2–6.4 mm/s. Moreover, the extraction efficiency decreases with increased volume flow rate ratio, total flow velocity, and the inner diameter of the main channel. The volumetric mass transfer coefficient decreases with increased volume flow rate ratio and the inner diameter of the main channel, increases with increased total flow velocity [41].

8.5.2 HYDRODYNAMICS AT MICRO/MINI SCALE

Hydrodynamics is the key role in controlling the mass transfer and kinetics in any type of chemical reaction. In mini–fluidic channels, specific flow patterns are observed for two immiscible fluids, including stratified flow, wavy annular flow and slug/plug flow. Usually, the flow patterns at mini/micro scales are well defined, and thus the interfacial areas of the flows are uniform, allowing simple and precise modeling of numerous chemical processes. Among the various flow patterns, slug flow offers promising advantages in mini/microfluidic liquid–liquid extraction because of the large interfacial area-to-volume ratio and an increased mass transfer coefficient due to liquid recirculation within each slug [16]. This enhances the rate of extraction in the mini-fluidic contacting system and microfluidics system. In order to define the reaction conditions and to design an optimum mini/micro-fluidic reactor for a solvent extraction processes, detailed behavior of flow pattern at micro/mini scales is necessary. Moreover, well-controlled hydrodynamics has many advantages such as decrease of pressure drop, enhancement of mass transfer and mixing and product separation from reaction medium.

As the mini-fluidic contacting system has the dimension in the range of 0.5 mm < ID < 3 mm, interfacial tension and viscous forces dominate the effects of gravity and inertia forces. The Weber, Capillary and Bond numbers are significant in characterizing multiphase phenomena in such mini scale flows. The Bond number is often used to justify whether the flow is dominated by interfacial tension force or gravity force, thereby defining the point at which geometry is "mini" or "micro" scale. However, it depends on the nature of binary liquid systems, but it affects the formation of flow patterns in the mini-channel. There are two types of liquid–liquid two-phase flows that can result when working with liquids at the micro and mini-fluidic scales: slug flow and parallel flow.

The capillary number can be used to predict the presence of such flow. A capillary number less than 1 indicates that the forces of interfacial tension dominate viscous forces, resulting in a decrease in the interfacial area. Depending on device geometry, a decrease in the interfacial area could lead to slug flow or parallel flow. A capillary number greater than 1 often results in parallel flow due to the viscous fluid's resistance to shear which extends the interface down the length of the channel. At low bond numbers this results in the eventual formation of slug flow, while at high bond numbers stratified flow can result. The Weber number is used to predict when inertial forces become so significant that they lead to instabilities at the interface of two-phase flow, which can also contribute to a transition from slug to stratified flow. A Weber number greater than 1 is indicative of flow instabilities. Stratified flow is quite rare within these systems due to the increased dominance of surface forces at these scales, which tends to result in the formation of slug flow. In the case of slug flow, two mechanisms are known to be responsible for the mass transfer between two immiscible liquids: internal circulation within each slug, and the concentration gradients between adjacent. In the case of stratified flow, the flow is laminar and the transfer of molecules between the two phases is supposed to occur only by diffusion [41, 45].

8.6   REAL TIME PERFORMANCE OF MINI/MICRO-FLUIDIC EXTRACTION OF OMEGA-3 PUFA

The liquid–liquid extraction of Omega-3 PUFA with silver ions at low-temperature favors advantages among several separation technologies. It is proved that extraction of the ethyl esters of Omega-3 PUFA with silver ions was an exothermic reaction that progressed faster at the lower temperature. Therefore, the solvent extraction of Omega-3 PUFA–et either pure components such as DHA/EPA-et or crude from fish oil with silver nitrate solutions has potential to overcome oxidation during the separation process. The extraction of Omega-3 PUFA in various micro/mini channels are performed with a variety of flow patterns.

8.6.1 EXTRACTION OF DOCOSAHEXAENOIC ACID (DHA) ETHYL ESTER IN SLUG FLOW BASED MICRO REACTOR

The separation of DHA that is dissolved in organic solvents was extracted with concentrated aqueous silver nitrate solution in slug flow based micro reactor. The uptakes of DHA-Et for the extraction system using a slug flow prepared by a microreactor are commented. The observed fast equilibration realized for the slug flow extraction system was due to the large specific interfacial area of the slug caused by the presence of wall film and the thin liquid film caused by the internal circulation caused by slug flow pattern. A glass chip having T-shape microchannel was used to create a contact between an aqueous and the organic phase to create a slug flow pattern in the micro reactor.

The syringe pump is used to connect the T-junction in the microchip in the configuration of slug flow micro reactor experimental system with the poly tetra fluoro ethylene (PTFE) tube of I.D. 5×10^{-4} m. From these syringe pumps, the aqueous and organic phases were fed at the same flow rate of 1.67×10^{-9} m^3/s, respectively. Behind the glass chip, a PTFE tube of 5×10^{-4} m I.D. was connected. A slug flow created in the glass chip was fed into the PTFE tube. During the slugs passed through the PTFE tube, DHA-Et was extracted from the organic phase to the aqueous phase by reaction between silver and Omega-3 PUFA.

8.6.2 CREATION OF SLUG FLOW PATTERN IN MICRO REACTOR

In this study, the slug flow produced by introducing and allowing contacting both ultrapure water and n-heptane was shown in Figure 8.8. The average length of each phase was found to be 2.54×10^{-3} m.

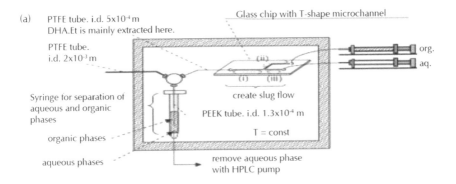

FIGURE 8.8 (a) Experimental setup for extraction of DHA-Et using slug flow prepared by glass chip with T-shape microchannel. Channel width: (i) 1.33×10^{-4} m, (ii) 9.8×10^{-5} m, and (iii) 1.31×10^{-4} m. Channel depth: 4.7×10^{-5} m.

8.6.3 EXTRACTION PERFORMANCE BY UPTAKE CURVE IN SLUG FLOW BASED MICRO REACTOR

As shown in Figures 8.9 and 8.10, the DHA-Et concentration in the organic phase reaches equilibrium after 10 s, e.g., equilibration was finished within 10 s. The ultra-fast equilibration was achieved by using slug flow pattern in the micro reactor.

By using a micro reactor, extraction of the ethyl ester of pure components of omega 3PUFA from an organic solution into an aqueous silver nitrate solution were investigated in Slug flow based a T-shape microchannel. They proved that slug flow produce large specific interfacial area between aqueous and organic phases, and offer ultra-fast extraction. Moreover, Kamio et al. [17] proved that slug flow is more advantageous from the point of handling and extraction rate than any another flow pattern like emulsion and stratified. Additionally, it confirmed that the extraction performance of slug flow was unchanged even the slug length are large. The equilibrium for solvent extraction has taken less than 20 s. In all of the experiments, the amount of DHA-Et or EPA-Et extracted was equal to that expected from the batch experiment where the equilibrium has reached to 5000 s.

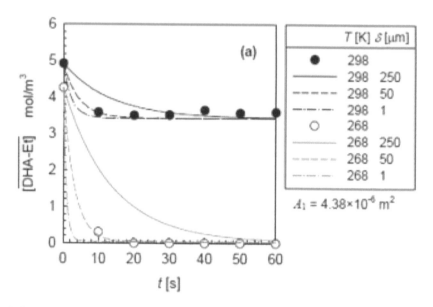

FIGURE 8.9 Uptake curves of DHA-Et extraction with the slug flow for $[Ag^+]_{aq,0} = 2000$ mol m^{-3}. Solid, broken, and dash-dotted lines are the calculated results for different values of δ. Interfacial area: (a) Columnar area, which is equivalent to the surface area of the unit slug ($A_1 = 4.38 \times 10^{-6}$ m^2).

Additionally, the extraction at low temperature offers a drastic improvement of extraction ratio with the low concentration of silver ion. Therefore, it can easily reduce consumption of silver nitrate by using a micro channel. This is preferable for the practical operation with considerable scaling up.

8.6.4 SLUG FLOW BASED MINI-FLUIDIC EXTRACTION PROCESS

In this case, the practical implications of extraction of Omega-3 PUFA from raw 18/12EE fish oils with concentrated silver nitrate solutions were evaluated in 1/16" slug flow mini-fluidic reactor channels. The extraction performance and hydrodynamic characteristics in idealized mixtures is evaluated by an experimental system have been constructed to contact 18/12EE fish oils with concentrated silver-salt solutions (50 wt. %) in a controlled manner and determine if previously idealized system

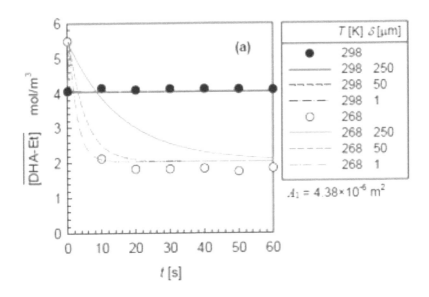

FIGURE 8.10 Uptake curves of DHA-Et extraction with the slug flow for $[Ag^+]_{aq,0} = 1000$ mol m^{-3}. Solid, broken, and dash-dotted lines are the calculated results for different values of δ. Interfacial area: (a) Columnar area which is equivalent to the surface area of the unit slug ($A_1 = 4.38 \times 10^{-6}$ m^2).

performance is observed when commercial fish oils are used. Silver-based solvent liquid–liquid extraction was carried out in 1/16″ ID food-grade Tygon tubing submersed in a temperature-controlled refrigerated reservoir. The organic phase was compared directly to the feedstock for each experiment to identify changes in the oil composition, while the aqueous solution was further processed in an attempt to elute and recover the extracted PUFA's. The deviation of flow patterns and physical property characteristics of the system were explored.

8.6.5 DEVELOPMENT OF SLUG FLOW BASED MINI-FLUIDIC REACTOR

In this work, The LLE experiments were performed with raw fish oil ethyl esters using concentrate aqueous silver nitrate solution under conditions for which slug flow was expected. In solvent-based extraction of Omega-3

PUFA, freshly prepared concentrated aqueous silver nitrate solution is used. The chemical used are silver nitrate (ACS grade – 99% Purity), sodium nitrate (Assay – 99%), 95% ethyl alcohol, deionized water, semi refined 18/12 fish oil ethyl esters from DSM Ocean Nutrition, Dartmouth, NS, Canada, and Nitrogen gas which is used to blanket and prevent the oxidation of fish oil and $AgNO_3$. Semi refined fish oil ethyl ester was provided from DSM, derived from anchovy (*Engraulisringens)* and Sardine (*Sardinopssagaxsagax*) on 29th June 2013. The slug flow mini-fluidic flow reactor was constructed for LLE of EPA and DHA ethyl esters from semi refined 18/12 EE fish oil ethyl esters.

The simplified mini-fluidic experimental setup is shown in Figure 8.11, consisting of a 1/16th inch ID Tygon mini-fluidic channel submerged in a cooled reservoir controlled to 10°C using an external refrigerated circulating bath. The solutions from the reservoir were pumped using a double syringe pump (2 NE 4000), whereby a 60 C.C. syringe was used for the silver nitrate solution and a 10 C.C. syringe was used for the 18/12 EE fish oil. By setting a dispense at rate of 5 ml/min for the 60 C.C. syringe, the silver salt solution flow rate was 5 ml/min and the oil flow rate was 1.47 ml/min, thus maintaining an approximate salt to oil solution flow ratio of ~3.4:1 [16, 17]. The fish oil ethyl ester and silver nitrate solution are precooled in a 1.5 m length of tubing prior to being contacted together in a "Y" junction, after which the immiscible fluids were allowed to contact for a set residence time before being sampled via syringe. Sampling ports were fitted into the immersion vessel by creation of holes through the side, minimizing the time which the fluids spent outside of the refrigerated environments. The holes through the vessel walls were below the water line inside the vessel, and were sealed by silicon caulking. Water inside the cooler was circulated using a submersible pump, with a copper line run through from the refrigerated bath recirculation loop. In the LLE experiments, 50 wt. % concentrated silver nitrate solution was used as extraction solvent for extraction of Omega-3 PUFA from fish Oil EE. Silver nitrate is toxic and corrosive, necessitating minimum exposure to avoid immediate or any significant side effects other than the purple skin stains, but with more exposure, noticeable side effects or burns may result (Fisher Scientific, 2014).

Figure 8.12 shows dual syringe pump to control flow into the mini-fluidic system. Stepwise procedure for mini-fludic solvent extraction is shown in Figure 8.13. Post processing overview of LLE extraction in mini–fluidic experimental set up is detailed in Figure 8.14. The Figure 8.15 presents the yield of Omega-3 PUFA versus residence time (mins) in case of mini-fluidic reactor. Figure 8.16 shows the stratification of flow at Y-junction. Additionally, the properties of fish oil and silver nitrate solution would deviate the anticipated flow pattern.

The results are provided in plot of equilibrium concentration, which reached in 36 secs in mini-fluidic contacting system. These values were thus likely due to problems in handling the samples in the vial during nitrogen evaporation of hexane or recovering the organic fractions between syringes. The remaining values had typical recoveries of 75–78 wt. % Omega-3 (on average 28.9% of the initial oil mass) with concentrations of

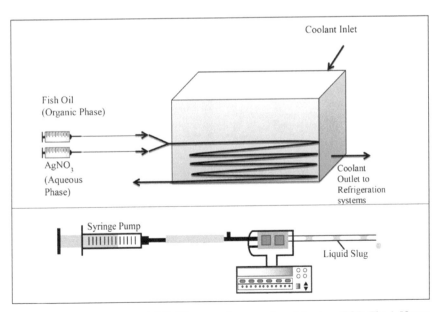

FIGURE 8.11   Slug flow mini-fluidic extraction experimental set up. TOP: The 1.58 mm ID mini-fluidic channel is immersed in subcooled water (10°C) in the bath, which is circulated by submersible pump. BOTTOM: the flow patterns produced also depends on the contactor (mixing system) in the experimental systems, so that the various flow patterns will be anticipated like slug flow/drop let flow/stratified flow. The experimental set up was designed for slug flow pattern.

~78 wt. % total Omega 3. However, an unanticipated flow pattern within the mini-fluidic channels could have potentially affected the relative flow rates of the two species (Fish Oil and Silver nitrate solution) in question (and the contact times), creating additional uncertainty in the yield. Despite the change in flow pattern from slug to stratified, the time scale for extraction coincides with the previous work of Kamio et al. [16,17]. In their work, pure DHA or EPA dissolved in an organic solvent reached equilibrium with a silver salt solution in 20 s with a silver ion concentration of 1000 mol/m^3 at 268 K, and 10 s with a silver ion concentration of 2000 mol/m^3 at 288 K. For the purpose of process feasibility, looking at shorter residence times was not required as most of the batch-wise processes would operate for longer than 15 min due to filling and emptying considerations, and a 36 second residence time within a continuous process provides an upper bound of what might be needed [5, 6].

Components of Experimental Setup

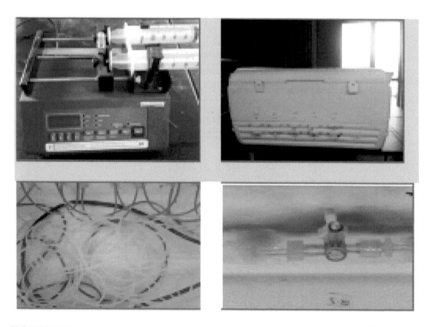

FIGURE 8.12 Dual syringe pump from Longer Instruments used to control flow into the mini-fluidic system; Immersion vessel for cooling the mini-fluidic channels; Tygon mini-fluidic channel; and sample port.

Flow patterns in Mini-fluidic Channel

- 18/12 Fish Oils EE (Organic Phase)~1.5 ml/min
- 50%wt.AgNO$_3$ (Aqueous Phase)~5 ml/min
- Temperature = 10±0.5°C
- Residence times varies from 0.6 to 7.3 mins
- Phase inversion observed at "Y" Junction
- Stratification of flow has been observed.
- Samples are collected at specified location.
- Gravity settling has been allowed.

FIGURE 8.13   Stepwise procedure for mini-fludiic solvent extraction.

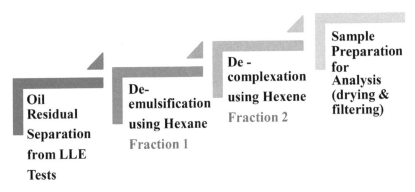

FIGURE 8.14   Post processing overview of LLE extraction in mini–fluidic experimental set up.

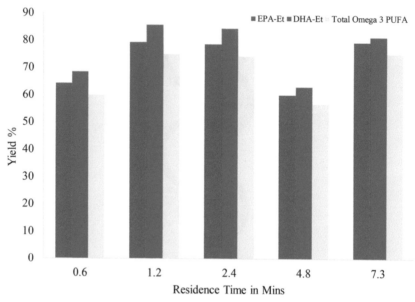

FIGURE 8.15   Yield of Omega-3 PUFA vs. residence time (mins) – mini-fluidic reactor.

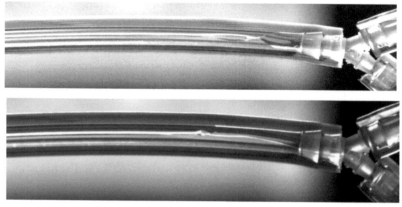

FIGURE 8.16   Stratification of Flow at Y Junction. Additionally, the properties of fish oil and silver nitrate solution would deviate the anticipated flow pattern.

8.6.5.1 Deviation of Flow Patterns

A notable difference between this study and previous designed mini-fluidic studies was the formation of a stratified flow profile in the mini-fluidic channels where a slug flow profile was expected. The stratified flow is characterized by a complete separation of the liquid–liquid interface, where the relative residence times of both phases may vary depending on the relative cross-section occupied by each phase. It is quite odd to observe stratified flow under these conditions, as one of the characteristic properties of minifluidic channels is the predominance of surface forces over gravity forces, which should limit the formation of stable stratified flow. The dimensionless parameters relevant to flow at this scale are provided in Table 8.6. In addition to that, the flow patterns will be affected the effects of tube wettability and fluid properties. At present, it is important to note the high Bond number present under these conditions with interfacial tension IFT of this particular system, as the use of commercial fish oils seems result in a sufficient reduction in IFT to cause a change in flow dynamics in the system. For now, a comprehensive discussion on the flow patterns is mentioned below and various forces involved in flow pattern are commented [5, 6].

The "Y" mixing section in experimental set up is used for creation of slug flow pattern, which is anticipated. However, raw fish oils contained numerous compounds such as saturated and unsaturated fatty acids, which reduce the interfacial tension at the interface between fish oil and silver nitrate, resulting in a stratified flow within the channels used in this study.

In addition to this, the interfacial reaction between fish oil ethyl ester and silver nitrate systems might cause reduction of interfacial tension between them. This system is an anomaly within the context of minifluidics, as very few fluids form stable stratified flow in millimeter-scale geometries. Despite this flow pattern being present, the overall extraction performance did not appear to suffer from the change in flow dynamics. Additionally, the relative area for mixed flow, stratified flow and slug flow in the extraction process are 290 m^2/m^3, 806.25 m^2/m^3 and 1520 m^2/m^3, respectively. From this information, the relative area for slug flow and stratified flow is higher than the idealized stirred tank system and provided better contact area for mass transfer.

In the case of Fish Oil Water system, at lower volumetric flow rates (Figure 8.17), slug flow pattern appeared in the mini-channel and Bond number for this system < 1. Therefore, Interfacial tension force dominates gravity force in fluid path. As a result, the flow becomes slug. In the case

TABLE 8.6 Influence of Various Dominating Force at "Y" Junction

Dimensionless number	Definition	Formula	Value
Bond number	Gravity force/ Interfacial tension	$Bo = \dfrac{\Delta \rho g d^2_H}{\sigma}$	54.94
Capillary number	Viscous force/ Interfacial tension	$Ca = \dfrac{\mu u}{\sigma}$	0.45
Reynolds number	Inertial force/viscous force	$Re = \dfrac{d_H \rho u}{\mu}$	48.35
Weber number	Inertial force/ Interfacial tension	$We = \dfrac{d_H \rho u^2}{\sigma}$	21.78

of Fish Oil – Silver nitrate system, the flow was stratified due to reduction of interfacial tension and then gravity force dominates interfacial tension force. Bo for this system is > 1.

For Ca<<1, the flow is dominated by capillary forces and interfacial tension forces (Figure 8.18). However, in the case of fish oil silver nitrate solution system, Ca<<1, the flow still is stratified due to reduction of interfacial tension at interface and high density of silver nitrates solution when compared to fish oil ethyl esters.

Even though the extraction performance is consistent with previously reported literature, the flow pattern is deviated from expected. The addition of alkanes into the fish oil phase could elevate the interfacial tension between fish oil phase and silver nitrate solution to bring the slug flow pattern in the mini-channel. These trends are important for future processing considerations, both in recognizing that the low surface tension between fish oil and silver nitrate will facilitate contacting and mass transfer, while possibly making separation difficult. The addition of hexane or, preferably, hexene, could be used to help in separating the two phases in the design of either mini-fluidic technology for liquid–liquid extraction of EPA/DHA from fish oil ethyl ester or on a more general basis in de-emulsification [5, 6].

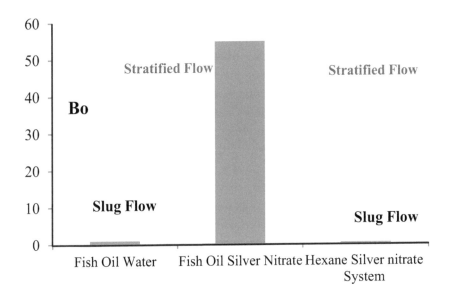

FIGURE 8.17 Role of Bond Number in hydrodynamics and its effects for formation of flow patterns.

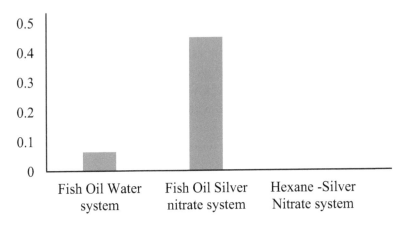

FIGURE 8.18 Role of Capillary Number in flow patterns – on the micro/mini scale, the relative effect of viscous force and interfacial tension forces across the interface between two immiscible fluids.

8.7   CONCLUSIONS AND FUTURE DIRECTIONS

The micro/mini-fluidic extraction of Omega-3 PUFA either pure component dissolved in organic solvents or concentrated form from raw fish oil with concentrated silver nitrate solution was investigated using slug and stratified flow pattern. The preliminary results suggest that silver-based solvent extraction is satisfactory at mini/micro scale. However, there are some possible recommendations to improve the existing process. In the extraction process, the de-complexation process currently uses dilution with water to elute bound ethyl esters, making recycle of the solvent stream difficult. To improve processing economics associated with solvent-based extractions, it would be ideal to de-complex the bound EPA and DHA without irreversibly modifying the silver nitrate solution (i.e., by combining a temperature increase with a secondary stripping agent). The experiments to determine the extent of EPA and DHA eluted at temperatures from 10 to 60°C when in contact with hexane and hexane would be performed.

Temperature appears to have a dramatic effect on extraction. While the results of the 10°C results are promising, equilibrium has been reached in less than 36 seconds in the mini-fluidic reactor, suggesting that even shorter flow paths could be used. Given the increase in viscosity at these lower temperatures, minimizing channel length is important for limiting energy dissipation. The Tygon tubing currently used is also not suitable for long-term use, softening over time with increased exposure to Fish Oil ethyl ester which was transesterified. This will necessitate a transition to a more process-based plate and frame design constructed of a suitable material. There are also indications that hexane/hexene addition both before or after the initial contacting could be used within a processing strategy to control flow patterns and, if coupled with wettability-based separation, be used to limit external vessel requirements. Based on these observations, there are a number of areas which like to explore in more detail:

- A different material of construction (i.e., stainless steel) significantly used for construction of mini-fluidic contacting system could affect the flow pattern observed (i.e., transition from stratified flow to slug flow due to wettability differences).
- To recycle the silver nitrate solution after recovery of silver from chemical precipitation method, the addition of food based antioxidant

addition to the aqueous solution could improve stability, and will the antioxidants remain in the aqueous solution or be partitioned between the residual oil and concentrated ethyl esters.

- To investigate the equilibrium concentration of EPA/DHA affected by both the silver nitrate solution concentration, ionic strength and operating temperature. The current aqueous: organic volumetric feed ratio is approximately 3.3:1, which does require excess solvent. To design the optimum ratio for a mini-fluidic flow-based system for these applications.

- To evaluate the fundamental mechanism by which the silver ions and EPA/DHA bond. The two compounds in question have multiple double-bonds which are capable of complexing with the ethyl esters, raising the question of how these bonds form, the number which is required to make the compounds soluble in organic solution, and how the extraction and de-complexation conditions can be manipulated to transition between solubility and insolubility with minimal energy input.

- To develop models for each stage of silver based solvent extraction of Omega-3 PUFA for better conceptual design and process design and cost estimation and to evaluate detailed investigation of physical properties of Omega-3 PUFA for process development and scale up studies for LLE [38].

8.8 SUMMARY

The fractionation of fats/oils or fatty acids from a variety of feedstock could be achieved by enchanting advantage of reactive solubility difference of triglycerides or fatty acids in a mono or binary solvent systems which are immiscible in nature. In this case, solvent extraction plays an important role in the separation of fatty acids of interest in oil/fats using numerous aqueous/organic solvent. One of its applications was the purification of fatty acids ethyl ester based on the formation of the complex between double bonds between PUFA-Et and silver ions by solvent extraction was recently reported as a novel process. However, In this case, separation by extraction was performed exclusively via reversible fast chemical reaction between silver nitrate solution and the double bond of Omega-3 PUFA.

The novel extraction methods to increase overall EPA and DHA yield is required to replace various conventional processes. In order to produce pharmaceutical and food grade Omega-3 PUFA, efficient extraction and concentration methods have been developed and are continually reviewed for improvement in separation performance. With the successful implementation of a number of conventional technologies at the industrial scale, the challenge now faced is to develop more cost–effective procedures with significant yields to produce Omega-3 PUFA concentrates to meet the growing demand. Past decade, the performing extraction at micro/mini scales is an emerging process intensification technology to enhance the mixing performance that augments to increase interfacial areas in liquid– liquid extraction. In addition to that, the flow patterns generated in micro/mini-fluidic technology have been proved to offer high mass transfer areas and consistent extraction performance, while minimizing solvent inventory requirements compared to other conventional extraction systems. The performance of mini/microfluidic technology in Liquid–Liquid extraction of EPA and DHA from commercial fish oil has been described and commented flow patterns in the mini-channel with possible recommendations in this book chapter.

ACKNOWLEDGEMENT

Author expresses his gratitude to his thesis advisor Dr. Adam A. Donaldson, Assistant Professor, Chemical Engineering, Department of Process Engineering and Applied Science, Dalhousie University, Halifax, NS, Canada for providing graduate research scholarship for pursuing MASc Program at Dalhousie University, Halifax, NS, Canada.

KEYWORDS

- bond number
- capillary number
- EPA/DHA extraction
- extraction performance
- fish oil
- flow patterns/hydrodynamics
- interfacial tension
- liquid–liquid extraction
- mini-fluidic flow reactor
- mini-fluidic process design
- mini-Fluidics
- miniaturization
- omega 3 PUFA
- process intensification
- residence time/contacting time
- reynolds number
- silver nitrate solvent
- slug flow
- solvent extraction
- stratification of flow
- stratified flow
- tygon mini-channel
- weber number
- "Y" Junction

REFERENCES

1. Kapoor, R., & Patil, U. K. (2011). Importance and production of omega-3 fatty acids from natural sources-Mini Review. *International Food Research Journal,* 18, 493–499.

2. James, R. Fair, & Jimmy L. Humphrey (1983). Liquid–liquid extraction processes. Proceedings Fifth Industrial Energy Technology Conference, Volume II, Houston, TX, April 17–20, pp. 846–856.

3. Wanasundara, U. N., & Shahidi, F. (1998). Production of omega-3 PUFA concentrates. *Trends in Food Science and Technology, 9,* 231–240.

4. Lembke, P. (2013). Production Techniques for Omega-3 Concentrates. In: *Omega-6/3 Fatty Acids,* De Meester, F., Watson, R. R., & Zibadi, S. (Eds.), pp. 353–364, Humana Press.

5. Donaldson, A. (2013). *Silver-Based Liquid/Liquid Extraction of EPA/DHA from Fish Oil Ethyl Esters: Feasibility Analysis of Mini-Fluidic Technology.* Technical Report. Halifax, NS: DSM Nutritional Products.

6. Kirubanandan, S. (2015). *Mini-Fluidic Silver Based Solvent Extraction of EPA/DHA from Fish Oil.* MASc Thesis, Dalhousie University, Halifax, NS, Canada.

7. Yazawa, K. (1995). Purification of highly unsaturated fatty acids, AA, EPA, DHA. In: *Highly Unsaturated Fatty Acids, edited by M.Kayama,* Kouseisyakouseikaku Co., Ltd., Tokyo, Japan, pp. 1–10.

8. Fagan, P., & Wijesundera, C. (2013). Rapid isolation of omega-3 long-chain polyunsaturated fatty acids using monolithic high performance liquid chromatography columns. *Journal of Separation Science, 36*(11), 1743–1752.

9. Breivik, H., Libnau, F. O., & Thorstad, O. (2012). Process for concentrating omega-3 fatty acids. Google Patents.

10. Li, M., Pittman Jr, C. U., & Li, T. (2009). Extraction of polyunsaturated fatty acid methyl esters by imidazolium-based ionic liquids containing silver tetrafluoroborate-extraction equilibrium studies. *Talanta, 78*(4–5), 1364–1370.

11. Li, M., & Li, T. (2008). Enrichment of Omega-3 polyunsaturated fatty acid methyl esters by ionic liquids containing silver salts. *Separation Science and Technology, 43*(8), 2072–2089.

12. Dewar, M. J. S. (1951). A review of π complex theory. *Bull. Soc. Chim. France, 18,* C71–C79.

13. Calder, P. C. (2013). Omega-3 polyunsaturated fatty acids and inflammatory processes: nutrition or pharmacology. *British Journal of Clinical Pharmacology, 75*(3), 645–662.

14. Connor, W. E. (2000). Importance of $n-3$ fatty acids in health and disease. *The American Journal of Clinical Nutrition, 71*(1), 171S-175S.

15. Covington, M. B. (2004). Omega-3 fatty acid. *Maryland American Family Physician, 70*(1), 133–140.

16. Kamio, E., Seike, Y., Yoshizawa, H., Matsuyama, H., & Ono, T. (2011). Micro-fluidic extraction of docosahexaenoic acid ethyl ester: Comparison between Slug flow and emulsion. *Industrial and Engineering Chemistry Research, 50*(11), 6915–6924.

17. Kamio, E., Seike, Y., Yoshizawa, H., & Ono, T. (2010). Modeling of extraction behavior of docosahexaenoic acid ethyl ester by using slug flow prepared by micro-reactor. *AIChE Journal, 56*(8), 2163–2172.

18. Charpentier, J. C. (2005). Process intensification by miniaturization. *Chemical Engineering and Technology, 28,* 255–258.

19. Kashid, M. N. et al. (2011). *Chemical Engineering Science, 66,* 3876–3897.

20. Oroskar, A. R. (2002). *Proceedings of the 5th International Conference on Micro reaction Technology*, 27–30 May 2001, Strasbourg, (Berlin, Heidelberg, New York: Springer), p. 153.
21. Kashid, M. N., Gerlach, I., Goetz, S., Franzke, J., Acker, J. F., Platte, F., & Turek, S. (2005). Internal circulation within the liquid slugs of a liquid–liquid slug-flow capillary microreactor. *Industrial and Engineering Chemistry Research, 44*(14), 5003–5010.
22. Kashid, M. N., Harshe, Y. M., & Agar, D. W. (2007). Liquid–liquid slug flow in a capillary: an alternative to suspended drop or film contactors. *Industrial and Engineering Chemistry Research, 46*(25), 8420–8430.
23. Benz, K., Jäckel, K. P., Regenauer, K. J., Schiewe, J., Drese, K., Ehrfeld, W., & Löwe, H. (2001). Utilization of micromixers for extraction processes. *Chemical Engineering and Technology, 24*(1), 11–17.
24. Burns, J. R., & Ramshaw, C. (2001).The intensification of rapid reactions in multi-phase systems using slug flow in capillaries. *Lab on a Chip, 1*(1), 10–15.
25. Canas, B. J., & Yurawecz, M. P. (1999). Ethyl carbamate formation during urea complexation for fractionation of fatty acids. *Journal of the American Oil Chemists' Society, 76*(4), 537–537.
26. Crexi, V., Monte, M., Monte, M., & Pinto, L. A. (2012). Polyunsaturated fatty acid concentrates of carp oil: Chemical hydrolysis and urea complexation. *Journal of the American Oil Chemists'Society, 89*(2), 329–334.
27. Cunnane, S. C., Plourde, M., Stewart, K., & Crawford, M. A. (2007). Docosahexae-noic acid and shore-based diets in hominin encephalization: A rebuttal. *American Journal of Human Biology, 19*(4), 578–581.
28. Domart, C., Miyauchi, D. T., & Sumerwell, W. N. (1955). The fractionation of marine-oil fatty acids with urea.*Journal of the American Oil Chemists'Society, 32*(9), 481–483.
29. Dummann, G., Quittmann, U., Gröschel, L., Agar, D. W., Wörz, O., & Morgensch-weis, K. (2003). The capillary-microreactor: a new reactor concept for the intensification of heat and mass transfer in liquid–liquid reactions. *Catalysis Today, 79–80*, 433–439.
30. Fisher Scientific, Inc. (2014). "MSDS of Silver sulfate". *https://www.fishersci.ca.* Fair Lawn, New Jersey.
31. Meester, F. (2013). Introduction: The economics of Omega-6/3. In: *Omega-6/3 Fatty Acids*. De Meester, F., Watson, R. R., Zibadi, S. (Eds.), pp. 3–11, Humana Press.
32. Jovanović, J., Rebrov, E. V., Nijhuis, T. A., Kreutzer, M. T., Hessel, V., & Schouten, J. C. (2011). Liquid–liquid flow in a capillary microreactor: hydrodynamic flow patterns and extraction performance. *Industrial and Engineering Chemistry Research, 51*(2), 1015–1026.
33. Okubo, Y., Maki, T., Aoki, N., Hong Khoo, T., Ohmukai, Y., & Mae, K. (2008). Liquid–liquid extraction for efficient synthesis and separation by using micro spaces. *Chemical Engineering Science, 63*(16), 4070–4077.
34. Ratnayake, W. M. N., Olsson, B., Matthews, D., & Ackman, R. G. (1988). Preparation of Omega-3 PUFA concentrates from fish oils via urea complexation. *Lipid/Fett, 90*(10), 381–386.

35. Rubio-Rodríguez, N., Beltrán, S., Jaime, I., de Diego, S. M., Sanz, M. T., & Carballido, J. R. (2010). Production of omega-3 polyunsaturated fatty acid concentrates: A review. *Innovative Food Science and Emerging Technologies, 11*(1), 1–12.
36. Seike, Y., Kamio, E., Ono, T., & Yoshizawa, H. (2007). Extraction of ethyl ester of polyunsaturated fatty acids by using Slug flow prepared by microreactor. *Journal of Chemical Engineering of Japan, 40*(12), 1076–1084.
37. Shahidi, F., & Wanasundara, U. N. (1998). Omega-3 fatty acid concentrates: nutritional aspects and production technologies. *Trends in Food Science and Technology, 9*(6), 230–240.
38. Sharratt, P. N., Wall, K., & Borland, J. N. (2003). Generating innovative process designs using limited data. *Journal of Chemical Technology and Biotechnology, 78*(2–3), 156–160.
39. Shibasaki, A., Irimoto, Y., Kim, M., Saito, K., Sugita, K., Baba, T., & Sugo, T. (1999). Selective binding of docosahexaenoic acid ethyl ester to a silver-ion-loaded porous hollow-fiber membrane. *Journal of the American Oil Chemists' Society, 76*(7), 771–775.
40. Steinke, M. E., & Kandlikar, S. G. (2004). Single-phase heat transfer enhancement techniques in micro channel and mini-channel flows, microchannels and mini-channels. June 17–19, New York, NY, Paper No. ICMM 2004–2328.
41. Taha, T., & Cui, Z. F. (2004). Hydrodynamics of slug flow inside capillaries. *Chemical Engineering Science, 59*(6), 1181–1190.
42. Teramoto, M., Matsuyama, H., Ohnishi, N., Uwagawa, S., & Nakai, K. (1994). Extraction of ethyl and methyl esters of polyunsaturated fatty acids with aqueous silver nitrate solutions. *Industrial and Engineering Chemistry Research, 33*(2), 341–345.
43. Thorsen, T., Roberts, R. W., Arnold, F. H., & Quake, S. R. (2001). Dynamic pattern formation in a vesicle-generating microfluidic device. *Physical Review Letters, 86*(18), 4163–4166.
44. Traynham, J. G., & Sehnert, M. F. (1956). Ring size and reactivity of cyclic Olefins: Complexation with aqueous silver ion. *J. Am. Chem. Soc., 78*, 4024–4027.
45. Vir, A. B., Fabiyan, A. S., Picardo, J. R., & Pushpavanam, S. (2014). Performance comparison of liquid–liquid extraction in parallel micro flows. *Industrial and Engineering Chemistry Research, 53*(19), 8171–8181.
46. Kubota, F., Goto, M., Nakashio, F., & Hano, T. (1997). Separation of Poly unsaturated fatty acids with silver nitrate using a hollow-fiber membrane extractor. *Separation Science and Technology, 32*, 1529–1541.

CHAPTER 9

ANTI-OXIDANT AND ANTI-BACTERIAL PROPERTIES OF EXTRACTS: TERMINALIA CHEBULA AND TERMINALIA BELLERICA

HARSHA PATEL, HEMALI PADALIA, MITAL J. KANERIA, YOGESH VAGHASIYA, and SUMITRA CHANDA

CONTENTS

9.1   INTRODUCTION

Bacterial pathogens have shown a remarkable ability to adapt to environmental parameters. In particular, the increasing use of antibiotics over the past few decades has led to the emergence and spread of various mechanisms of antimicrobial drug resistance among bacterial pathogen [36]. *Staphylococcus aureus* is a versatile bacterial opportunist responsible for a wide spectrum of infections [9]. Methicillin-resistant *Staphylococcus aureus* (MRSA) is an important cause of morbidity and mortality in hospital acquired infections; however, in recent years, there has also been a worldwide increase of community-acquired MRSA infections [43]. Development of MRSA is an important step in the evolution of this pathogen [28]. So there is an urgent need to control MRSA infections by improved antibiotic usage or to find out alternative source especially from medicinal plants [11].

In recent years, food and other biological materials contain natural antioxidants which are safe, have high nutritional and therapeutic value. Humans are surrounded and exposed daily to a multitude of toxic products, xenobiotics and even food, which are known to have a broad range of *reactive oxygen species* (ROS) generating mutagenic compounds. There is also increasing evidence indicating that ROS and free radical mediated reactions are known to be implicated in many cell disorders and in the development of many diseases including cardiovascular diseases, atherosclerosis, cataracts, cancer and chronic inflammation [2]. Antioxidants are substances that delay the oxidation process, inhibit the polymerization chain initiated by free radicals and other subsequent oxidizing reactions [26]. This concept is fundamental to biomedical, nutraceutical, food chemistry and phytochemical sciences, where synthetic antioxidants like butylated hydroxy toluene (BHT) have long been used to preserve quality of food by protecting against oxidation-related deterioration [12]. However, toxicity and potential health hazards of synthetic antioxidants forced the scientists to find out natural and safer antioxidants [25].

The importance of medicinal plants is being highlighted as a source of natural antibacterial and antioxidant agent [13, 48]. *Terminalia* is well known for medicinal properties, a genus of mostly deciduous trees. Different species of Terminalia are reported in most of the Indian forest [56]. Triphala a combination of *T. chebula*, *T. bellerica* and *Emblica officinalis* is extensively used in *Indian System of Medicine* for treating different kind of diseases and to promote health, immunity and longevity [32]. *T. chebula* and *T. bellerica* belong to the family

Combretaceae, commonly known as harade and baheda, respectively in Gujarat, India [20, 31]. Dried ripe fruit of *T. chebula* has been traditionally used to treat various ailments in Asia [50]. *T. chebula* has been reported to exhibit a variety of biological activities including anticancer [55], antidiabetic [21], antimutagenic [6], antimicrobial [5], antiviral activities [7], antioxidant activities [33] and antianaphylactic [58]. *T. chebula* extracts also showed inhibitory activity against HIV [1]. *T. bellerica* is also reported for several pharmacological effects including antibacterial, antimalarial, antifungal, anti-HIV, antioxidant, and anti-mutagenic, anticancer effects [4, 8, 19, 30, 47, 60].

In the present study, methanol extracts of the fruit rind and seed coat of *T. chebula* and *T. bellerica* were investigated for anti-MRSA and antioxidant activities.

9.2   MATERIAL AND METHODS

9.2.1   REAGENTS

The reagents used for the study were 2,2-diphenyl-1-picrylhydrazyl (DPPH), Nitroblue tetrazolium (NBT), Phenazine methosulfate (PMS), Nicotinamide adenine dinucleotide reduced (NADH), gallic acid, ascorbic acid, $FeSO_4$, quercetin, Folin-Ciocalteu's reagent, aluminum chloride, potassium acetate, ferric chloride, 2,4,6- tripyridyl-s-triazine (TPTZ), Tris-HCl, sodium acetate, 2,2-azino-bis-(3-ethylbenzothiazoline-6-sulfonic acid) (ABTS), potassium persulfate ($K_2S_2O_8$), Brain heart infusion broth, Muller Hinton Agar No. 2, hexane, methanol, and dimethylsulphoxide (DMSO). All the Chemicals were obtained from Hi-Media, Merck and Sisco Research Laboratories Pvt. Ltd., India. All reagents were of analytical grade.

9.2.2   PLANT MATERIAL

Fruit rind and seed coat of both plants *T. chebula* (PSN292) and *T. bellerica* (PSN290) was purchased in April, 2010 from the local market. The plant parts were separately homogenized to fine powder and stored in air tight bottles.

9.2.3   EXTRACTION

The dried powder of each of the plant parts was defatted with hexane and then extracted in methanol for 24 h on a rotary shaker by cold percolation method [51, 59]. Ten grams of dried powder was placed in 100 ml of solvent in a conical flask, plugged with cotton wool, and then kept on a rotary shaker at 190–220 rpm for 24 h. Then the extract was filtered with 8 layers of muslin cloth. The filtrate was centrifuged at 5000 g for 10 min, the supernatant was collected, and the solvent was evaporated. Each solvent extract was concentrated to dryness and then stored at 4°C in an air-tight bottle. The methanol extract was used for antibacterial and antioxidant studies.

9.2.4   PREPARATION OF EXTRACTS FOR ANTIMICROBIAL ASSAY

Plant extracts were separately dissolved in 100% dimethylsulphoxide (DMSO) for antibacterial study. The concentration of the each of the extracts used for antibacterial activity was 15 mg/ml.

9.2.5   ANTIBACTERIAL ASSAY

A loop full of each strain was inoculated in 25 ml of Brain heart infusion broth in a conical flask and then incubated at room temperature on a rotary shaker for 24 h in order to activate the test bacteria. The final cellular concentration was 1×10^8 cfu/ml. Muller Hinton agar was used to determine antibacterial susceptibility. Antibacterial assay was performed using the agar well diffusion method [49]. Media and test bacterial cultures were poured into petri dishes (Hi-Media). Each test strain (200 µl) was inoculated into the media when the temperature was 40–42°C. Care was taken to ensure proper homogenization. After media were solidified a well was made in the plates with the help of a cup-borer (8.5 mm). The well was filled with 100 µl of extract (1.5 mg/well) and the plates were incubated overnight at 37°C. Bacterial growth was determined according to the diameter of the zone of inhibition. The experiments were performed 3 times and the mean values are presented. For each bacterial strain,

DMSO was used as a negative control and vancomycin, gentamycin and oxacillin were used as positive controls.

9.2.6 ANTIOXIDANT TESTING ASSAYS

9.2.6.1 Superoxide Anion Radical Scavenging Assay [54]

The reaction mixture (3.0 ml) contained 0.5 ml Tris–HCl buffer (16 mM, pH 8.0), 0.5 ml of NBT (0.3 mM), 0.5 ml NADH (0.936 mM) solution and 1.0 ml of different concentrations of the extracts. The reaction was initiated by adding 0.5 ml PMS solution (0.12 mM) to the mixture. The reaction mixture was incubated at 25°C for 5 min and the absorbance was measured at 560 nm against a blank sample. Gallic acid was used as a positive control [52].

9.2.6.2 DPPH (2,2-diphenyl-1-picryl hydrazyl) Free Radical Scavenging Assay [41]

The reaction mixture consisting of DPPH in methanol (0.3 mM, 1.0 ml), 1.0 ml methanol and different concentrations of the extracts (1.0 ml) was incubated for 10 min in dark, after which the absorbance was measured at 517 nm. The radical scavenging activity was expressed in terms of the amount of antioxidants necessary to decrease the initial DPPH absorbance by 50% (IC_{50}). The IC_{50} value for each sample was determined graphically by plotting the percentage disappearance of DPPH purple color as a function of the sample concentration. Ascorbic acid was used as a positive control [52].

9.2.6.3 ABTS (2,2-azino-bis-(3-ethylbenzothiazoline-6-sulfonic acid)) Radical Scavenging Activity [53]

ABTS was dissolved in water to a 7 mM concentration. ABTS was produced by reacting ABTS stock solution with 2.45 mM potassium persulfate (final concentration) and allowing the mixture to stand in the dark at room temperature for 16 h before use. The radical was stable in this form for more than 2 days when stored in dark at room temperature. The ABTS solution was diluted with methanol to an absorbance of 0.850 ± 0.05 at 734 nm. 3.0 ml of this ABTS solution was added to 1.0 ml of different concentrations of the extract and incubated for 4 min at room temperature. Absorbance was measured spectrophotometrically at 734 nm. For control, 1.0 ml of methanol was used in place of extract. Ascorbic acid was used as a positive control [3, 52]. The radical scavenging activity of the each of the tested samples is also expressed as the IC_{50} value.

9.2.6.4   Ferric Reducing Antioxidant Power (FRAP) [10]

FRAP reagent was prepared by mixing 10 volumes of 300 mmol/L acetate buffer, pH 3.6, with 1 volume of 10 mmol/L TPTZ (2,4,6- tripyridyl-s-triazine) in 40 mmol/L hydrochloric acid and with 1 volume of 20 mmol/L ferric chloride. Aqueous solutions of known Fe-II concentration, in range of 100–1000 μmol/L ($FeSO_4 \times 7H_2O$) were used for calibration. All solutions were used on the day of preparation. Freshly prepared FRAP reagent (3.0 ml) was warmed to 37°C and a reagent blank reading was taken at 593 nm. Subsequently, 0.1 ml of sample was added to the FRAP reagent. The reaction mixture was incubated at 37°C. The absorbance at 10 min after starting the reaction (adding the sample) was selected as final reading (A sample). Results of FRAP were expressed as mM $FeSO_4$ equivalents per gram of extracted compounds [52].

9.2.6.5   Total Phenol Content [42]

The plant extract (0.5 ml) and 0.5 N Folin-Ciocalteu's reagent (0.1 ml) was mixed and the mixture was incubated at room temperature for 15 min. Then 2.5 ml saturated sodium carbonate solution was added and further incubated for 30 min. at room temperature and the absorbance was measured at 760 nm. Total phenol values are expressed in terms of gallic acid equivalent (mg/g of extracted compound).

9.2.6.6   Flavonoid Content [14]

The reaction mixture (3 ml) contained 1.0 ml of sample (1 mg/ml) 1.0 ml methanol and 0.5 ml of (1.2%) aluminum chloride and 0.5 ml (120 mM) potassium acetate was incubated at room temperature for 30 min. The absorbance of all the samples was measured at 415 nm. Quercetin was used as a positive control [22, 29]. Flavonoid content is expressed in terms of quercetin equivalent (mg/g of extracted compound).

9.2.7 STATISTICAL ANALYSIS

Correlation analysis was done by Pearson's correlation coefficient at $P > 0.05$ significance level.

9.3 RESULTS

9.3.1 EXTRACTIVE YIELD

The extractive yield varied among seed coat and fruit rind (Table 9.1). *T. chebula* fruit rind showed highest extractive yield (40%) while *T. bellerica* seed coat showed lowest extractive yield (3.7%).

9.3.2 ANTI-MRSA ACTIVITY

All the extracts of *Terminalia chebula* and *Terminalia bellerica* showed potent anti-MRSA activity (Figure 9.1). All the MRSA strains were resistant

TABLE 9.1 Extractive Yield, Free Radical Scavenging Property (SO, DPPH and ABTS), Total Phenol Content, Flavonoid Content and FRAP Activity of Methanol Extracts of *T. bellerica* and *T. chebula*

Plants	Extractive yield (%)	IC$_{50}$ values (µg/ml)			FRAP (mmol/g)	Phenol (mg/g)	Flavonoid (mg/g)
		SO	DPPH	ABTS			
TBS	3.7	190	26.5	9.5	38133	203.398	7.463
TBF	34.1	106.5	15.5	7.6	57520	211.341	13.881
TCS	4.1	90	18	6.8	81227	217.482	29.831
TCF	40.0	83	14	6	70187	232.631	23.239
GA	—	185	—	—	—	—	—
AA	—	—	11.4	6.5	—	—	—

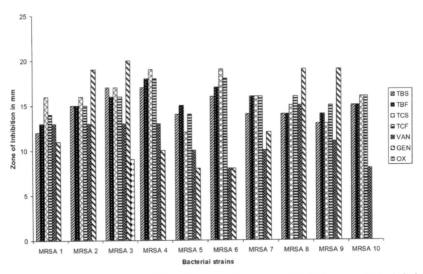

FIGURE 9.1 Antibacterial activity of methanol extracts of *T. bellerica* and *T. chebula* against methicillin resistant *S. aureus.*

to oxacillin and intermediate/susceptible to vancomycin and gentamycin. The anti-MRSA activity of the plant extracts was comparable with standard antibiotics studied. Seed coat extract of *T. chebula* showed higher anti-MRSA activity than other three extracts.

9.3.3 SUPEROXIDE SCAVENGING ACTIVITY

The scavenging effect of the methanol extracts with the superoxide radical is in the following order: *T. chebula* (Fruit rind); IC_{50} value – 83 µg/ml > *T. chebula* (Seed coat); IC_{50} value – 90 µg/ml > *T. bellerica* (Fruit rind); IC_{50} value – 106.5 µg/ml > *T. bellerica* (Seed coat); IC_{50} value – 190 µg/ml. *T. chebula* (Fruit rind) showed best superoxide radical scavenging activity, with a very low IC_{50} value, indicating its high potential to scavenge the super oxide free radicals. The other three extracts also showed low IC_{50} value, either less than standard Gallic acid (Table 9.1) or almost equal (185 µg/ml) to it substantiating the potentiality of fruit rind and seed coat of these two *Terminalia* species as good superoxide anion scavengers.

9.3.4 DPPH FREE RADICAL SCAVENGING ACTIVITY

In the present study, the scavenging effect of the methanol extracts with the DPPH radical is in the following order: *T. chebula* (Fruit rind); IC_{50} value – 14 µg/ml > *T. bellerica* (Fruit rind); IC_{50} value – 15.5 µg/ml > *T. chebula* (Seed coat); IC_{50} value – 18 µg/ml > *T. bellerica* (Seed coat); IC_{50} value – 26.5 µg/ml. Fruit rind extracts showed better DPPH scavenging property than seed coat extracts in both the plants. The methanol extract of *T. chebula* exerted greater DPPH scavenging activity than that of methanol extract of *T. bellerica* (Table 9.1). Overall, the extracts showed good DPPH scavenging activity comparable with standard ascorbic acid (11.4 µg/ml) (Table 9.1).

9.3.5 ABTS FREE RADICAL SCAVENGING ACTIVITY

In this assay, *T. chebula* and *T. bellerica* extracts were evaluated by comparing their ABTS scavenging capacities, using a spectrophotometric method. The scavenging effect of the methanol extracts with the ABTS radical is in the following order: *T. chebula* (Fruit rind); IC_{50} value – 6 µg/ml > *T. chebula* (Seed coat); IC_{50} value – 6.8 µg/ml > *T. bellerica* (Fruit rind); IC_{50} value – 7.6 µg/ml > *T. bellerica* (Seed coat); IC_{50} value – 9.5 µg/ml. All the tested extracts showed significant ABTS scavenging capacity (Table 9.1). Methanol extract of *T. chebula* showed higher ABTS free radical scavenging activity than *T. bellerica*. Fruit rind extracts showed better ABTS free radical scavenging activity than seed coat extracts in both the plants, which was comparable with standard ascorbic acid (6.5 µg/ml) (Table 9.1).

9.3.6 FRAP ASSAY

The FRAP reducing capacity of the methanol extracts is in the following order: *T. chebula* (Seed coat); 81,226.67 mmol/g > *T. chebula* (Fruit rind); 70,186.66 > *T. bellerica* (Fruit rind); 57,520 mmol/g > *T. bellerica* (Seed coat); 38,133.33 mmol/g. These results clearly demonstrated that all the extracts possessed antioxidant capacities, especially methanol extract of seed coat of *T. chebula*; it showed strong FRAP reducing capacity (81,226.67 mmol/g) (Table 9.1).

9.3.7 TOTAL PHENOL AND FLAVONOIDS

Total Phenolic content of the methanol extracts is in the following order: *T. chebula* (Fruit rind); 232.63 mg/g > *T. chebula* (Seed coat); 217.48 mg/g > *T. bellerica* (Fruit rind); 211.34 mg/g > *T. bellerica* (Seed coat); 203.39 mg/g. Flavonoid content of the methanol extracts is in the following order: *T. chebula* (Seed coat); 29.83 mg/g > *T. chebula* (Fruit rind); 23.24 mg/g > *T. bellerica* (Fruit rind); 13.88 mg/g > *T. bellerica* (Seed coat); 7.46 mg/g. All the studied extracts contained higher total phenolic content (Table 9.1). Amongst the extracts studied, *T. chebula* fruit rind extract showed higher total phenolic content while *T. chebula* seed coat showed higher flavonoid content.

9.4 DISCUSSION

There are many reports validating that the extractive yield varied with different solvents [29, 37]. However, in the present study, different parts of the same plant extracted in the same solvent also showed different extractive yield. Methanol extract of *T. chebula* fruit rind showed highest yield, so that extract may be the major source for antibacterial and antioxidants.

Staphylococcus aureus is one of the most prominent and widespread human pathogens, causing skin and tissue infections, deep abscess formation, pneumonia, endocarditis, osteomyelitis, toxic shock syndrome and bacteremia. Methicillin-resistant *Staphylococcus aureus* (MRSA) has been recognized as one of the main pathogenic causes of nosocomial infections throughout the world [18, 44]. The search for new ways to treat MRSA infections stimulates the investigation of natural compounds as an alternative treatment of these infections especially from medicinal plants. Present study showed all the extracts showed potent activity against MRSA strains and can be the alternative source than antibiotics.

The results indicated a concentration-dependent radical scavenging activity for all the samples tested (Table 9.1). Superoxide is biologically quite toxic and is deployed by the immune system to kill invading microorganisms. It is oxygen- centered radical with selective reactivity. It is also produced by a number of enzyme systems in auto oxidation reactions and by nonenzymatic electron transfers that univalently reduce molecular oxygen. The biological toxicity of superoxide is due to its capacity to inactivate iron–sulfur cluster containing enzymes, which are critical in a wide variety of metabolic pathways, thereby liberating free iron in the cell, which can undergo Fenton-chemistry and generate the highly reactive hydroxyl radical [24].

DPPH is usually used to evaluate antioxidant activity due to its ability to scavenge free radicals or donate hydrogen [17]. In this work, DPPH scavenging properties was evaluated using six different concentrations for each extracts and repeating experiments in triplicate. Results are reported as concentration required for 50% radical inhibition (IC_{50}, expressed as l g of extract per milliliter of solvent); higher antiradical activity corresponds to lower IC_{50} values [39].

ABTS method monitors the decay of the radical cation ABTS produced by the oxidation of ABTS caused by the addition of antioxidants [39]. The ABTS scavenging assay, which employs a specific absorbance (734 nm) at a wavelength remote from the visible region and requires a short reaction time, can be used in both organic and aqueous solvent systems [62] and can also be an index reflecting the antioxidant activity of the test samples [61]. ABTS radical cation decolorization assay is an excellent tool for determining the antioxidant activity of hydrogen-donating anti-oxidants (scavengers of aqueous phase radicals) and of chain breaking antioxidants (scavenger of lipid peroxyl radicals) [35].

The FRAP assay treats the antioxidants contained in the samples as reductant in a redox linked colorimetric reaction, and the value reflects the reducing power of antioxidants [15]. The reducing potential of plant extracts was determined using iron (III) to iron (II) reduction assay. In this assay, the yellow color of the test solution changes to various shades of green and blue depending on the reducing power of extracts or compounds. The reducing capacity of a compound may serve as a significant indicator of its potential antioxidant activity [24]. The presence of reductant in solution causes reduction of Fe^{3+}/Ferricyanide complex to the ferrous form. Therefore, the Fe^{2+} can be monitored by measurement of the formation of Perl's Prussian blue at 700 nm [63].

Plant phenolics comprise a great diversity of compounds, which can be classified into different groups, based on the number of phenol rings that they contain and on the structural elements that bind these rings to one another [40]. The flavonoids, which contain hydroxyl groups, show anti-oxidant activity through scavenging or chelating processes have considerable effects on human nutrition and health [38].

There are also other reports showing that different extracts/fractions/isolated compounds showed potent antioxidant activity of *T. chebula* and *T. bellerica* with different antioxidant assays either *in vivo* or *in vitro* [16, 34], but the comparison of different parts of these plants for antibacterial and antioxidants was not reported. Thus, this chapter reported comparative evaluation of seed coat and fruit rind extracts of both the plants for *in vitro* antibacterial and antioxidant capacities. Naik et al. [45] has reported DPPH scavenging property of aqueous extract of *T. chebula* (fruit) which showed IC_{50} value 12 ± 2 μg/ml. In our study, similar result was observed with methanol extract. Nampoothiri et al. [46] has found different extracts of *T. bellerica* showed potential DPPH and SO radical scavenging property which was more than the present study. It may be because of different extraction procedure. Hazra et al. [27] has reported high antioxidant capacity of 70% methanol extract of *T. chebula* and *T. bellerica*. Thus 70% methanol extract is more capable to scavenge free radicals than methanol and aqueous extracts of *T. chebula* and *T. bellerica*.

In this study, higher correlation between FRAP and flavonoid content was observed (Table 9.2), that supports previous studies [23]. Significant correlations were also observed between antioxidant capacities determined by different systems ($p < 0.05$), indicating that these four methods have

Table 9.2 Correlation Analysis of Different Antioxidant Parameters, Total Phenol and Flavonoid Content

Assay	Correlation analysis				
	ABTS	**FRAP**	**SO**	**Phenol**	**Flavonoid**
DPPH	0.908 ($p < 0.05$)	0.723	0.945 ($p < 0.05$)	0.784	0.612
ABTS		0.889 ($p < 0.05$)	0.967 ($p < 0.01$)	0.928 ($p < 0.05$)	0.843 ($p < 0.1$)
FRAP			0.909 ($p < 0.05$)	0.709	0.985 ($p < 0.01$)
SO				0.805 ($p < 0.1$)	0.834 ($p < 0.1$)
Phenol					0.697

satisfactory correlations for the examination of antioxidants. Significant correlation was observed between ABTS-DPPH, SO-DPPH, SO-ABTS, SO-FRAP, Phenol-ABTS and Flavonoid-FRAP (Table 9.2). Significance level between FRAP-DPPH and FRAP-ABTS was also near to the critical values for Pearson's correlation coefficient. Lower significance may be because of number of fewer samples. Many reports showed close relationship between phenolic compounds and antioxidant activity [57]. Present results also showed correlation between total phenolic contents and antioxidant activity that was near to the critical values for Pearson's correlation coefficient.

9.5   CONCLUSIONS

The fruits of Indian medicinal plant *Terminalia* species are known for their pharmacological activity and in this chapter it has been shown that the extracts can be used as an effective antioxidant and antibacterial. Fruit rind extract of *T. chebula* showed good antioxidant activity while seed coat extract of *T. chebula* showed good anti-MRSA activity. It is interesting to note that the extracts are not pure compounds and in spite of it, good results were obtained which only suggests the potency of these extracts. A high correlation between antioxidant activity and total phenolic/flavonoid content indicated that phenolic/flavonoid compounds could be one of the main components responsible for free radical scavenging activity. The extracts also showed antibacterial activity against MRSA and that was comparable with standard antibiotics used. Taken collectively, these results lead to the conclusion that the methanol extracts of *T. chebula* and to some extent *T. bellerica* have powerful antioxidant and anti-MRSA components which may be helpful in controlling complications during degenerative and infectious diseases.

9.6   SUMMARY

Terminalia species have been frequently used as medicinal food to treat various ailments. In the present study, fruit rind and seed coat of two *Terminalia* species viz. *Terminalia chebula* and *Terminalia bellerica* were evaluated for their antibacterial and antioxidant potentials. The antibacterial activity was evaluated against multidrug resistant methicillin-resistant *Staphylococcus aureus* (MRSA) and antioxidant capacity was evaluated using superoxide, 2,2-diphenyl-1-picrylhydrazyl (DPPH), 2,2-azino-bis-(3-ethylbenzthiazoline-6-sulphonic acid) (ABTS) radical scavenging assays and ferric reducing/antioxidant power (FRAP) assay. Quantitative estimation of total phenol and flavonoid content was also done. The extraction was done in methanol by cold percolation method. The antibacterial activity was measured by agar well diffusion method. Oxacillin, vancomycin and gentamycin were used as positive controls. The methanol extracts of both plants showed good anti-MRSA activity which was comparable with that of standard antibiotics studied. The methanol extract of fruit rind of *T. chebula* showed highest antioxidant activity as well as higher total phenol content. Significant correlation between antioxidant assays was also observed in all the extracts. Based on all these results, it is concluded that *T. chebula* and *T. bellerica* methanol extracts could act as potent anti-MRSA and antioxidant agent.

ACKNOWLEDGEMENTS

The authors are thankful to Prof. S. P. Singh, Head, Department of Biosciences, Saurashtra University, Rajkot, India for providing excellent research facilities. The authors are thankful to Dr. Parijat Goswami (GCRI, Ahmedabad) for help in collection of clinical isolates. Authors (Dr. Harsha Patel, Dr. Mital J. Kaneria and Ms. Hemali Padalia) are thankful to University Grant Commission (UGC), New Delhi, India for financial assistance.

KEYWORDS

- ABTS
- agar well diffusion method
- anti-MRSA
- antioxidant activity
- cold percolation method
- correlation analysis

- **DPPH**
- *emblica officinalis*
- **extractive yield**
- **FRAP**
- **gentamycin**
- **IC$_{50}$ values**
- **methanol extracts**
- **oxacillin**
- **ROS**
- **SO**
- *terminalia bellerica*
- *terminalia chebula*
- **TFC**
- **TPC**
- **vancomycin**

REFERENCES

1. Ahn, M. J., Kim, C. Y., Lee, J. S., Kim, T. G., Kim, S. H., Lee, C. K., Lee, B. B., Shin, C. G., Huh, H., & Kim, J. (2002). Inhibition of HIV-1 integrase by galloyl glucoses from *Terminalia chebula* and flavonol glycoside gallates from *Euphorbia pekinensis. Planta Medica, 68,* 457–459.
2. Ali, S. S., Kasoju, N., Luthra, A., Singh, A., Sharanabasava, H., Sahu, A., & Bora, U. (2008). Indian medicinal herbs as sources of antioxidants. *Food Research International, 41,* 1–15.
3. Alzoreky, N., & Nakahara, N. (2001). Antioxidant activity of some edible Yemeni plants evaluated by ferryl myoglobin/ABTS assays. *Food Science and Technology Research, 7,* 144–146.
4. Aqil, F., & Ahmad, I. (2007). Antibacterial properties of traditionally used Indian medicinal plants. *Methods Findings in Experimental Clinical Pharmacology, 29,* 79–92.
5. Aqil, F., Khan, M. S., Owais, M., & Ahmad, I. (2005). Effect of certain bioactive plant extracts on clinical isolates of beta-lactamase producing methicillin resistant *Staphylococcus aureus. Journal of Basic Microbiology, 45,* 106–114.
6. Arora, S., Kaur, K., & Kaur, S. (2003). Indian medicinal plants as a reservoir of protective phytochemicals. *Teratogenesis, Carcinogenesis and Mutagenesis, 1,* 295–300.

7. Badmaev, V., & Nowakowski, M. (2000). Protection of epithelial cells against influenza A virus by a plant derived biological response modifier Ledretan-96. *Phytotherapy Research, 14*, 245–249.

8. Bajpai, M., Pande, A., Tewari, S. K., & Prakash, D. (2005). Phenolic contents and antioxidant activity of some food and medicinal plants. *International Journal of Food Science and Nutrition, 56*, 287–291.

9. Beaume, M., Hernandez, D., Docquier, M., Delucinge-Vivier, C., Descombes, P., & Francois, P. (2011). Orientation and expression of methicillin-resistant *Staphylococcus aureus* small RNAs by direct multiplexed measurements using the nCounter of NanoString technology. *Journal of Microbiological Methods, 84*, 327–334.

10. Benzie, I. F., & Strain, J. J. (1996). The ferric reducing ability of plasma (FRAP) as a measure of "antioxidant power": the FRAP assay. *Analytical Biochemistry, 239*, 70–76.

11. Buenz, E. J., Bauer, B. A., Schnepple, D. J., Wahner-Roedler, D. L., Vandell, A. G., & Howe, C. L. (2007). A randomized Phase I study of *Atuna racemosa*: a potential new anti-MRSA natural product extract. *Journal of Ethnopharmacology, 114*, 371–376.

12. Cespedes, C. L., Valdez-Morales, M., Avila, J. G., El-Hafidi, M., Alarcon, J., & Paredes-Lopez, O. (2010). Phytochemical profile and the antioxidant activity of Chilean wild black-berry fruits, *Aristotelia chilensis* (Mol) Stuntz (Elaeocarpaceae). *Food Chemistry, 119*, 886–895.

13. Chanda, S., Dudhatra, S., & Kaneria, M. (2010). Antioxidative and antibacterial effects of seeds and fruit rind of nutraceutical plants belonging to the Fabaceae family. *Food and Function,* 1, 308–315.

14. Chang, C., Yang, M., Wen, H., & Chern, J. (2002). Estimation of total flavonoid content in *Propolis* by two complementary colorimetric methods. *Journal of Food and Drug Analysis, 10*, 178–182.

15. Chen, R., Liu, Z., Zhao, J., Chen, R., Meng, F., Zhang, M., & Ge, W. (2011). Antioxidant and immunobiological activity of water-soluble polysaccharide fractions purified from *Acanthopanax senticosu. Food Chemistry, 127*, 434–440.

16. Cheng H. Y., Lin T. C., Yu K. H., Yang C. M., & Lin C. C. (2003). Antioxidant and free radical scavenging activities of *Terminalia chebula. Biological and Pharmaceutical Bulletin, 26*, 1331–1135.

17. Da Porto, C., Calligaris, S., Celotti, E., & Nicoli, M. C. (2000). Antiradical properties of commercial cognacs assessed by the DPPH˙ test. *Journal of Agricultural and Food Chemistry, 48*, 4241–4245.

18. Eileen, C. J. S., Elizabeth, M. W., Neale, W., Glenn, W. K., & Simon, G. (2007). Antibacterial and modulators of bacterial resistance from the immature cones of *Chamaecyparis awsoniana. Phytochemistry, 68*, 210–217.

19. Elizabeth, K. M. (2006). Antimicrobial activity of *Terminalia bellerica. Indian Journal of Clinical Biochemistry, 20*, 150–153.

20. Gandhi, N. M., & Nair, C. K. (2005). Radiation protection by *Terminalia chebula*: some mechanistic aspects. *Molecular and Cellular Biochemistry, 277*, 43–48.

21. Gao, H., Huang, Y. N., Gao, B., & Kawabata, J. (2008). Chebulagic acid is a potent alpha-glucosidase inhibitor. *Bioscience, Biotechnology and Biochemistry, 72*, 601–603.

22. Ghasemi, K., Ghasemi, Y., & Ebrahimzadeh, M. A. (2009). Antioxidant activity, phenol and flavonoid contents of 13 *Citrus* species peels and tissues. *Pakistan Journal of Pharmaceutical Sciences, 22*, 277–281.

23. Ghasemzadeh, A., & Jaafar, H. Z. (2011). Effect of CO_2 enrichment on synthesis of some primary and secondary metabolites in ginger (*Zingiber officinale* Roscoe). *International Journal of Molecular Sciences, 12*, 1101–1114.

24. Gulcin, I., Huyut, Z., Elmastas, M., & Aboul-Enein, H. Y. (2010). Radical scavenging and antioxidant activity of tannic acid. *Arabian Journal of Chemistry, 3*, 43–53.

25. Gulcin, I., Mshvildadze, V., Gepdiremen, A., & Elias, R. (2006). Screening of anti-radical and antioxidant activity of monodesmosides and crude extract from *Leontice smirnowii* tuber. *Phytomedicine, 13*, 343–351.

26. Halliwell, B., & Aruoma, O. I. (1991). DNA damage by oxygen-derived species. Its mechanism and measurement in mammalian systems. *FEBS Letters, 281*, 9–19.

27. Hazra, B., Sarkar, R., Biswas, S., & Mandal, N. (2010). Comparative study of the antioxidant and reactive oxygen species scavenging properties in the extracts of the fruits of *Terminalia chebula, Terminalia belerica* and *Emblica officinalis*. *BMC Complementary and Alternative Medicine, 10*, 20.

28. Ho, P. L., Chow, K. H., Lo, P. Y., Lee, K. F., & Lai, E. L. (2009). Changes in the epidemiology of methicillin-resistant *Staphylococcus aureus* associated with spread of the ST45 lineage in Hong Kong. *Diagnostic Microbiology and Infectious Disease, 64*, 131–137.

29. Kaneria, M., Baravalia, Y., Vaghasiya, Y., & Chanda, S. (2009). Determination of antibacterial and antioxidant potential of some medicinal plants from Saurashtra region, India. *Indian Journal of Pharmaceutical Sciences, 71*, 406–412.

30. Kaur, S., Michael, H., Arora, S., Harkonen, P. L., & Kumar, S. (2005). The *in vitro* cytotoxic and apoptotic activity of Triphala-an Indian herbal drug. *Journal of Ethnopharmacology, 97*, 15–20.

31. Khare, C. P. (2007). *Indian Medicinal Plants – An Illustrated Dictionary*. Springer Science, New York, pp. 653–654.

32. Kumar, M. S., Kirubanandan, S., Sripriya, R., & Sehgal, P. K. (2008). Triphala promotes healing of infected full-thickness dermal wound. *Journal of Surgical Research, 144*, 94–101.

33. Lee, H. S., Jung, S. H., Yun, B. S., & Lee, K. W. (2007). Isolation of chebulic acid from *Terminalia chebula* Retz. and its antioxidant effect in isolated rat hepatocytes. *Archives in Toxicology, 81*, 211–218.

34. Lee, H. S., Won, N. H., Kim, K. H., Lee, H., Jun, W., & Lee, K. W. (2005). Antioxidant effects of aqueous extract of *Terminalia chebula in vivo* and *in vitro*. *Biological and Pharmaceutical Bulletin, 28*, 1639–1644.

35. Leong, L. P., & Shui, G. (2002). An investigation of antioxidant capacity of fruits in Singapore markets. *Food Chemistry, 76*, 69–75.

36. Levy, S. B., & Marshall, B. (2004). Antibacterial resistance worldwide: causes, challenges and responses. *Nature Medicine, 10*, S122–S129.

37. Lin, L., Cui, C., Wen, L., Yang, B., Luo, W., & Zhao, M. (2011). Assessment of *in vitro* antioxidant capacity of stem and leaf extracts of *Rabdosia serra* (MAXIM.) HARA and identification of the major compound. *Food Chemistry, 12*, 54–59.

38. Liu, J., Wang, C., Wang, Z., Zhang, C., Lu, S., & Liu, J. (2011). The antioxidant and free-radical scavenging activities of extract and fractions from corn silk (*Zea mays* L.) and related flavone glycosides. *Food Chemistry, 126*, 261–269.

39. Locatelli, M., Travaglia, F., Coisson, J. D., Martelli, A., Stevigny, C., & Arlorio, M. (2010). Total antioxidant activity of hazelnut skin (*Nocciola Piemonte* PGI): Impact of different roasting conditions. *Food Chemistry, 119*, 1647–1655.

40. Manach, C., Scalbert, A., Morand, C., Remesy, C., & Jimenez, L. (2004). Polyphenols: food sources and bioavailability. *American Journal of Clinical Nutrition, 79*, 727–747.

41. McCune, L. M., & Johns, T. (2002). Antioxidant activity in medicinal plants associated with the symptoms of diabetes mellitus used by the indigenous peoples of the North American boreal forest. *Journal of Ethnopharmacology, 82*, 197–205.

42. McDonald, S., Prenzler, P. D., Antolovich, M., & Robards, K. (2001). Phenolic content and antioxidant activity of olive extracts. *Food Chemistry, 73*, 73–84.

43. Miller, L. G., & Kaplan, S. L. (2009). *Staphylococcus aureus*: a community pathogen. *Infectious Disease Clinics of North America, 23*, 35–52.

44. Murthy, A., De Angelis, G., Pittet, D., Schrenzel, J., Uckay, I., & Harbarth, S. (2010). Cost-effectiveness of universal MRSA screening on admission to surgery. *Clinical Microbiology and Infection, 16*, 1747–1753.

45. Naik, G. H., Priyadarsini, K. I., Naik, D. B., Gangabhagirathi, R., & Mohan, H. (2004). Studies on the aqueous extract of *Terminalia chebula* as a potent antioxidant and a probable radioprotector. *Phytomedicine, 11*, 530–538.

46. Nampoothiri, S. V., Prathapan, A., Cherian, O. L., Raghu, K. G., Venugopalan, V. V., & Sundaresan, A. (2011). *In vitro* antioxidant and inhibitory potential of *Terminalia bellerica* and *Emblica officinalis* fruits against LDL oxidation and key enzymes linked to type 2 diabetes. *Food and Chemical Toxicology, 49*, 125–131.

47. Padam, S. K., Grover, I. S., & Singh, M. (1996). Antimutagenic effects of polyphenols isolated from *Terminalia bellerica* myroblan in *Salmonella typhimurium*. *Indian Journal of Experimental Biology, 34*, 98–102.

48. Pereira, E. M., Gomes, R. T., Freire, N. R., Aguiar, E. G., Brandao, M. G., & Santos, V. R. (2011). *In vitro* antimicrobial activity of Brazilian medicinal plant extracts against pathogenic microorganisms of interest to dentistry. *Planta Medica, 77*, 401–404.

49. Perez, C., Paul, M., & Bazerque, P. (1990). An antibiotic assay by the agar well diffusion method. *Acta Biologica Medica Experimentalis, 15*, 113–115.

50. Perry, L. M. (1980). *Medicinal Plants of East and South-east Asia Attributed Properties and Use*. The MIT, Cambridge, pp. 80.

51. Rakholiya, K., Kaneria, M., & Chanda, S. (2014). Inhibition of microbial pathogens using fruit and vegetable peel extracts. *International Journal of Food Science and Nutrition, 65*, 733–739.

52. Rakholiya, K., Kaneria, M., Nagani, K., Patel, A., & Chanda, S. (2015). Comparative analysis and simultaneous quantification of antioxidant capacity of four *Terminalia* species using various photometric assays. *World Journal of Pharmaceutical Science, 4*, 1280–1296.

53. Re, R., Pellegrini, N., Proteggente, A., Pannala, A., Yang, M., & Rice-Evans, C. (1999). Antioxidant activity applying an improved ABTS radical cation decolorization assay. *Free Radical Biology and Medicine, 26*, 1231–1237.

54. Robak, J., & Gryglewski, R. J. (1988). Flavonoids are scavengers of superoxide anions. *Biochemical Pharmacology, 37*, 837–841.

55. Saleem, A., Husheem, M., Harkonen, P., & Pihlaja, K. (2002). Inhibition of cancer cell growth by crude extract and the phenolics of *Terminalia chebula* retz. fruit. *Journal of Ethnopharmacology, 81*, 327–336.

56. Sarwat, M., Das, S., & Srivastava, P. S. (2011). Estimation of genetic diversity and evaluation of relatedness through molecular markers among medicinally important trees: *Terminalia arjuna, T. chebula* and *T. bellerica. Molecular Biology Reports, 38*, 5025–5036.

57. Shi, Y. X., Xu, Y. K., Hu, H. B., Na, Z., & Wang, W. H. (2011). Preliminary assessment of antioxidant activity of young edible leaves of seven *Ficus* species in the ethnic diet in Xishuangbanna, South-west China. *Food Chemistry, 128*, 889–894.

58. Shin, T. Y., Jeong, H. J., Kim, D. K., Kim, S. H., Lee, J. K., Chae, B. S., Kim, J. H., Kang, H. W., Lee, C. M., Lee, K. C., Park, S. T., Lee, E. J., Lim, J. P., Kim, H. M., & Lee, Y. M. (2001). Inhibitory action of water soluble fraction of *Terminalia chebula* on systemic and local anaphylaxis. *Journal of Ethnopharmacology, 74*, 133–140.

59. Vaghasiya, Y., & Chanda, S. V. (2007). Screening of methanol and acetone extracts of 14 Indian medicinal plants for antimicrobial activity. *Turkish Journal of Biology, 31*, 243–248.

60. Valsaraj, R., Pushpangadan, P., Smitt, U. W., Adsersen, A., Christensen, S. B., Sittie, A., Nyman, U., Nielsen, C., & Olsen, C. E. (1997). New anti-HIV-1, antimalarial, and antifungal compounds from *Terminalia bellerica. Journal Natural Products, 60*, 739–742.

61. Wang, H., Gan, D., Zhang, X., & Pan, Y. (2010). Antioxidant capacity of the extracts from pulp of *Osmanthus fragrans* and its components. *LWT – Food Science and Technology, 43*, 319–325.

62. Wu, L. C., Hsu, H. W., Chen, Y. C., Chiu, C. C., Lin, Y. I., & Ho, J. A. (2006). Antioxidant and antiproliferative activities of red pitaya. *Food Chemistry, 95*, 319–327.

63. Zou, Y., Lu, Y., & Wei, D. (2004). Antioxidant activity of a flavonoid-rich extract of *Hypericum perforatum* L. *in vitro. Journal of Agricultural and Food Chemistry, 52*, 5032–5039.

PART IV

ANTIMICROBIAL ACTIVITIES IN FOOD

CHAPTER 10

IN VITRO ANTIMICROBIAL ACTIVITY: SALVADORA SPECIES

JAGRUTI SONAGARA, KALPNA D. RAKHOLIYA,
HEMALI PADALIA, SUMITRA CHANDA, and MITAL J. KANERIA

CONTENTS

10.1   INTRODUCTION

The frequency of life threatening infections caused by pathogenic microorganisms has increased worldwide and is becoming an important cause of morbidity and mortality in immune compromised patients all over the world [5, 6]. This ongoing emergence of multi drug resistant bacteria and the infectious diseases caused by them are serious global problem. Apart from causing infectious diseases, the pathogenic microorganisms also cause food spoilage and food poisoning. The introduction and increasing use of antibiotics for antibacterial therapy has initiated a rapid development and expansion of antibiotic resistance in human pathogens due to antimicrobial misuse and decreased the treatment options [15, 26, 30]. The discovery of antibiotics was an essential part in combating bacterial infections that once ravaged humankind. Different antibiotics exercise their inhibitory activity on different pathogenic organisms. The widespread indiscriminate use has led to many medical problems including hypersensitivity, allergic reaction and immunity suppression [3, 5, 23, 24]. Therefore, it is necessary to search for strategies that are accessible, simple in application and nontoxic.

Traditionally common people use crude extracts of different parts of plants as curative agents. Plant extracts have also been used in the treatment of infectious diseases caused by resistant microbes. In fact, herbal medicines have received much attention as sources of lead compounds since they are considered to be time tested and relatively safe for both human use and environment friendly [5, 12, 25, 29]. They are also

TABLE 10.1   Plant Description of *Salvadora oleoides* Decne. and *Salvadora persica* L.

Parameter	Plant name	
	Salvadora oleoides **Decne.**	*Salvadora persica* **L.**
Family	Salvadoraceae	Salvadoraceae
Habit	Shrub or small tree, 6–9 m height under favorable condition	Evergreen shrub or small tree to 6–7 m
Distribution	Arid and semiarid region	Arid and semiarid region
Vernacular name	Khakan, Pilava, Pilu, Mityal, Meethajal	Kharijai, Piludi, Kharajal
Constituents	β-sitosterol, glucosides, rutin, terpenoids, phospholipids, dihydro-isocoumarin [1]	Alkaloid, resin, trimethyl amine, β-sotisterol, di-benzyl thiourea, rutin, thioglucoside, chlorine, potash, sulphar [16]
Action /uses	Leaves is used open wound, blood purifiers, cooling agent. Stem is anthelmintic, diuretic property, fever, asthma, cough, leprosy, rheumatism [19]	Stem is used gastric trouble, vesicant, stimulant. Leaves is used cough, piles, tumors, rheumatism, asthma [21]
Photo		

economical, easily available and affordable. Lastly, natural products, either as pure compounds or as standardized plant extracts, provide unlimited opportunities for new drug leads because of the unmatched availability of chemical diversity. All parts or any part of the plant such as bark, leaves, peel, seed, and stem can potentially possess antimicrobial properties [7, 9, 10, 18, 27]. Plants produce secondary metabolites which inhibit bacteria, fungi, viruses and pests.

However, almost no comparative studies have so far been reported on antimicrobial properties of different solvent extracts of leaf and young stem and woody stem of *Salvadora oleoides* Decne. and *Salvadora persica* L. Therefore, this chapter presents the study on the comparative assessment of antimicrobial activity of different solvent extracts of both *Salvadora* species using different *in vitro* assays (Table 10.1).

10.2 MATERIALS AND METHODS

10.2.1 PLANT COLLECTION

The leaf, young stem (YS), woody stem (WS) of *S. oleoides* and *S. persica* were collected in September 2014 from Mudi, Surendranagar, Gujarat, India and in August 2014 from Balachadi, Jamnagar, Gujarat, India, respectively. Parts of the plant were washed thoroughly with tap water, shade dried and homogenized to fine powder and stored in air tight bottles.

10.2.2 PLANT EXTRACTION

The dried powder of the plant part was extracted individually by cold percolation method [17, 28] using different organic solvents like Petroleum ether (PE), Ethyl acetate (EA), Acetone (AC), Methanol (ME), Aqueous (AQ). Ten grams of dried powder was taken in 150 ml petroleum ether in a conical flask, plugged with cotton wool and then kept on shaker at 120 rpm for 24 h. After 24 h, it was filtrated through eight layers of muslin cloth, centrifuged at 5000 rpm in a centrifuge for 15 min and the supernatant was collected and the solvent was evaporated using a rotary vacuum evaporator to dryness. This dry powder was then taken individually in

150 ml of each solvent (ethyl acetate, acetone, methanol and water) and was kept on a shaker at 120 rpm for 24 h. Then the procedure followed in each case was the same as above, and the residues were stored in air tight bottles at 4°C.

10.2.3   ANTIMICROBIAL ACTIVITY

10.2.3.1   Microorganisms Tested

The microbial strains studied were obtained from National Chemical Laboratory (NCL), Pune. They consisted of 4 Gram positive bacteria: *Bacillus cereus* ATCC1778 (BC), *Bacillus subtilis* (ATCC6833) (BS), *Corynebacterium rubrum* ATCC14898 (CR), *Staphylococcus aureus* ATCC29737 (SA); 4 Gram negative bacteria: *Escherichia coli* ATCC25922 (EA), *Klebsiella pneumonia* NCIM2719 (KP), *Pseudomonas aeruginosa* ATCC9027 (PA), *Salmonella typhimarium* ATCC23564 (ST); and 4 Fungi: *Candida albicans* ATCC2091 (CA), *Candida epicola* NCIM3102 (CE), *Candida glabrata* NCIM3448 (CG), *Cryptococcus neoformans* ATCC34664 (CN). The bacteria and fungi were maintained on nutrient agar and MGYP medium, respectively at 4°C and were subcultured before use. The microorganisms studied are clinically important causing several infections and spoilage.

10.2.3.2   Susceptibility Test by Agar Well Diffusion Assay

In vitro antimicrobial activity of different solvent extracts was studied against different pathogenic microbial strains by agar well diffusion assay [20, 26]. Mueller Hinton agar and Sabouraud dextrose agar media was used for antibacterial and antifungal susceptibility test, respectively. Molten Mueller Hinton agar/Sabouraud dextrose agar (40–42°C) were seeded with 200 µl of inoculums (1×10^8 cfu/ml) and poured into Petri dishes. The media were allowed to solidify and wells were prepared in the seeded agar plates with the help of a cup borer (8.5 mm). Different extracts were dissolved in 100% DMSO at a concentration of 20 mg/ml, from this 100 µl of different extracts were added into the sterile 8.5 mm diameter well. The plates were incubated at 37°C and 28°C for 24 and 48 h for bacteria and fungi, respectively. DMSO was used as a negative control. Antimicrobial activity was assayed by measuring the diameter of the zone of inhibition formed around the well in millimeters. The experiment was carried out in triplicate and the average values are presented with ± SEM.

10.3 RESULTS AND DISCUSSION

Researchers for many decades have been trying to develop new broad-spectrum antibiotics for treating the infectious diseases caused by pathogenic microorganisms. Prolonged usage of these broad-spectrum antibiotics has led to the emergence of drug resistance. There is a tremendous need for novel antimicrobial agents from different sources. Plants are rich in a wide variety of secondary metabolites such as tannins, alkaloids, terpenoids, flavonoids, etc. These secondary metabolites have been found *in vitro* to have antimicrobial property and they may serve as an alternative, effective, cheap and safe antimicrobials for the treatment of microbial infections [2].

Antimicrobial activity of leaf, YS and WS of *S. oleoides* and *S. persica* was evaluated against 12 medically important microbial strains. All the extracts of the different parts of both plants in different solvents exhibited different level of antimicrobial activity against different selected bacterial and fungal strain. The antimicrobial activity was determined by measuring the zone of inhibition in mm. the antimicrobial activity of leaf, YS and WS of *S. oleoides* and *S. persica* against microorganisms is shown in Figures 10.1–10.6.

All the five different solvent extracts of *S. oleoides* showed activity against all the Gram-positive bacteria (Figure 10.1). Maximum activity of different solvent extracts was exhibited against BC followed by SA. BS was not inhibited by PE and AQ extracts of leaf (Figure 10.1A), AC, ME and AQ extracts of YS (Figure 10.1B) and ME and AQ extracts of WS (Figure 10.1C), while CR was not inhibited by AC extract of leaf (Figure 10.1A) and EA, ME and AQ extracts of WS (Figure 10.1C).

Different solvent extracts of *S. oleoides* showed poor antimicrobial activity against Gram negative bacteria and fungi as compare to Gram positive bacteria (Figures 10.1–10.3). PA was inhibited by AC and ME extracts of leaf (Figure 10.2A) and YS (Figure 10.2B), while EC was inhibited by

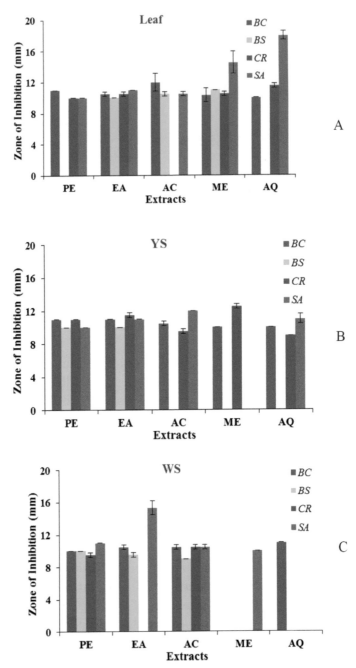

FIGURE 10.1 Antibacterial activity of different solvent extracts of *S. oleoides* (A) Leaf, (B) YS, and (C) WS against Gram positive bacteria.

PE extract of YS (Figure 10.2B) and AQ extract of WS (Figure 10.2C). None of the other extracts were able to inhibit Gram negative bacteria. Maximum activity of different solvent extracts was exhibited against CN followed by CA, while none of the extracts were able to inhibit CE and CG (Figure 10.3).

All the five different solvent extracts of *S. persica* showed activity against all the Gram positive bacteria (Figure 10.4). Maximum activity of different solvent extracts was exhibited against SA. BS was not inhibited by PE and AQ extracts of leaf (Figure 10.4A) and ME and AQ extracts of WS (Figure 10.4C), while CR was not inhibited by AC extract of leaf (Figure 10.4A), AQ extract of YS (Figure 10.4B) and EA and AQ extracts of WS (Figure 10.4C).

Different solvent extracts of *S. persica* showed poor antimicrobial activity against Gram negative bacteria and fungi as compare to Gram positive bacteria (Figures 10.4–10.6). Maximum activity of different solvent extracts was exhibited against ST followed by PA, while none of the extracts were able to inhibit EC and KP (Figure 10.5). Maximum activity of different solvent extracts was exhibited against CA followed by CN, while none of the extracts were able to inhibit CG and CE (Figure 10.6).

From the above results, it can be stated that *S. persica* showed better antimicrobial activity than *S. oleoides*. The implied that the Gram positive bacteria were more susceptible to the extract than the Gram negative bacteria. This difference may be due structural differences in cell wall of these bacteria. The Gram-negative cell wall is complex and multilayered structure; it has an outer phospholipid membrane carrying the structural lipopolysaccharide components, which makes a barrier to many environmental substances including synthetic and natural antibiotics [11]. The Gram-positive bacteria contain a single outer peptidoglycan layer, which is not an effective permeability barrier [4, 8, 14, 22, 31, 32]. Garveya et al. [13] also reported that Gram-negative bacteria have innate multidrug resistance to many antimicrobial compounds owing to the presence of efflux pumps.

The results of this study may be useful as potential bio-control agents and afford efficient and safe antimicrobics which will certainly contribute to the ongoing search for new antimicrobial agents to fight against infectious diseases caused by antibiotic resistant strains.

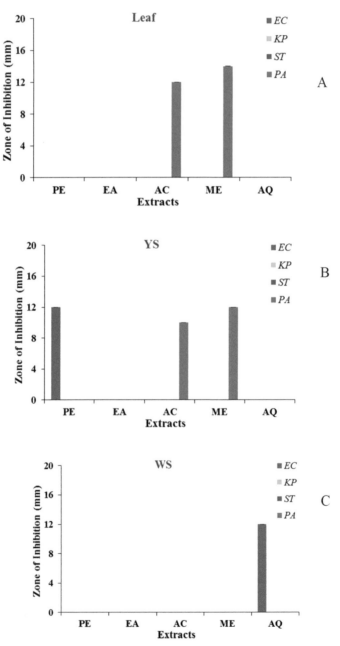

FIGURE 10.2 Antibacterial activity of different solvent extracts of *S. oleoides* (A) Leaf, (B) YS, and (C) WS Gram negative bacteria.

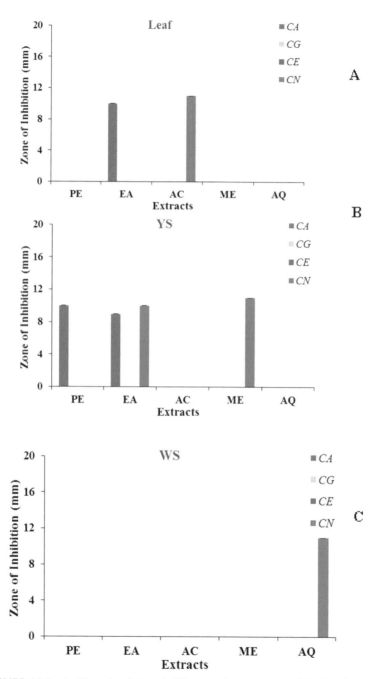

FIGURE 10.3 Antifungal activity of different solvent extracts of *S. oleoides* (A) Leaf, (B) YS, and (C) WS against fungi.

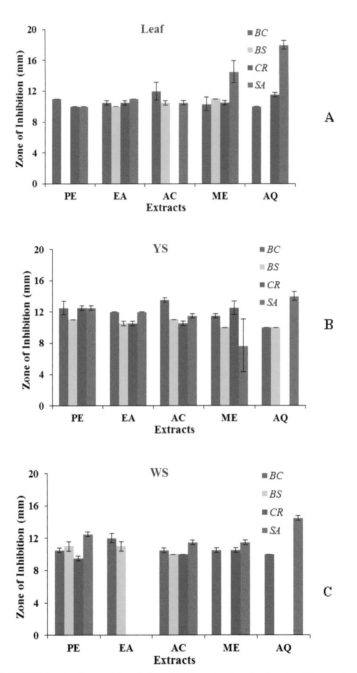

FIGURE 10.4 Antibacterial activity of different solvent extracts of *S. persica* (A) Leaf, (B) YS, and (C) WS against Gram positive bacteria.

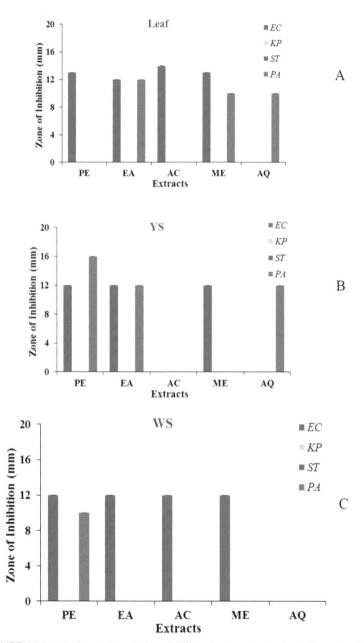

FIGURE 10.5 Antibacterial activity of different solvent extracts of *S. persica* (A) Leaf, (B) YS, and (C) WS against Gram negative bacteria.

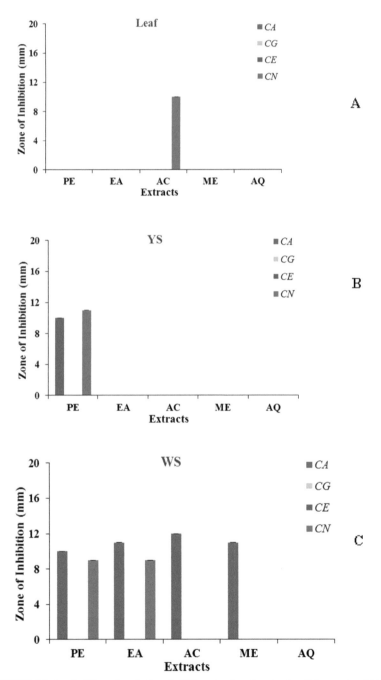

FIGURE 10.6 Antifungal activity of different solvent extracts of *S. persica* (A) Leaf, (B) YS, and (C) WS against fungi.

10.4   CONCLUSIONS

The present work has demonstrated the antimicrobial potential of *Salvadora oleoides* and *S. persica* extracted in various solvents. This is the first report on the comparative antimicrobial activity of different solvent extracts of *S. oleoides* and *S. persica* leaf, young stem and woody stem. From the result of the present study, it can be concluded that Gram positive bacteria were more susceptible than Gram negative bacteria as well as fungi towards different solvent extracts. Most susceptible microorganisms were both Gram positive bacteria, e.g., *B. cereus* and *S. aureus* from the tested pathogenic strains. Different solvent extracts of *S. persica* showed better antimicrobial activity than *S. oleoides*. Hence it can be concluded that *S. persica* may be a good source of new therapeutic agents. However, further research is required to identify individual components responsible for the activity.

10.5   SUMMARY

Plants are important sources of biologically active natural products, which differ widely in terms of structure. They are the original source of a variety of compounds used by the pharmaceutical industry as medicines. Much research has been and is being done on plants in popular use, with the objective of identifying natural products with therapeutic potential. From this point of view, two species of *Salvadora* were selected for the comparative assessment of antimicrobial activity.

In the present study, antimicrobial potential of two different *Salvadora* species was evaluated and compared. The leaf, young stem (YS), woody stem (WS) of *S. oleoides* and *S. persica* were collected and were individually extracted by cold percolation method using Petroleum ether, Ethyl acetate, Acetone, Methanol, Aqueous. Antimicrobial activity was done by agar well diffusion assay against 4 Gram positive bacteria [*Bacillus cereus* ATCC1778, *Bacillus subtilis* (ATCC6833), *Corynebacterium rubrum* ATCC14898, *Staphylococcus aureus* ATCC29737]; 4 Gram negative bacteria [*Escherichia coli* ATCC25922, *Klebsiella pneumonia* NCIM2719, *Pseudomonas aeruginosa* ATCC9027, *Salmonella typhimarium* ATCC23564] and 4 Fungi [*Candida albicans* ATCC2091, *Candida epicola* NCIM3102, *Candida glabrata* NCIM3448, *Cryptococcus neoformans* ATCC34664].

The microorganisms studied are clinically important causing several infections, food borne diseases, spoilages, skin infection and it is essential to overcome them through some active therapeutic agents. *S. persica* showed better antimicrobial activity than *S. oleoides*. The implied that the Gram positive bacteria were more susceptible to the extract than the Gram negative bacteria. *S. persica* could be a good source of bioactive phytochemicals with antimicrobial agent against pathogenic microbial strain. Nevertheless further studies are warranted for enlightening the antimicrobial mechanism of the components present in *S. persica* extracts responsible for biological potentiality.

KEYWORDS

- acetone
- agar well diffusion method
- antibacterial activity
- antifungal activity
- antimicrobial agents
- aqueous
- bioactivity
- cold percolation extraction
- ethyl acetate
- fungi
- gram negative bacteria
- gram positive bacteria
- infectious diseases
- leaf
- methanol
- microbial infection
- pathogenic microbes
- petroleum ether
- phyto constituents

- *salvadora oleoides*
- *salvadora persica*
- **woody stem**
- **young stem**

REFERENCES

1. Arora, M., Siddiqui, A. A., Paliwal, S., & Mishra, R. (2013). Pharmacognostical and phytochemical investigation of *Salvadora oleoides* decne stem. *International Journal of Pharmacy and Pharmaceutical Sciences, 5*, 128–130.
2. Bag, A., Bhattacharyya, S. K., Pal, N. K., & Chattopadhyay, R. R. (2012). *In vitro* antibacterial potential of *Eugenia jambolana* seed extracts against multidrug-resistant human bacterial pathogens. *Microbiological Research, 167*, 352–357.
3. Cakir, A., Kordali, S., Kilic, H., & Kaya, E. (2005). Antifungal properties of essential oil and crude extracts of *Hypericum linarioides Bosse. Biochemical Systematics and Ecology, 33*, 245–256.
4. Chan, E. W. C., Lim, Y. Y., & Omar, M. (2007). Antioxidant and antibacterial activity of leaves of *Etlingera* species (*Zingiberaceae*) in Peninsular Malaysia. *Food Chemistry, 104*, 1586–1593.
5. Chanda, S., & Rakholiya, K. (2011). Combination therapy: Synergism between natural plant extracts and antibiotics against infectious diseases. In: Mendez-Vilas, A. (Ed.) *Science against Microbial Pathogens: Communicating Current Research and Technological Advances*, Formatex Research Center, Spain, pp. 520–529.
6. Chanda, S., Baravalia, Y., Kaneria, M., & Rakholiya, K. (2010). Fruit and vegetable peels – strong natural source of antimicrobics. In: Mendez-Vilas, A. (Ed.) *Current Research, Technology and Education Topics in Applied Microbiology and Microbial Biotechnology*, Formatex Research Center, Spain, pp. 444–450.
7. Chanda, S., Kaneria, M., & Baravalia, Y. (2012). Antioxidant and antimicrobial properties of various polar solvent extracts of stem and leaves of four *Cassia* species. *African Journal of Biotechnology, 11*, 2490–2503.
8. Chanda, S., Rakholiya, K., & Nair, R. (2011). Antimicrobial activity of *Terminalia catappa* L. leaf extracts against some clinically important pathogenic microbial strains. *Chinese Medicine, 2*, 171–177.
9. Chanda, S., Rakholiya, K., & Parekh, J. (2013). Indian medicinal herb: Antimicrobial efficacy of *Mesua ferrea* L. seed extracted in different solvents against infection causing pathogenic strains. *Journal of Acute Disease, 2*, 277–281.
10. Chanda, S., Rakholiya, K., Dholakia, K., & Baravalia, Y. (2013). Antimicrobial, antioxidant and synergistic property of two nutraceutical plants: *Terminalia catappa* L., & *Colocasia esculenta* L. *Turkish Journal of Biology, 37*, 81–91.
11. Da Silva, M., Iamanaka, B. T., Taniwaki, M. H., & Kieckbusch, T. G. (2013). Evaluation of the antimicrobial potential of alginate and alginate/chitosan films containing potassium sorbate and natamycin. *Packaging Technology and Science, 26*, 479–492.

12. Fazly Bazzaz, B. S., Khajehkaramadin, M., & Shokooheizadeh, H. R. (2005). *In vitro* antibacterial activity of *Rheum ribes* extract obtained from various plant parts against clinical isolates of Gram-negative pathogens. *Iranian Journal of Pharmaceutical Research, 2*, 87–91.

13. Garveya, M. I., Rahman, M. M., Gibbons, S., & Piddock, L. J. V. (2011). Medicinal plant extracts with efflux inhibitory activity against Gram-negative bacteria. *International Journal of Antimicrobial Agents, 37*, 145–151.

14. Kaneria, M., Baravalia, Y., Vaghasiya, Y., & Chanda, S. (2009). Determination of antibacterial and antioxidant potential of some medicinal plants from Saurashtra region, India. *Indian Journal of Pharmaceutical Sciences, 71*, 406–412.

15. Kaneria, M., Rakholiya, K., & Chanda, S. (2014). *In vitro* antimicrobial and anti-oxidant potency of two nutraceutical plants of Cucurbitaceae. In: Gupta, V. K. (Ed.) *Traditional and Folk Herbal Medicine: Recent Researches*, Daya Publication House, India, pp. 221–248.

16. Mohammed, S. (2013). Comparative study of *in vitro* antibacterial activity of miswak extracts and different toothpastes. *American Journal of Agricultural and Biological Sciences, 8*, 82–88.

17. Parekh, J., & Chanda, S. (2007). *In vitro* antibacterial activity of the crude methanol extract of *Woodfordia fruticosa* Kurz. flower (Lythraceae). *Brazilan Journal of Microbiology, 38*, 204–207.

18. Parekh, J., Jadeja, D., & Chanda, S. (2005). Efficacy of aqueous and methanol extracts of some medicinal plants for potential antibacterial activity. *Turkish Journal of Biology, 29*, 203–210.

19. Patel, M., Pandya, S. S., & Rabari, A. (2013). Anti-inflammatory activity of leaf extracts of *Salvadora oleoides* (decne.). *International Journal of Pharma and BioSciences, 4*, 985–993.

20. Perez, C., Paul, M., & Bazerque, P. (1990). An Antibiotic assay by the agar well diffusion method. *Acta Biologiae Medecine Experimentalis, 15*, 113–115.

21. Prasad, S., Anthonmma, Jyothirmayi, N., Sowjanya, K., Shariotte, V., Priyanka, A., & Mounika, S. (2011). *In vitro* assay of herbaceous extracts of *Salvadora persica* L. against some pathogenic microbes. *Research Journal of Pharmaceutical, Biological and Chemical Sciences, 2*, 860–863.

22. Rakholiya, K., & Chanda, S. (2012). *In vitro* interaction of certain antimicrobial agents in combination with plant extracts against some pathogenic bacterial strains. *Asian Pacific Journal of Tropical Biomedicine, 2*, S876–S880.

23. Rakholiya, K., Kaneria, M., & Chanda, S. (2013). Medicinal plants as alternative sources of therapeutics against multidrug-resistant pathogenic microorganisms based on their antimicrobial potential and synergistic properties. In: *Fighting Multidrug Resistance with Herbal Extracts, Essential Oils and their Components,* Gupta, V. K. (Ed.). Rai, M., & Kon, K. (Eds.). Elsevier, USA, pp. 165–179.

24. Rakholiya, K., Kaneria, M., & Chanda, S. (2014). Mango Pulp: A potential source of natural antioxidant and antimicrobial agents. In: *Medicinal Plants: Phytochemistry, Pharmacology and Therapeutics,* Daya Publication House, India, pp. 253–284.

25. Rakholiya, K., Kaneria, M., & Chanda, S. (2014). Inhibition of microbial pathogens using fruit and vegetable peel extracts. *International Journal of Food Sciences and Nutrition, 65*, 733–739.

26. Rakholiya, K., Kaneria, M., & Chanda, S. (2015). *In vitro* assessment of novel antimicrobial from methanol extracts of matured seed kernel and leaf of *Mangifera indica* L. (Kesar Mango) for inhibition of *Pseudomonas* and their synergistic potential. *American Journal of Drug Discovery and Development, 5*, 13–23.

27. Rakholiya, K., Kaneria, M., Desai, D., & Chanda, S. (2013). Antimicrobial activity of decoction extracts of residual parts (seed and peels) of *Mangifera indica* L. var. Kesar against pathogenic and food spoilage microorganism. In: *Microbial Pathogens and Strategies for Combating them: Science, Technology and Education.* Mendez-Vilas, A. (Ed.). Formatex Research Center, Spain. pp. 850–856.

28. Rakholiya, K., Kaneria, M., Nagani, K., Patel, A., & Chanda, S. (2015). Comparative analysis and simultaneous quantification of antioxidant capacity of four *Terminalia* species using various photometric assays. *World Journal of Pharmaceutical Science, 4*, 1280–1296.

29. Rakholiya, K., Vaghela, P., Rathod, T., & Chanda, S. (2014). Comparative study of hydroalcoholic extracts of *Momordica charantia* L. against foodborne pathogens. *Indian Journal of Pharmaceutical Sciences, 76*, 148–156.

30. Trigui, M., Hsouna, A. B., Tounsi, S., & Jaoua, S. (2013). Chemical composition and evaluation of antioxidant and antimicrobial activities of Tunisian *Thymelaea hirsuta* with special reference to its mode of action. *Industrial Crops and Products, 41*, 150–157.

31. Tyagi, A. K., & Malik, A. (2011). Antimicrobial potential and chemical composition of *Eucalyptus* oil in liquid and vapor phase against food spoilage microorganisms. *Food Chemistry, 126*, 228–235.

32. Yagi, S., Chretien, F., Duval, R. E., Fontanay, S., Maldini, M., Piacente, S., Henry, M., Chapleur, Y., & Laurain-Mattar, D. (2012). Antibacterial activity, cytotoxicity and chemical constituents of *Hydnora johannis* roots. *South African Journal of Botany, 78*, 228–234.

CHAPTER 11

ANTIMICROBIAL PROPERTIES OF LEAF EXTRACT: POLYALTHIA LONGIFOLIA VAR. PENDULA UNDER IN-VITRO CONDITIONS

KALPNA RAKHOLIYA, YOGESH BARAVALIA, and SUMITRA CHANDA

CONTENTS

11.1   INTRODUCTION

Many infectious diseases are known to be treated with herbal remedies throughout the history of mankind. Even today, plant materials continue to play a major role in primary health care as therapeutic remedies in many developing countries. Human infections, parasite and microbial, constitute a serious problem, especially in tropical and subtropical developing countries, despite advances in medical science and discovery of new antibiotics [25].

During the last few years, antimicrobial properties of extracts and natural products have been intensively investigated as the demand for safe drugs has increased due to the misuse of antibiotics and an increase in immuno deficiency [17, 26]. There is continuous and urgent need to discover new antimicrobial compounds with diverse chemical structures and novel mechanisms of action because there has been an alarming increase in the incidence of new and reemerging infectious diseases.

Gram positive cocci and in particular *Staphylococcus* species are predominant among the organisms responsible for infective complications following surgical vascular grafts or the implantation of prosthetic devices [11]. *Staphylococcus aureus* is a particularly virulent organism that causes a broad array of health conditions including: abscesses, wound infection, pneumonia, toxic shock syndrome and other diseases [10]. It rapidly develops resistance to many antimicrobial agents. *Staphylococcus epidermidis* is the most common cause of nosocomial bacteraemia and is the principal organism responsible for infections of implanted prosthetic medical devices such as prosthetic heart valves, artificial joints, and cerebrospinal fluid shunts [29]. Infections caused by *S. epidermidis* are often persistent and relapsing. *Bacillus cereus* is a spore forming pathogen associated with various opportunistic clinical infections [7]. It has also been associated with food borne diseases caused by toxins. *Micrococcus luteus* is considered as an emerging nosocomial pathogen in immune-compromised patients.

Frequency of *Candida albicans* infections has risen dramatically since the advent of antibiotics and the development of drug-resistant *C. albicans* is a major concern worldwide [24]. A rapid and effective response to challenge pathogens is essential for the survival of all living organisms. The need for efficient agents increases with the expanding number of immune-deficient patients and with the emergence of bacterial and fungal pathogens resistant to current therapies [20]. To overcome the alarming problem of microbial resistance, the discovery of novel active compounds against new targets is a matter of urgency.

"*Polyalthia longifolia var. pendula (Annonaceae) is a tall evergreen tree found throughout India and Sri Lanka. Polyalthia longifolia (False Ashoka) is a lofty evergreen tree commonly planted due to its effectiveness in alleviating noise pollution. It exhibits symmetrical pyramidal growth with willowy weeping pendulous branches and long narrow lanceolate leaves with undulate margins. The tree is known to grow over 30 ft. in height. Polyalthea is derived from a combination of Greek words meaning 'many cures' with reference to the medicinal properties of the tree while Longifolia, in Latin, refers to the length of its leaves. Polyalthia longifolia is sometimes incorrectly identified as the Ashoka tree (Saraca indica) because of the close resemblance of both trees. The Polyalthia is allowed to grow naturally (without trimming the branches out for decorative reasons) grows into a normal large tree with plenty of shade,"* according to https://en.wikipedia.org/wiki/Polyalthia_longifolia. Many medicinal uses of this tree and the parts of the plant have been reported [8, 9, 13, 19, 21, 31].

This chapter presents the research study to evaluate the antimicrobial efficacy of different solvent extracts of *P. longifolia* leaf and also to evaluate the bactericidal effect and drug stability of the potent extract.

11.2   MATERIAL AND METHODS

11.2.1   PLANT MATERIAL

Polyalthia longifolia (Sonn.) Thw. variety *pendula* (Annonaceae) leaves were collected in July, 2006 from Rajkot, Gujarat (Western India) and identified by comparison with specimens (PSN4) available at the Herbarium of the Department of Biosciences, Saurashtra University, Rajkot, Gujarat, India. The leaves were washed, cleaned and air dried. The dried leaves were crushed in a homogenizer to fine powder and stored in air tight containers.

11.2.2   SOXHLET EXTRACTION

Successive extraction method was used for extraction. Hexane (PHE), toluene (PTE), chloroform (PCE), acetone (PAE) and methanol (PME) were used for the extraction. The dried powder of *P. longifolia* leaf (10 g) was extracted with 150 ml of solvent for different time duration by using soxhlet equipment. The extractive yield was different with different solvents. Maximum extractive yield was in methanol (10.68%) followed by toluene (9.25%) and minimum was in hexane (0.49%) followed by chloroform (2.98%).

11.2.3   TEST MICROORGANISMS

The microbial strains are identified strains and were obtained from National Chemical Laboratory (NCL), Pune, India. The microorganisms studied were five Gram positive bacteria viz. *Staphylococcus epidermidis* NCIM2493, *Micrococcus luteus* ATCC10240, *Staphylococcus aureus* ATCC25923-1, *Bacillus cereus* ATCC29737, *Staphylococcus aureus* ATCC29737-2; and five Gram negative bacteria viz. *Pseudomonas aeruginosa* ATCC27853, *Salmonella typhimurium* ATCC23564, *Klebsiella aerogenes* NICM2098, *Escherichia coli* ATCC25922, *Citrobacter freundii* ATCC10787. The fungal strains studied were *Candida albicans* ATCC2091, *Cryptococcus luteolus* ATCC32044, *Candida tropicalis* ATCC4563 and 6 clinical isolates. Bacterial cultures were grown on nutrient broth (Hi-Media) at 37°C for 24 h and the fungal cultures were grown on Sabouraud dextrose broth (Hi-Media) at 28°C for 48 h. All the microbial cultures of bacteria and Yeast were maintained on nutrient agar slants and MGYP slants, respectively at 4°C.

11.2.4   ANTIBACTERIAL ASSAY

The antibacterial assay was performed by agar disc diffusion method [6, 33]. The molten Mueller Hinton Agar (Hi-Media) was inoculated with the 200 μl of the inoculum and poured into the sterile Petri plates (Hi-Media). The disc (0.7 cm) (Hi-Media) was saturated with 20 μl of the test compound, allowed to dry and was introduced on the upper layer of the seeded agar plate. The plates were incubated for 24 h. Microbial growth was determined by measuring the diameter of zone of inhibition. For each bacterial strain controls were maintained where pure solvents were used instead of the extract. The results were obtained by measuring the zone diameter. The experiment was replicated thrice and the mean values are presented. The results were compared with the standard Cefotaxime sodium (20 μl/disc).

11.2.5   BACTERICIDAL ACTIVITY

The bacterial cultures were grown in N-broth. The cultures were grown for at least three generations to the late logarithmic phase (Optical density of 0.8 at 600 nm) [3]. The bacteria in the mid exponential phase of growth were treated with the test samples (drug) in 12.5, 25 and 50 μg concentrations. The change in optical density was measured spectrophoto-metrically. Growth monitoring was done after treatment with the test sample after 1, 2 and 5 h. Positive control and negative control was kept along with test sample for comparison. Graph was plotted for OD at 600 nm versus concentration (μg/ml).

11.2.6 DRUG STABILITY TEST

Long-term stability of the extracts for potency was checked by Drug Stability Assay. The potency of PHE, PCE and PTE was tested against *Staphylococcus epidermidis*, *Micrococcus luteus*, *Bacillus cereus*, *Klebsiella aerogenes* and *Staphylococcus aureus* at a dose level of 5 mg/ml for 49 days (7 weeks). Stock solution was prepared in DMSO. The stock sample was used for 0, 7, 14, 21, 28, 35, 42, and 49th days (7 weeks) to check the drug stability. Graph is plotted for activity against Days.

11.3 RESULTS

11.3.1 ANTIBACTERIAL SCREENING

The antibacterial activity of all five extracts (PHE, PTE, PCE, PAE and PME) against 10 bacterial strains was evaluated and the results are presented in Table 11.1. Each extract showed a different trend of potency towards the bacterial strains investigated (Table 11.1). Maximum antibacterial activity was shown against *B. cereus* when all the 5 extracts were considered and maximum activity was shown by PHE inhibited 6 strains while PME inhibited only *B. cereus*; PAE inhibited *M. luteus*, *K. aerogenes* and *B. cereus*. PCE was not active only against *S. aureus*-1. PTE and PHE showed similar level of potency.

11.3.2 ANTIBACTERIAL ACTIVITY

Three different concentrations (50 μg, 100 μg and 200 μg) of PHE, PTE and PCE were evaluated for antibacterial activity against 5 bacterial strains (*S. aureus*-2, *M. luteus*, *S. epidermidis*, *B. cereus* and *K. aerogenes*). The results are shown in Table 11.2. A clear concentration effect was

TABLE 11.1 Antibacterial Activity of *Polyalthia longifolia* var. *pendula* Leaf Extracts

Extracts (100 µg)	Inhibition zone (mm)*									
	SA-1	SA-2	ML	SE	BC	EC	PA	CF	ST	KA
PHE	10 ± 0.5	11 ± 0.7	18 ± 1.1	14 ± 0	23 ± 0.5	—	—	—	—	17 ± 0
PTE	9 ± 0	11 ± 0.5	16 ± 0.4	13 ± 0.1	22 ± 0.1	—	—	—	—	16 ± 0
PCE	—	9 ± 0.2	16 ± 0.2	11 ± 0.4	20 ± 1.0	—	—	—	—	13 ± 0
PAE	—	—	12.5 ± 0.2	—	12 ± 0.2	—	—	—	—	9 ± 0
PME	—	—	—	—	10 ± 0.2	—	—	—	—	—
CS	34 ± 0	24 ± 2.7	4 ± 1.1	28 ± 0.2	18 ± 0.2	15 ± 0.2	19 ± 1.7	32 ± 0.8	28 ± 0	15 ± 0.2

* Values are Mean ± S.E.M ($n = 3$).

TABLE 11.2 Antibacterial Activity of PHE, PTE and PCE at Different Concentrations

Drug concentration	Inhibition zone (mm)*				
	SA-2	SE	ML	BC	KA
PHE					
200 µg	9 ± 0	13 ± 1	20 ± 1	19 ± 0.8	15 ± 0.2
100 µg	14 ± 1	12 ± 1	13 ± 1	19 ± 0.2	14 ± 0.2
50 µg	—	12 ± 0.8	12 ± 0.2	17 ± 0.5	12 ± 0.2
PTE					
200 µg	13 ± 0.2	13 ± 0.1	18 ± 0	20 ± 5	15 ± 0
100 µg	14 ± 0.8	13 ± 1	14 ± 1	18 ± 0.2	14 ± 0.1
50 µg	—	12 ± 0.2	12 ± 0.8	15 ± 0.2	14 ± 0.8
PCE					
200 µg	13 ± 0.5	9 ± 0	15 ± 2	18 ± 0	14 ± 0.4
100 µg	10 ± 0.2	10 ± 0.2	10 ± 0.5	16 ± 0.7	12 ± 0
50 µg	—	4 ± 2	13 ± 1	16 ± 0.9	10 ± 0.1
CS 100 µg	23 ± 0.5	28 ± 0	39 ± 0.5	13 ± 0.1	15 ± 0

* Values are Mean ± S.E.M ($n = 3$).

observed in all the 3 extracts which was comparable to that of standard antibiotic cefotaxime sodium (Table 11.2). All the extracts showed better activity than cefotaxime sodium against *B. cereus* and almost same trend was shown against *K. aerogenes* though the level of activity was less. The action of *P. longifolia* leaf extracts on *B. cereus* and *K. aerogenes* is noteworthy (Table 11.2). *B. cereus* is a human pathogen whose infections are almost the most difficult to treat with conventional antibiotics. The susceptibility of these bacteria to various extracts of this plant may be a pointer to its potentiality as a drug that can be used against these organisms.

11.3.3 BACTERICIDAL ACTIVITY

The *in vitro* antibacterial potential (bactericidal effect) of all the 3 extracts (PHE, PTE and PCE) of *P. longifolia* leaf are shown in Figures 11.1–11.4. Cefotaxime selectively interfere with synthesis of the bacterial cell wall – a structure that mammalian cell do not possess. The cell wall is a polymer called peptidoglycan that consists of glycan units joined by each other by a peptide cross link. To be maximally effective, inhibitors of cell wall synthesis require actively proliferating microorganisms; they have little or no effect on bacteria that are not growing and dividing. The bactericidal activity of PHE, PTE and PCE demonstrated their role as cell wall synthesis inhibitor. PHE, PTE and PCE are showing comparable bactericidal activity as that of Cefotaxime sodium. The efficacy of the extracts for potency against Gram positive bacteria suggests their mode of action as inhibitor of bacterial cell wall.

11.3.4 DRUG STABILITY TEST

The concentration 100 µg of each extract, e.g., PHE, PTE and PCE was subjected to drug stability test. The results of drug stability test of the 3 extracts against 5 bacteria are shown in Figures 11.5 and 11.6. The activity was checked for a period of 49 days (7 weeks). The respective drugs were prepared on day 0, and the same drug was used for 49 days, e.g., their antibacterial potency was evaluated against the same 5 bacteria as above.

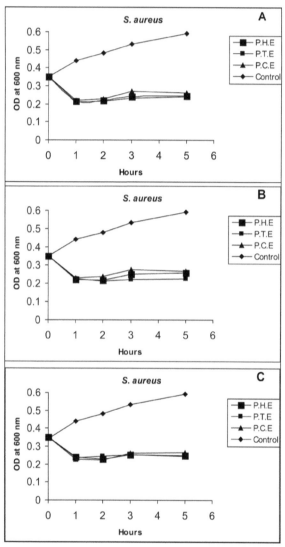

FIGURE 11.1   Bactericidal activity of different extracts of *P. longifolia* against *S. aureus* with different concentrations: A: 12.5 µg/disc, B: 25 µg/disc and C: 50 µg/disc.

As can be seen from the figures, it is evident that no degradation of the drug occurred, till the period of 49 days (7 weeks) and potency was consistent (Figures 11.5 and 11.6). Even after 49 days, the same antibacterial activity was evident, proving the drug to be stable up to 49 days at 4–8°C.

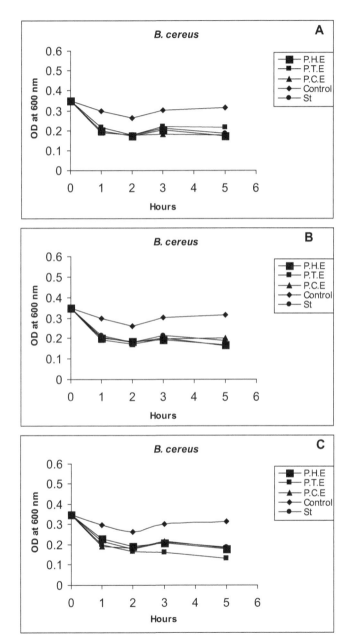

FIGURE 11.2   Bactericidal activity of different extracts of *P. longifolia* against *B. cereus* with different concentrations: A: 12.5 µg/disc, B: 25 µg/disc and C: 50 µg/disc.

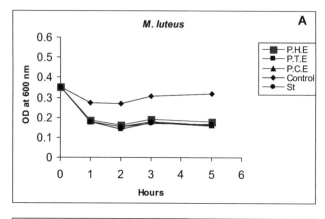

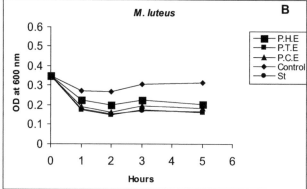

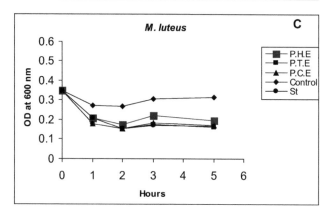

FIGURE 11.3   Bactericidal activity of different extracts of *P. longifolia* against *M. luteus* with different concentrations: A: 12.5 µg/disc, B: 25 µg/disc and C: 50 µg/disc.

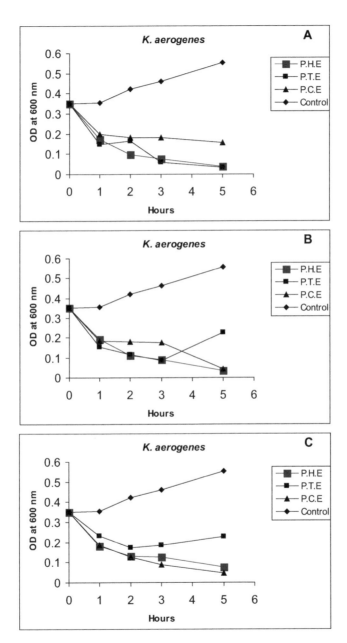

FIGURE 11.4 Bactericidal activity of different extracts of *P. longifolia* against *K. aerogenes* with different concentrations (A: 12.5 µg/disc, B: 25 µg/disc and C: 50 µg/disc).

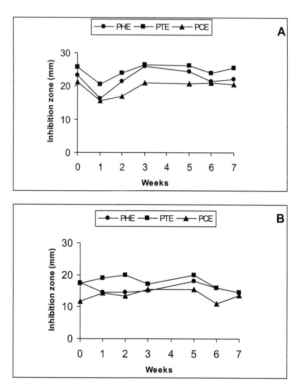

FIGURE 11.5 Drug stability of different extracts of *P. longifolia* leaf at the concentration of 100 µg/disc (Top – A: *B. cereus* and bottom – B: *K. aerogenes*).

11.4   DISCUSSION

It is well known that plants have been a valuable source of products to treat a wide range of medical problems, including ailments caused by microbial infection. Numerous studies have been carried out in different parts of the globe to extract plant products for screening antibacterial activity [1, 4, 5, 22, 27, 28, 30, 34]. All these extracts were inactive against most of the Gram negative strains investigated. This is in agreement with previous reports that plant extracts are more active against Gram positive bacteria than Gram negative bacteria [14, 16, 23]. PHE, PTE and PCE showed poor antifungal activity against three identified strains and six clinically isolates (data is not shown). Hence further experiments were carried out with bacterial strains only.

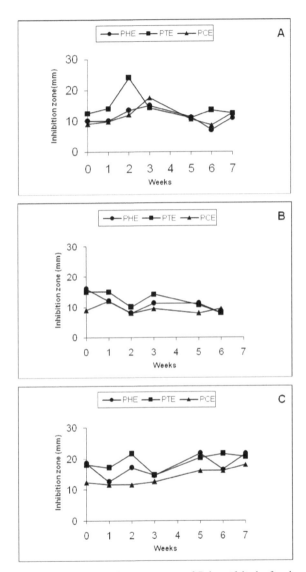

FIGURE 11.6 Drug stability of different extracts of *P. longifolia* leaf at the concentration of 100 μg/disc (A: *S. aureus*, B: *S. epidermidis* and C: *M. luteus*).

Truiti et al. [32] reported that the bactericidal effect of *C. nutans* extract was more apparent against Gram positive bacteria than Gram negative bacteria. Several reports have indicated that catechins in green tea show antibacterial or bactericidal activities against Gram positive and Gram negative bacteria [15, 18]. The mechanism of these activities has been explained by the suggestion that catechins exert an effect on the bacterial membrane, changing its fluidity and resulting in the leakage of intracellular materials [15]. The present work has highlighted the antibacterial effects of *P. longifolia* leaf on clinically important pathogens. Some antibiotics have become almost obsolete because of the problem of drug resistance [12] and that the consequence of drug resistance implies that new drugs must be sought for and to treat diseases for which known drugs are no longer useful. The three drugs PHE, PTE and PCE showed positive results for the presence of tannins, steroids, glycosides and alkaloids (Table 11.3). The antibacterial activity can be due to alkaloids, tannins, steroids, saponins found in crude extract and fractions. These phytochemical groups are known to possess antimicrobial compounds [2].

The above-mentioned compounds, which are present in *P. longifolia* leaf may influence antibacterial activity in an effective manner and may have multiple actions against bacteria. Elucidation of these processes is necessary to define the exact mechanism of active principles present in *P. longifolia* leaf.

TABLE 11.3 Preliminary Qualitative Phytochemical Analysis of Different Extracts of *Polyalthia longifolia* Var. *pendula*

Phytochemicals	TEST	PCR	PHE	PTE	PCE	PAE	PME
Flavonoids	Shinoda's	–	–	–	–	+++	++
Saponins	Froth formation	+++	–	–	–	–	+++
Tannins	Ferric chloride	+++	++	–	–	–	+++
Steroids (Terpenoids)	Liebermann-Buchard's	–	+	+++	+++	–	–
Cardiac Glycosides	Keller-Kiliani	+++	–	+++	–	–	–
Alkaloids	Dragendroff's	+++	–	–	+++	+++	+++
	Wagner's	+++	–	–	–	–	+++
	Mayer's	–	–	–	–	–	–

11.5 CONCLUSIONS

The potential for developing antimicrobials from higher plants appears rewarding as it will lead to the development of a phytomedicine to act against microbes. Plant based antimicrobials have enormous therapeutic potential as they can serve the purpose without any side effects that are often associated with synthetic antimicrobials. Continued further exploration of plant-derived antimicrobials is the need of the hour. *P. longifolia* might possess a novel antimicrobial molecule, which has an effect mainly against Gram positive and to a certain extent on Gram negative bacteria. Further research is needed in order to obtain information about the chemical composition of *P. longifolia*, as well as to reveal their mode of action on microbial cells.

11.6 SUMMARY

The leaves of *Polyalthia longifolia* (Sonn.) Thw. var. *pendula* (Annonaceae) were extracted with hexane (PHE), toluene (PTE), chloroform (PCE), acetone (PAE) and methanol (PME) successively. The antibacterial activity was evaluated by agar disc diffusion method at 100 µg/disc concentration against five Gram positive bacteria and five Gram negative bacteria Cefotaxime sodium (100 µg/disc) was used as standard drug. Antifungal activity was evaluated against three fungal strain and six clinical isolates. PHE, PTE and PCE showed strong antimicrobial activity against 5 bacterial strains; hence these 3 extracts were selected for antibacterial activity at a concentration of 50, 100 and 200 µg/disc. Bactericidal activity (12.5 µg, 25 µg and 50 µg) and drug stability test (100 µg/disc) was also checked. Plant extracts showed poor antifungal activity. The most susceptible bacterial strain was *B. cereus*. PHE, PTE and PCE showed comparable bactericidal activity as that of cefotaxime against *B. cereus* and *M. luteus*. All the 3 extracts showed almost same antibacterial activity till 7 weeks. The results of the present work indicate that *P. longifolia* leaf extracts may be an ideal candidate for further research for antibacterial products.

KEYWORDS

- acetone
- annonaceae
- antimicrobial activity
- *bacillus cereus*
- bactericidal activity
- cefotaxime
- chloroform
- drug stability
- fungal strains
- gram negative bacteria
- gram positive bacteria
- hexane
- *klebsiella aerogenes*
- leaves
- methanol
- *micrococcus luteus*
- *polyalthia longifolia*
- qualitative phytochemical analysis
- soxhlet extraction
- *staphylococcus aureus*
- *staphylococcus epidermidis*
- toluene

REFERENCES

1. Ait-Ouazzou, A., Loran, S., Arakrak, A., Laglaoui, A., Rota, C., Herrera, A., Pagan, R., & Conchello, P. (2012). Evaluation of the chemical composition and antimicrobial activity of *Mentha pulegium, Juniperus phoenicea*, and *Cyperus longus* essential oils from Morocco. *Food Research International, 45,* 313–319.
2. Ameri, A., Ghadge, C., Vaidya, J. G., & Deokule, S. S. (2011). Anti-*Staphylococcus aureus* activity of *Pisolithus albus* from Pune, India. *Journal of Medicinal Plant Research, 5,* 527–532.

3. Aronovitch, J., Godianger, D., & Czapski, G. (1991). Bactericidal activity of catecholamine copper complexes. *Free Radical Research, 12,* 179–488.
4. Bag, A., Bhattacharyya, S. K., Pal, N. K., & Chattopadhyaya, R. R. (2012). *In vitro* antibacterial potential of *Eugenia jambolana* seed extracts against multidrug-resistant human bacterial pathogens. *Microbiological Research, 167,* 352–357.
5. Baravalia, Y., Kaneria, M., Vaghasiya, Y., Parekh, J., & Chanda, S. (2009). Antioxidant and antibacterial activity of *Diospyros ebenum* Roxb. leaf extracts. *Turkish Journal of Biology, 33,* 159–164.
6. Bauer, A. W., Kirby, W. M. M., Sherris, J. C., & Turck, M. (1966). Antibiotic susceptibility testing by a standardized single disk method. *American Journal of Clinical Pathology, 45,* 493–496.
7. Boyd, R. F. (1995). *Basic Medical Microbiology,* 5th ed., Little, Brown and company, Boston, pp. 310–314.
8. Chanda, S., & Nair, R. (2010). Antimicrobial activity of *Polyalthia longifolia* (Sonn.) Thw. var. *pendula* leaf extracts against 91 clinically important pathogenic microbial strains. *Chinese Medicine, 1,* 31–38.
9. Chanda, S., Baravalia, Y., & Kaneria, M. (2011). Protective effect of *Polyalthia longifolia* var. *pendula* leaves on ethanol and ethanol/HCl induced ulcer in rats and its antimicrobial potency. *Asian Pacific Journal of Tropical Medicine, 4,* 673–679.
10. Daka. D. (2011). Antibacterial effect of garlic (*Allium sativum*) on *Staphyloccus aureus*: An *in vitro* study. *African Journal of Biotechnology, 10,* 666–669.
11. De Lalla, F. (1999). Antimicrobial chemotherapy in the control of surgical infectious complications. *Journal of Chemotherapy, 11,* 440–445.
12. Ekpendu, T. O., Akshomeju, A. A., & Okogun, J. J. (1994). Anti-inflammatory antimicrobial activity. *Letters in Applied Microbiology, 30,* 379–384.
13. Faizi, S., Mughal, N. R., Khan, R. A., Khan, S. A., Ahmad, A., Bibi, N., & Ahmed, S. A. (2003). Evaluation of the antimicrobial property of *Polyalthia longifolia* var. *pendula*: isolation of a lactone as the active antibacterial agent from the ethanol extract of the stem. *Phytotherapy Research, 17,* 1177–1181.
14. Fernandez-Agullo, A., Pereira, E., Freire, M. S., Valentao, P., Andrade, P. B., Gonzalez-Alvarez, J., & Pereira, J. A. (2013). Influence of solvent on the antioxidant and antimicrobial properties of walnut (*Juglans regia* L.) green husk extracts. *Industrial Crops and Products, 42,* 126–132.
15. Ikigai, H., Nakae, T., Hara, Y., & Shimamura, T. (1993). Bactericidal catechins damage the lipid bilayer. *Biochimia Biophys Acta, 1147,* 132–136.
16. Kaneria, M., & Chanda, S. (2013). Evaluation of antioxidant and antimicrobial capacity of *Syzygium cumini* L. leaves extracted sequentially in different solvents. *Journal of Food Biochemistry, 37,* 168–176.
17. Kaneria, M., Baravalia, Y., Vaghasiya, Y., & Chanda, S. (2009). Determination of antibacterial and antioxidant potential of some medicinal plants from Saurashtra region, India. *Indian Journal of Pharmaceutical Sciences, 71,* 406–412.
18. Kimura, T., Hocino, N., Yamaji, A., Hayakawa, F., & Ando, T. (1998). Bactericidal activity of Catechin-copper (II) complexes on *Escherichia coli* ATCC11775 in the absence of hydrogen peroxide. *Letters in Applied Microbiology, 27,* 328–330.
19. Kishore, N., Dubey, N. K., Tripathi, R. D., & Singh, S. K. (1982). Fungitoxic activity of leaves of some higher plants. *National Academy Science Letters, 6,* 9–10.

20. Lugardon, K., Raffner, R., Goumon, Y., Corti, A., Delmas, A., Bulet, P., Aunis, D., & Metz-Boutigue, M. H. (2000). Antibacterial and antifungal activities of vasostatin-1, the N-terminal fragment of chromogranin A. *Journal of Biological Chemistry, 275*, 10745–10753.
21. Marthanda, M., Subramanyam, M., Hima-Bindu, M., & Annapurna, J. (2005). Antibacterial activity of Clerodane diterpenoids from *Polyalthia longifolia* seeds. *Fitoterapia, 76*, 3–4.
22. Martins, M., Amorim, E. L. C., Sobrinho, T. J. S. P, Saraiva, A. M., Pisciottano, M. N. C., Aguilar, C. N., Teixeira, J. A., & Mussatto, S. I. (2013). Antibacterial activity of crude methanolic extract and fractions obtained from *Larrea tridentata* leaves. *Industrial Crops and Products, 4*, 306–311.
23. Mathabe, M. C., Nikolova, R. V., & Nyazema, N. Z. (2006). Antibacterial activities of medicinal plants used for the treatment of diarrhea in Limpopo Province, South African. *Journal of Ethnopharmacology, 105*, 286–293.
24. Nolte, F. S., Parkinson, T., & Falconer, D. J. (1997). Isolation and characterization of Flucanazole- and amphotericin B-resistant *Candida albicans* from blood of two patients with leukemia. *Antimicrobial Agents Chemotherapy, 41*, 196–199.
25. Portillo, A., Vila, R., Freixa, B., Adzet, T., & Canigueral, S. (2001). Antifungal activity of Paraguayan plants used in traditional medicine. *Journal of Ethnopharmacology, 76*, 93–98.
26. Rakholiya, K., & Chanda, S. (2012). *In vitro* interaction of certain antimicrobial agents in combination with plant extracts against some pathogenic bacterial strains. *Asian Pacific Journal of Tropical Biomedicine*, S876–S880.
27. Rakholiya, K., Kaneria, M., & Chanda, S. (2014). Inhibition of microbial pathogens using fruit and vegetable peel extracts. *International Journal of Food Science and Nutrition, 65*, 733–739.
28. Rakholiya, K., Kaneria, M., & Chanda, S. (2015). *In vitro* assessment of novel antimicrobial from methanol extracts of matured seed kernel and leaf of *Mangifera indica* L. (Kesar Mango) for inhibition of *Pseudomonas* spp. and their synergistic potential. *American Journal of Drug Discovery and Development, 5*, 13–23.
29. Rupp, M. E., & Archer, G. L. (1994). Coagulase-negative staphylococci pathogens associated with medical progress. *Clinical Infectious Diseases, 19*, 231–245.
30. Silvan, J. M., Mingo, E., Hidalgo, M., de Pascual-Teresa, S., Carrascosa, A. V., & Martinez-Rodriguez, A. J. (2013). Antibacterial activity of a grape seed extract and its fractions against *Campylobacter* spp. *Food Control, 29*, 25–31.
31. Tanna, A., Nair, R., & Chanda, S. (2009). Assessment of antiinflammatory and hepatoprotective potency of *Polyalthia longifolia* var. *pendula* leaf in wistar albino rats. *Journal of Natural Medicine, 63*, 80–85.
32. Truiti, M. C., Sarragiotto, M. H., Filho, B. A. D. A., Nakamura, C. V., & Filho, B. P. D. (2003). *In vitro* antibacterial activity of 7-O-β-D-glucopyranosyl-nutanocoumarin from *Chaptalia nutans* (Asteraceae). *Mem Institute Oswald Cruz Rio de Janeiro, 98*, 283–286.
33. Vaghasiya, Y., & Chanda, S. V. (2007). Screening of methanol and acetone extracts of 14 Indian medicinal plants for antimicrobial activity. *Turkish Journal of Biology, 31*, 243–248.
34. Vaghasiya, Y., Patel, H., & Chanda, S. (2011). Antibacterial activity of *Mangifera indica* L. seeds against some human pathogenic bacterial strains. *African Journal of Biotechnology, 10*, 15788–15794.

PART V

ACTIVE CONSTITUENTS OF FOODS

ISOLATION, VALIDATION AND CHARACTERIZATION OF MAJOR BIOACTIVE CONSTITUENTS FROM MANGO RIPE SEED

KALPNA D. RAKHOLIYA, MITAL J. KANERIA,
and SUMITRA V. CHANDA

CONTENTS

12.1   INTRODUCTION

In healthy individuals, there is an equilibrium between the natural antioxidative defense system and the reactive oxygen species (ROS), generated from both living organisms and exogenous sources. When the equilibrium is disrupted, the ROS can induce oxidative and cause cellular damages to various biomolecules [38]. As a result, the excess of free radicals can damage both the structure and function of cell membrane in a chain reaction leading to degenerative diseases [40, 85].

Free radicals not only cause human disease but also cause lipid oxidation in food system. Oxidation of lipids, which is the main cause of quality deterioration in many food systems, may lead to the development of undesirable off-flavors and formation of some toxic compounds and may lower the quality and nutritional value of foods. The principal focus of ethno pharmacological research today is related to the discovery of new antioxidants [17, 35, 92]. Antioxidants are also added to food products, especially to lipids and lipid containing foods, can increase the shelf life by retarding the process of lipid peroxidation, which is one of the major reasons for deterioration of food products during processing and storage. Synthetic phenolic antioxidants, such as butylated hydroxyanisole (BHA), butylated hydroxytolune (BHT) and tert-butylhydroxyquinone (TBHQ) effectively inhibit lipid oxidation but they have restricted use in foods as these synthetic antioxidants are suspected to be carcinogenic.

Nowadays, food industries need new food ingredients obtained from natural sources, to develop novel functional foods or nutraceuticals. Fruits and vegetables, apart from being good sources of vitamins, minerals, and fiber, are also rich sources of potentially bioactive compounds known as phytochemicals [11]. The biological compounds from agro-waste may not only increase the stability of foods, by preventing lipid peroxidation and microbial growth, but in humans or animals may also protect biomolecules and supra-molecular structures from oxidative damage and infectious diseases caused by microbes [11, 18, 59]. It has now been reported that byproducts and waste materials of fruits, including peels, seeds and stems contain high levels of health enhancing substances due to large amounts of polyphenols that can be extracted from the byproducts to provide nutraceuticals [10, 65, 76].

Mangifera indica L. (Anacardiaceae) is one of the important tropical fruits in the world and India. Mango is considered as a king of fruits in Indian delicacy. There are many traditional medicinal uses for the bark, pulp, roots, seed kernel and leaves of Mango throughout the globe [58, 63, 67].

This chapter presents the research study on isolation, validation and characterization of major bioactive constituents by various techniques from mango ripe seed.

12.2 MATERIAL AND METHODS

12.2.1 CHEMICALS AND REAGENTS

Nitroblue tetrazolium (NBT), phenazine methosulfate (PMS), nicotinamide adenine dinucleotide reduced (NADH), gallic acid, ascorbic acid, ferrous sulfate ($FeSO_4$), Folin-Ciocalteu's reagent, aluminum chloride ($AlCl_3$), potassium acetate, ferric chloride ($FeCl_3$), Tris buffer, sodium acetate, trichloroacetic acid (TCA), potassium persulfate, thiobarbituric acid (TBA), Hydrogen peroxide (H_2O_2), hydrochloric acid (HCl), petroleum ether, acetone, methanol, glacial acetic acid, 2,2-diphenyl-1-picrylhydrazyl (DPPH), quercetin, 2,4,6- tripyridyl-5-triazine (TPTZ), potassium ferricyanide ($K_3Fe(CN)_6$), sodium carbonate, 2-deoxy-D-ribose, di-Potassium hydrogen phosphate (K_2HPO_4), Potassium biphosphate (KH_2PO_4), Potassium hydroxide (KOH), Ethylenediamine tetraacetic acid (EDTA), Chloramphenicol (CH), ceftazidime (CF), 2-(4-Iodo phenyl)-3-(4-nitro phenyl)-5-phenyltetrazolium chloride (INT) and 2,2′-Azino-bis-(3-ethyl)benzothiazoline)-6-sulfonic acid diammonium salt (ABTS) were obtained from Hi-media, Merck and SRL. Water was purified with a Milli-Q system (Millipore, Bedford, USA). All solvents and chemical used were of analytical grade.

12.2.2 PLANT MATERIAL

The ripe seeds of Kesar variety Mango were collected in May, 2010 from Gujarat, India and washed thoroughly with tap water. The peels were removed from the seeds manually for further extraction, shade dried, homogenized to fine powder and stored in air tight bottle.

12.2.3 INDIVIDUAL COLD PERCOLATION EXTRACTION METHOD

The dried powder ripe seed was individually extracted by cold percolation method [55, 62]. The powder was first defatted with petroleum ether and then extracted in methanol. Ten grams of dried powder was taken in 100 ml of petroleum ether in a conical flask, plugged with cotton wool, and then kept on a rotary shaker at 120 rpm for 24 h. After 24 h, it was filtered through eight layers of muslin cloth, centrifuged at 5,000 rpm in a centrifuge (Remi Centrifuge, India) for 15 min and the supernatant was collected and the solvent was evaporated using a rotary vacuum evaporator (Equitron, India) to dryness. Petroleum ether was evaporated from the powder. This dry powder was then taken individually in 100 ml of methanol and was kept on a rotary shaker at 120 rpm for 24 h. Then the procedure followed was same as above, and the residues were weighed to obtain the extractive yield of extracts and were stored in air tight bottles at 4°C.

12.2.4 ISOLATION OF METHANOL EXTRACTS FROM RIPE MANGO SEED

Fractionation of the methanol extract was done by solvent-solvent partition [63, 83]. Five grams of methanol extracts was dissolved in hot methanol (200 ml). Slight precipitation obtained was discarded as methanol insoluble matter. The methanol soluble fraction was filtered and collected. It was concentrated to about 50 ml volume and ethyl acetate was added to it till faint turbidity was obtained. Then it was allowed to settle down in a refrigerator. The settled gelatinous reddish mass and supernatant was separated and collected separately. The supernatant was further concentrated and ethyl acetate step was repeated till reddish gelatinous mass obtained. All the settled mass was collected together and dissolved in methanol. It was concentrated further to dryness and designated as RSM I. The collected supernatant was concentrated further to near dryness and then dissolved in methanol. Then chloroform was added to it and cooled. Light yellow waxy sediment was separated and light buff colored supernatant was collected. This supernatant was concentrated further to dryness and designated as RSM II.

12.2.5 ANTIOXIDANT ASSAYS

12.2.5.1 DPPH Free Radical Scavenging Assay (DPPH)

The free radical scavenging activity of RSM and its fractions was measured by using DPPH by the modified method of McCune and Johns [49].

The reaction mixture (3.0 ml) consisted of 1.0 ml DPPH in methanol (0.3 mM), 1.0 ml methanol and 1.0 ml of different concentrations (2 to 60 µg ml^{-1}) of RSM and fractions diluted by methanol, was incubated for 10 min, in dark, after which the absorbance was measured at 517 nm using a UV-VIS Spectrophotometer (Shimadzu, Japan), against a blank sample. Ascorbic acid (2 to 16 µg ml^{-1}) was used as positive control [66, 73]. Percentage of inhibition was calculated using the following formula:

$$\% \text{ Inhibition} = [1 - (A/B)] \times 100 \tag{1}$$

where, B = absorbance of blank (DPPH + methanol), and A = absorbance of sample (DPPH + methanol + sample).

12.2.5.2   Superoxide Anion Radical Scavenging Assay (SO)

The superoxide anion radical scavenging activity was measured by the method as described by Robak and Gryglewski [72]. Superoxide radicals are generated by oxidation of NADH and assayed by the reduction of NBT. The reaction mixture (3.0 ml) consisted of 1.0 ml of different concentrations (20 to 500 µg ml^{-1}) of RSM and its fractions diluted by distilled water, 0.5 ml Tris-HCl buffer (16 mM, pH 8), 0.5 ml NBT (0.3 mM), 0.5 ml NADH (0.936 mM) and 0.5 ml PMS (0.12 mM). The superoxide radical generating reaction was started by the addition of PMS solution to the mixture. The reaction mixture was incubated at 25°C for 5 min, and then the absorbance was measured at 560 nm using a UV-VIS Spectrophotometer (Shimadzu, Japan), against a blank sample. Gallic acid (50 to 225 µg ml^{-1}) was used as a positive control [13, 66]. Gallic acid was used as a positive control. Percentage of inhibition was calculated using Eq. (1).

12.2.5.3   Hydroxyl Radical Scavenging Assay (OH)

The hydroxyl radical scavenging radical scavenging activity was measured by studying the competition between deoxyribose and test compound for hydroxyl radicals generated by Fe^{+3}-Ascorbic acid-EDTA-H$_2$O$_2$ system (Fenton reaction) according to the method of Kunchandy and Rao [43].

The reaction mixture (1.0 ml) consisted of 100 μl 2-deoxy-D-ribose (28 mM in 20 mM KH_2PO_4-KOH buffer, pH 7.4), 500 μl of different concentrations (200 to 1000 μg ml^{-1}) of RSM and fractions diluted by distilled water, 200 μl EDTA (1.04 mM) and 200 μM $FeCl_3$ (1:1 v/v), 100 μl H_2O_2 (1.0 mM) and 100 μl ascorbic acid (1.0 mM), and was incubated at 37°C for 1 h. The 1.0 ml TBA (1%) and 1.0 ml TCA (2.8%) was added, and incubated at 100°C for 20 min. After cooling, the absorbance of pink color was measured at 532 nm using a UV-VIS Spectrophotometer (Shimadzu, Japan), against a blank sample. Gallic acid (20 to 200 μg ml^{-1}) was used as a positive control [25, 67]. Percentage of inhibition was calculated using Eq. (1).

12.2.5.4 ABTS Radical Cation Scavenging Assay

The ABTS radical cation scavenging activity was measured by the method as described by Re et al. [69]. ABTS radical cations are produced by reacting ABTS (7 mM) and potassium persulfate (2.45 mM) and incubating the mixture at room temperature in the dark for 16 h. The ABTS working solution obtained was further diluted with methanol to give an absorbance of 0.85 ± 0.20 at 734 nm. 1.0 ml of different concentrations (1 to 24 μg ml^{-1}) of RSM and fractions diluted by methanol was added to 3.0 ml of ABTS working solution. The reaction mixture was incubated at room temperature for 4 min, and then the absorbance was measured at 734 nm using a UV-VIS Spectrophotometer (Shimadzu, Japan), against a blank sample. Ascorbic acid (1 to 10 μg ml^{-1}) was used as a positive control [66, 93]. The assay was carried out in triplicate and the mean values with ± SEM are presented. Percentage of inhibition was calculated using Eq. (1).

12.2.5.5 Reducing Capacity Assessment (RCA)

The reducing capacity assessment was determined using the modified method of Athukorala et al. [5]. The 1.0 ml of different concentrations (20 to 180 μg ml^{-1}) of RSM and its fractions diluted by distilled water was mixed with 2.5 ml phosphate buffer (200 mM, pH 6.6) and 2.5 ml $K_3Fe(CN)_6$ (30 mM). The mixture was then incubated at 50°C for 20 min. Thereafter, 2.5 ml of TCA (600 mM) was added to the reaction mixture, and then centrifuged for 10 min at 3,000 rpm. The upper layer of solution (2.5 ml) was mixed with 2.5 ml distilled water and 0.5 ml $FeCl_3$ (6 mM), and the absorbance was measured at 700 nm using a UV-VIS Spectrophotometer (Shimadzu, Japan), against a blank sample. Ascorbic acid (20 to 180 μg ml^{-1}) was used as positive control [42, 67].

12.2.5.6 Ferric Reducing Antioxidant Power (FRAP)

The reducing ability was determined by FRAP assay of Benzie and Strain [8]. FRAP assay is based on the ability of antioxidants to reduce Fe^{3+} to Fe^{2+} in the presence of TPTZ, forming an intense blue Fe^{2+}-TPTZ complex with an absorption maximum at 593 nm. This reaction is pH-dependent (optimum pH 3.6). The RSM and its fractions (0.1 ml) was added to 3.0 ml FRAP reagent (10 parts 300 mM sodium acetate buffer at pH 3.6, 1 part 10 mM TPTZ in 40 mM HCl, and 1 part 20 mM $FeCl_3$), and the reaction mixture was incubated at 37°C for 10 min. And then, the absorbance was measured at 593 nm using a UV-VIS Spectrophotometer (Shimadzu, Japan), against a blank sample. The calibration curve was made by preparing a $FeSO_4$ (100 to 1000 μM ml^{-1}) solution in distilled water [66, 86]. The antioxidant capacity based on the ability to reduce ferric ions of sample was calculated from the linear calibration curve and expressed as M $FeSO_4$ equivalents per gram of extracted compounds.

12.2.6 QUANTITATIVE PHYTOCHEMICAL ANALYSIS

12.2.6.1 Determination of Total Phenol Content (TPC)

The amount of TPC, in methanol extract and their fractions, was determined by Folin-Ciocalteu's reagent method [50]. The extract (0.5 ml) and 0.1 ml Folin-Ciocalteu's reagent (0.5 N) were mixed, and the mixture was incubated at room temperature for 15 min. Then, 2.5 ml saturated sodium carbonate solution was added and further incubated for 30 min at room temperature, and the absorbance was measured at 760 nm using a UV-VIS Spectrophotometer (Shimadzu, Japan), against a blank sample. The calibration curve was made by preparing gallic acid (10 to 100 μg ml^{-1}) solution in distilled water [66]. Total phenol content is expressed in terms of gallic acid equivalent (mg g⁻¹ of extracted compounds).

12.2.6.2   Determination of Flavonoid Content (TFC)

The amount of TFC, in methanol extract and their fractions, was determination by aluminum chloride colorimetric method [19]. The reaction mixture (3.0 ml) consisted 1.0 ml extract (1 mg ml⁻¹), 1.0 ml methanol, 0.5 ml AlCl₃ (1.2%), and 0.5 ml potassium acetate (120 mM) and was incubated at room temperature for 30 min. The absorbance of all samples was measured at 415 nm using a UV-VIS Spectrophotometer (Shimadzu, Japan), against a blank sample. The calibration curve was made by preparing quercetin (5 to 60 µg ml⁻¹) solution in methanol [66]. The flavonoid content is expressed in terms of quercetin equivalent (mg g⁻¹ of extracted compounds).

12.2.7   PHENOLIC COMPOSITION BY REVERSED PHASE (RP) – HPLC

12.2.7.1   Sample Preparation and Separation Conditions

Ascorbic acid, Gallic acid, Quercetin and Rutin (10 mg each) were weighed accurately and transferred to separate 100 ml volumetric flasks. The standards (10 mg each) were dissolved in 100 ml methanol to prepare standard stock solution of 100 µg ml⁻¹ [75].

12.2.7.2   Separation Conditions

Phenolic compounds were analyzed using a Shimadzu HPLC system (LC-20A series, Tokyo, Japan), consisting of a binary pump and a diode-array detector (DAD), and equipped with a Shim-pack RP-C18 column (5 µm, 250 mm × 4.6 mm) (Shimadzu Co., Japan). Sample preparation for HPLC analysis was as follows: the extract was centrifuged at 10–000 rpm for 15 min, filtered using a Millipore filter (0.22 µm nylon membrane) at room temperature, and then injected into HPLC for analysis. Phenolic compounds in the samples were analyzed at 35°C with the following gradient elution program (solution A, 0.1% formic acid, and solution B, 100% methanol): 0–10 min, 0%–10% B; 10–25 min, 10%–20% B; 25–35 min, 20%–23% B; 35–45 min, 23%–28% B; 45–60 min, 28%–35% B; 60–75 min, 35%–50% B; 75–80 min, 50%–55% B; 80–85 min, 55%–75% B; 85–90 min, 75% B. The flow rate was 0.8 ml min^{-1} and the injection volume was 20 μl. Detection was monitored at 280 nm. The identification of phenolics was achieved through the following: comparison of the retention times of standard solutions with the retention times of compounds in the samples.

12.2.8   SPECTRAL ANALYSIS

12.2.8.1   Ultraviolet (UV) Visible Absorption

The RSM and its fractions were analyzed in UV-Visible range between 200–800 nm using UV-Visible Spectrophotometer (UV-1800, Shimadzu). This method is useful for analyzing organic compounds viz. ketones, dienes, etc.

12.2.8.2   Fourier Transform Infrared (FTIR) Spectroscopy

Fourier transform infrared (FTIR) spectroscopy is one of the powerful analytical techniques which offer the possibility of chemical identification. The technique is based on the simple fact that chemical substance shows selective absorption in infrared region. After absorption of IR radiations, the molecules vibrate, giving rise to absorption spectrum. It is an excellent method for the qualitative analysis because except optical isomers, the spectrum of compound is unique. It is most useful for the identification of purity and gross structural details. This method is useful in the field of natural products, forensic chemistry and in industrial analysis of competitive products. The IR spectra of the RSM and their fractions were scanned on FT-IR-Shimadzu-8400 over the frequency range from 4000–400 cm^{-1}. FTIR spectral analysis was done at Department of Pharmaceutical Sciences, Saurashtra University, Rajkot, India.

12.2.8.3   Mass Spectroscopy (MS)

It is an analytical technique that measures the mass to charge ratio of charged particles. Mass spectroscopy determination was carried out in Department of Chemistry, Saurashtra University, Rajkot, India.

12.2.8.4   HPTLC Finger Printing

Authors used CAMAG HPTLC system equipped with Linomat 5 applicator, TLC scanner 3, reprostar 3 with 12 bit CCD camera for photo documentation, controlled by WinCATS-4 software. All the solvents used for HPTLC analysis were obtained from MERCK. A total of 100 mg of RSM and its fractions were dissolved in 5 ml of methanol and used for HPTLC analysis as test solution. The samples (10 µl) were spotted in the bands of width 8 mm with a Camag micro liter syringe on precoated silica gel glass plate 60F-254. The sample-loaded plate was kept in TLC twin trough developing chamber (after saturated with Solvent vapor) with respective mobile phase and the plate was developed up to 83 mm in the respective mobile phase. The Toluene:Ethyl acetate:Glacial acetic acid (8:2:0.1) was employed as mobile phase for sample. Linear ascending development was carried out in 20 cm × 10 cm twin trough glass chamber saturated with the mobile phase and the chromate plate development with the same mobile phase to get good resolution of phytochemical contents. The optimized chamber saturation time for mobile phase was 30 min at room temperature.

The developed plate was dried by hot air to evaporate solvents from the plate. The plate was photo-documented at UV 366 nm and white light using photo documentation chamber. Finally, the plate was fixed in scanner stage and scanning was done at 366 nm. The plate was kept in photo-documentation chamber and captured the images under white light, UV light at 254 and 366 nm. Densitometric scanning was performed on Camag TLC scanner III and operated by CATS software (V 3.15, Camag). The HPTLC fingerprinting of the RSM and their fractions were carried out at Department of Pharmaceutical Science, Saurashtra University, Rajkot, India.

12.2.9 ANTIMICROBIAL STUDY

12.2.9.1 Test Microorganism and Growth Conditions

The microorganisms studied were obtained from National Chemical Laboratory (NCL), Pune, India. The microorganisms were maintained at 4°C. They consisted of 8 Gram positive bacteria: [(*Bacillus cereus* ATCC11778 (*BC*), *Corynebacterium rubrum* ATCC14898 (*CR*), *Listeria monocytogenes* ATCC19112 (*LM*), *Micrococcus flavus* ATCC10240 (*MF*), *Staphylococcus albus* NCIM2178 (*SAl*), *S. aureus* ATCC25923 (*SA1*) and *S. aureus* ATCC29737 (*SA2*), *S. epidermidis* ATCC12228 (*SE*)]; 13 Gram negative bacteria in which 8 bacteria belongs to Enterobacteriaceae family [*Citrobacter frundii* NCIM2489 (*CF*), *Enterobacter aerogenes* ATCC13048 (*EA*), *Escherichia coli* NCIM2931 (*EC*), *Klebsiella aerogenes* NCIM2098 (*KA*), *K. pneumoniae* NCIM2719 (*KP*), *Proteus mirabilis* NCIM2241 (*PMi*), *P. morganii* NCIM2040 (*PMo*), *Salmonella typhimurium* ATCC23564 (*ST*)] and 5 bacteria belongs to Pseudomonadaceae family [*Pseudomonas aeruginosa* ATCC27853 (*PA*), *P. pictorum* NCIB9152 (*PPi*), *P. putida* NCIM2872 (*PPu*), *P. syrigae* NCIM5102 (*PS*), *P. testosterone* NCIM5098 (*PT*)]. The bacteria was maintained on nutrient agar medium (Hi Media, India) while *L. monocytogenes* and *E. coli* were maintained on Brain heart infusion agar and Luria medium (Hi Media, India), respectively at 4°C and subcultured before use. The microorganisms studied are clinically important ones causing several infections, food borne diseases, spoilages, skin infection and it is essential to overcome them through some active therapeutic agents.

12.2.9.2 Preparation of the Extracts and/or Antibiotics for MIC and MBC Study

The extract and fractions dissolved in 100% of DMSO were first diluted to highest concentration (1250 μg ml^{-1}) to be tested, and then serial two-fold dilution was made in a concentration range from (0.605 to 1250 μg ml^{-1}). Chloramphenicol and ceftazidime were used as a positive control (0.0156 to 32 μg ml^{-1}).

12.2.9.3   Preparation of Bacterial Inocula for MIC and MBC Study

The inocula of the test organisms were prepared using the colony suspension method [28, 63, 74]. Colonies picked from 24 h old cultures grown on nutrient agar were used to make suspension of the test organisms in saline solution to give an optical density of approximately 0.1 at 600 nm. The suspension was then diluted 1:100 by transfer of 100 µl of the bacterial suspension to 9.9 ml of sterile nutrient broth before use to yield 6×10^5 CFU ml^{-1}.

12.2.9.4   Determination of the Minimum Inhibitory Concentrations (MIC)

The MIC was determined by the micro well dilution method [26, 63] with some modifications. This test was performed in sterile flat bottom 96 well micro test plates (Tarsons Products Pvt. Ltd.). 150 µl volume of Mueller-Hinton broth was dispensing into each wells and 20 µl of varying concentrations of the extract was added in decreasing order along with 30 µl of the test organism suspension. The final volume in each well was 200 µl (150 µl Mueller-Hinton broth, 30 µl of the test organism suspension, and 20 µl extract/antibiotic). Three control wells were maintained for each test batch. The positive control (CH/CF, MHB and test organism) and sterility control (MHB and DMSO) and organism control (MHB, test organism and DMSO). Plates were then incubated at 37°C for 24 hr. Experiments were carried out in duplicate. After incubation, 40 µl of INT solution (200 µg ml^{-1}) dissolved in sterile distilled water was added to each well [29]. The plates were incubated for further 30 min, and estimated visually for any change in color to pink indicating reduction of the dye due to bacterial growth. The highest dilution (lowest concentration) that remained clear corresponded to the MIC.

12.2.9.5   Determination of the Minimum Bactericidal Concentration (MBC)

MBC was determined from all wells showing no growth as well as from the lowest concentration showing growth in the MIC assay for all the samples.

Bacterial cells from the MIC test plate were subcultured on freshly prepared solid nutrient agar by making streaks on the surface of the agar. The plates were incubated at 37°C for 24 h overnight. Plates that did not show growth was considered to be the MBC for the extract used [2, 63]. The experiment was carried out in duplicate.

12.2.9.6 Determination of MIC Index

The MIC index (MBC/MIC) was calculated for extracts to determine whether an extract had bactericidal (MBC/MIC ≤4) or bacteriostatic (>4 MBC/MIC <32) effect on growth of bacteria [63, 79, 84].

12.2.10 STATISTICAL ANALYSIS

Each sample was analyzed individually in triplicate and the results are expressed as the mean value ($n = 3$) ± Standard Error of Mean (S.E.M.). The correlation coefficients between studies parameters were demonstrated by linear regression analysis.

12.3 RESULTS AND DISCUSSION

12.3.1 ANTIOXIDANT ACTIVITY

Since the antioxidant capacity of food is determined by a mixture of different antioxidants with different action mechanisms, among which synergistic interactions, it is necessary to combine more than one method in order to determine *in vitro* antioxidant capacity of foodstuffs [56, 60]. Therefore, six oxidant systems were selected in the present work.

12.3.1.1 DPPH Free Radical Scavenging Activity

DPPH radical is a stable organic free radical with an absorption band at 517 nm. It loses this absorption on accepting an electron or a free radical species, which results in a visually noticeable discoloration from purple to yellow. It can accommodate many samples in a short period and is sensitive enough to detect active ingredients at low concentrations [31]. In this study, RSM and its fractions showed DPPH scavenging activity in a concentration dependent manner whose profile varied among the different extracts (Figure 12.1a). The RSM showed highest DPPH radical scavenging activity as compared to fractions. The EC_{50} values were calculated and tabulated in Table 12.1 to facilitate the comparison of the free radical scavenging activities of RSM and its fractions. The EC_{50} value is defined as the concentration of the sample necessary to cause 50% inhibition, which is obtained by interpolation from linear regression analysis [89]. A lower EC_{50} value is associated with a higher radical scavenging

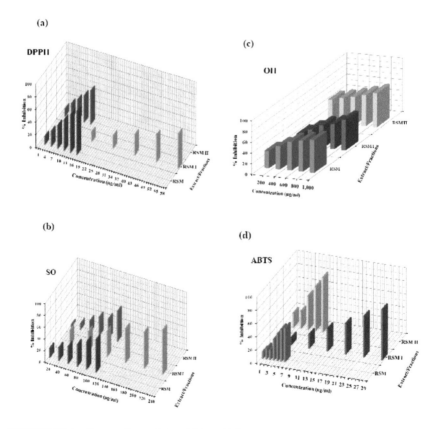

FIGURE 12.1 Scavenging activity: (a) DPPH (b) SO (c) OH and (d) ABTS of RSM and its fractions.

TABLE 12.1 EC_{50} Values of DPPH, SO, OH and ABTS Radical Cation Scavenging Activity and FRAP Values of RSM and Its Fractions

Extract/Fractions		EC_{50} Values ($\mu g\ ml^{-1}$)				FRAP ($M\ g^{-1}$)
		DPPH	SO	OH	ABTS	
RSM		12.33	106	560	6.2	55.95 ± 0.15
RSM I		55	140	880	11	41.12 ± 0.16
RSM II		13.8	118	580	6	48.69 ± 0.35
Standard	Ascorbic acid	11.4	NA	NA	6.5	NA
	Gallic acid	NA	185	140	NA	

activity. The EC_{50} value of RSM was 12.33 $\mu g\ ml^{-1}$ showed best and maximum DPPH radical scavenging activity which was almost similar to that of standard ascorbic acid ($EC_{50} = 11.4\ \mu g\ ml^{-1}$) followed by RSM II and RSM I and their EC_{50} values ware 13.8 and 55 $\mu g\ ml^{-1}$, respectively (Table 12.1). Christudas and Savarimuthu [22] reported that methanol extract of *S. cochinchinensis* exhibited a significant dose dependent inhibition of DPPH activity, with a 50% inhibition (EC_{50}) at a concentration of $620.30 \pm 0.14\ \mu g\ ml^{-1}$.

12.3.1.2 Superoxide Anion Radical Scavenging Activity

The superoxide anion radicals are derived in PMS-NADH-NBT system, where the decrease in absorbance at 560 nm with antioxidants [9, 41, 87] indicates the consumption of superoxide anion in the reaction mixture, thereby exhibiting a dose dependent increase in superoxide scavenging activity. The results obtained in the present work show that the choice of solvents of extraction cannot be arbitrary being, therefore, a very important step when searching components with capacity for scavenging free radicals. In addition, one solvent may be the most adequate for one plant species and not for another one. Figure 12.1b shows the percent inhibition of superoxide radical anion formed as determined by NBT method for RSM and its fractions. RSM, RSM I and RSM II had very good scavenging activity as compared to the standard gallic acid (Table 12.1). The EC_{50} value of RSM, RSM I and RSM II were 106, 140 and 118 $\mu g\ ml^{-1}$, respectively. The RSM exerted significantly higher activity in superoxide anion free radical assays, which can be explained by the fact that some of the compounds present in the extracts are good superoxide free radical scavengers.

12.3.1.3   Hydroxyl Radical Scavenging Activity

Hydroxyl radicals (OH˙) are generated by a mixture of an ascorbic acid, H_2O_2 and $FeCl_3$-EDTA. Ascorbic acid increases the rate of OH˙ generation by reducing iron and maintaining a supply of Fe^{2+}-EDTA. The generated OH˙ radical can degrade deoxyribose into a series of fragments that react with TBA upon heating and at low pH. This reaction forms a pink color that absorbs 532 nm light. When added to the reaction mixture, the test compounds compete with deoxyribose for OH˙ radicals and inhibit deoxyribose degradation. If an antioxidant is able to scavenge an OH˙ present in the system (e.g., polyphenols), it will compete with deoxyribose for OH˙ radicals and will inhibit the deoxyribose degradation [7, 36, 71]. RSM and its fractions showed a varied level of hydroxyl radical scavenging activity (Figure 12.1c). RSM showed better scavenging activity as compared to RSM I and RSM II (Table 12.1).

12.3.1.4   ABTS Radical Cation Scavenging Activity

The ABTS˙+ assay, which measures the ability of compounds to transfer labile H- atoms to radicals, is a common method for evaluating antioxidant activity. The abstraction of hydrogen by this stable free radical leads to bleaching of the absorption maxima at 743 nm and can easily be monitored spectrophotometrically [81]. The results are based on the ability of an antioxidant to decolorize the ABTS˙+. ABTS activity was quantified in terms of percentage inhibition of the ABTS˙+ by antioxidants in each sample. There was a significant variation in the percentage inhibition of the RSM and its fractions (Figure 12.1d). EC_{50} values of RSM II and RSM were 6 and 6.2 µg ml^{-1}, respectively and which were better than standard ascorbic acid used (Table 12.1). RSM and its fractions displayed a greater ABTS˙+ scavenging capacity than that of standard ascorbic acid (Table 12.1).

12.3.1.5   Reducing Capacity Assessment

The reducing power (RP) of a compound serves as a significant indicator of its antioxidant activity. Reducing power is a potent antioxidant defense mechanism. The two mechanisms that are available to affect this property are electron transfer and hydrogen atom transfer [1, 6]. The reduction of the Fe^{3+}/ferricyanide complex to the ferrous form (Fe^{2+}) due to antioxidants was monitored at 700 nm [80]. Figure 12.2a compares antioxidants based on their

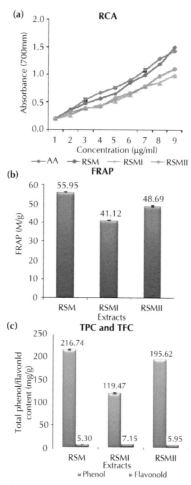

FIGURE 12.2   Antioxidant activity based on: (a) RCA (b) FRAP (c) TPC and TFC of RSM and its fractions.

ability to reduce ferric (Fe^{3+}) to ferrous (Fe^{2+}) ion through the donation of an electron, with the resulting ferrous ion (Fe^{2+}) formation. The reducing powers of the RSM and its fractions were excellent and increased steadily with the increase in concentration and showed some degree of electron-donating capacity in a linear concentration – dependent manner (Figure 12.2a). The reducing power obtained from RSM was significantly higher than the fractions. In Figure 2a, the reducing power followed the order: AA > RSM > RSM II > RSM I. The reducing power of a bioactive substance is closely related to its antioxidant ability [16, 54]. Considering the effective reducing power of the samples tested, the antioxidant compounds of all the extracts should function as good electron and hydrogen-atom donors and therefore be able to converting free radicals and reactive oxygen species to more stable products and terminating the free-radical chain reaction.

12.3.1.6 Ferric Reducing Antioxidant Power (FRAP)

FRAP assay, a redox-linked reaction which involves a single electron transfer mechanism, directly measures the antioxidants or reductants in a sample [45, 87]. It is a powerful tool, quick, simple and reproducible test, linearly related to the molar concentration of the sample antioxidants [24, 39]. FRAP method measures the ability of a compound to reduce Fe^{3+} to Fe^{2+} that depends on the reduction of a ferric tripyridyltriazine (Fe^{3+}-TPTZ) complex to a ferrous tripyridyltriazine (Fe^{2+}-TPTZ) by a reducing agent at low pH. The generated ferrous ion is a well-known pro-oxidant, due to its reactivity with H_2O_2 to produce OH·, the most *in vivo* harmful free radical. FRAP values as a measure of the antioxidant activity is shown in Table 12.1. The RSM possessed the highest FRAP value (55.95 M g^{-1}) followed by RSM II and RSM I (Figure 12.2b).

12.3.1.7 Total Phenol and Flavonoid Content

The main classes of natural antioxidant compounds in nature are flavonoids and phenolic acids in free or complex forms. These compounds have been identified and quantified in several fruits and vegetables. The TPC was more than the TFC in RSM and its fractions. RSM showed more TPC compared to RSM I and RSM II (Figure 12.2c).

Maximum flavonoid content was in RSM I (7.15 mg g^{-1}). Phenolics and flavonoids are important compounds responsible for antioxidant activity; however their amount and structure varied in different herbs when extracted with different solvents [21, 88]. Maximum phenol content was in methanol extract. Similar result was found by Ercetin et al. [27] richest total phenol (118.18 ± 10.29 mg g^{-1} extract) and flavonoid (74.14 ± 3.09 mg g^{-1} extract) contents were found in the methanol extract of flower of *C. arvensis*. While Thondre et al. [86] reported that acetone was superior to ethanol and acidified methanol extracts in extracting total phenolics from Barley β-glucan. Chanda and Kaneria [15] reported maximum total phenol content in acetone extract of *S. cumini*; the contradictory results may be because of the different extraction methods used by researchers. Liu and Yao [44] and Moyo et al. [52] suggested that difference in total phenolic content of the extracts might be attributed to the polarities of the solvents used.

12.3.1.8 Correlation of the Different Antioxidant Activities With TPC and TFC

Phenol compounds have been reported to be responsible for the antioxidant activity of plants [14, 70]. However, as mentioned by other authors [3, 12, 60, 64], this relationship is not always present or depends on the method used for evaluating the antioxidant ability. The correlation between total phenolic concentration and antioxidant activity has been widely studied in different foodstuffs, such as fruits and vegetables [33, 57, 82]. In this study, there was a distinct correlation between studied parameters (TPC, TFC and antioxidant activities) in RSM and its fractions. This correlation was demonstrated by linear regression analysis (Table 12.2). The correlations of the TPC and TFC against the different activities were satisfactory. Correlations are expressed through the following correlations coefficients: TPC versus DPPH, R = 0.983; TPC versus OH, R = 0.988; TPC versus SO, R = 0.989; TPC versus ABTS, R = 0.970; TPC versus RCA, R = 0.813; TPC versus FRAP, R = 0.954, while TFC versus DPPH, R = 0.950; TFC versus OH, R = 0.941; TFC versus SO, R = 0.800; TFC versus ABTS, R = 0.967; TFC versus RCA, R = 0.439; TFC versus FRAP, R = 0.697 (Table 12.2).

TABLE 12.2 Correlation Between TPC and TFC with Antioxidant Activities of RSM and Its Fractions

Correlation between	Linear regression equation	Correlation	
		R	Average
TPC vs. DPPH	$y = -0.466x + 109.6$	0.983	
TPC vs. SO	$y = -0.333x + 180.4$	0.989	
TPC vs. OH	$y = -3.463x + 1287.0$	0.988	0.950
TPC vs. ABTS	$y = -0.053x + 17.25$	0.970	
TPC vs. RCA	$y = 0.004x + 0.444$	0.813	
TPC vs. FRAP	$y = 0.138x + 24.07$	0.954	
TFC vs. DPPH	$y = 37.28x - 214.0$	0.950	
TFC vs. SO	$y = 22.36x - 23.28$	0.8	
TFC vs. OH	$y = 273.5x - 1095.0$	0.941	0.799
TFC vs. ABTS	$y = 4.442x - 20.99$	0.967	
TFC vs. RCA	$y = -0.191x + 2.434$	0.439	
TFC vs. FRAP	$y = -8.381x + 102.0$	0.697	

TPC: Total phenol content; TFC: Total flavonoid content; DPPH: DPPH free radical scavenging activity; OH: Hydroxyl radical scavenging activity; SO: Superoxide anion radical scavenging activity; ABTS: ABTS radical cation scavenging activity; RCA: Reducing capacity assessment; FRAP: Ferric reducing antioxidant power.

Different reports are found from the literature, whereby some authors suggested correlation between all these parameters [23, 37, 91], while other found no such relationship [82]. The results also revealed that the higher and/or lower values of antioxidant activities (evaluated using different assays) were detected from different plant parts and were not restricted towards certain part. The antioxidant activity from all assays which was highly correlated with total flavonoid content was DPPH and ABTS assay. Similar results were also reported by Lou et al. [46]. Al-Laith [4] reported that only FRAP was strongly correlated with total flavonoids, evaluated using different assays, e.g., DPPH and ABTS, not with total phenolics. A significant and linear relationship existed between the antioxidant activity and TPC and TFC of *M. indica* extracted thus indicating that phenolic compounds in the extracts are major contributed significantly to their antioxidant activity.

12.3.2 PRIMARY IDENTIFICATION OF PHENOLIC CONSTITUENTS BY RP-HPLC-DAD

Different phenolic compounds normally have specific chromatographic behavior (retention time, t_R) and spectral characteristics (λ_{max} and spectral shapes). Because of the diversity and complexity of natural mixtures of phenolic compounds in the extracts, it is difficult to characterize every compound and elucidate its structure. Therefore, only a preliminary identification of the major phenolic compounds was carried out in the present study. The different classes of phenolic compounds exhibit maximum absorbance at different wavelengths [34]. Peaks identification in the plant extract was determined by comparing their

TABLE 12.3 Retention Time (t_R) of Standard Phenolic Compounds Used

Peak no.	Retention time, tR (min)	Phenolic compounds
1	4.213	Ascorbic acid
2	10.946	Gallic acid
3	71.074	Quercetin
4	62.04	Rutin

chromatographic retention time; and the spectra obtained in the chromatographic peak with those from standards and their retention times are shown in Table 12.3. The four phenolic standards [gallic acid, ascorbic acid (phenolic acids), rutin, quercetin (flavonoids)] used in this study were eluted with t_R 10.946, 4.213, 62.04 and 71.074 min, respectively. Authors analyzed the major phenolic compounds of RSM, RSM I, RSM II using RP-HPLC with DAD by comparison with authentic phenolic standards and related published data [32, 48, 64, 78]. The studied four phenolic compounds were detected in crude RSM and the fractions obtained.

12.3.3   SPECTRAL ELUCIDATION

12.3.3.1   Ultraviolet Visible Absorption (UV)

UV spectroscopy method is very much useful for identification of unsaturated bonds present in a plant component which can be distinguished between conjugated and nonconjugated system. Using the principle of absorption maxima, the structure of compounds can be deduced [51]. The UV spectrum of RSM and its fractions are shown in Figure 12.3. The UV spectrum of RSM showed absorption maxima at 279, 216 and 242. The RSM I showed absorption maxima at 278 and 242 nm; while RSM II showed absorption maxima at 279, 218–242.5 and 262 nm.

12.3.3.2   FTIR Spectroscopy

FTIR spectra involve the energy transition between different vibrational energy levels. Mainly this can be used to find the functional group present in a pure or mixture of compounds because IR gives a strong absorption pattern at a particular frequency for a particular functional group. Sample are taken in KBr crystals and solids are grinded with KBr and pellets are made and spectra taken. IR spectrum was recorded in FT-IR-Shimadzu-8400 by KBr pellet method. FTIR spectra of RSM and their fractions are shown in Figure 12.4. The mid-infrared, approximately 4000–400 cm^{-1} (2.5–25 μm) was used to study the fundamental vibrations and associated rotational vibration spectrum. Interpretation of RSM and their fractions are given in Table 12.3.

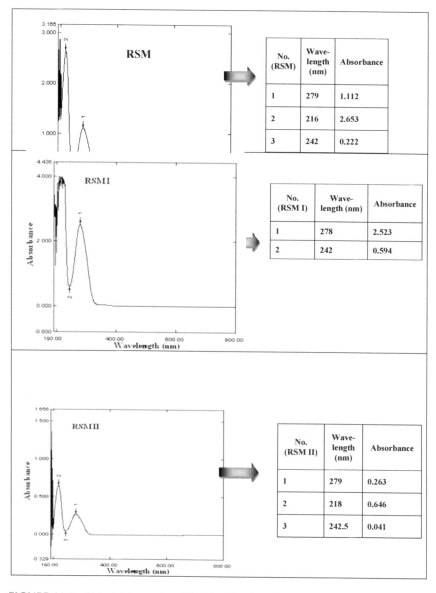

FIGURE 12.3   UV-visible spectra of RSM and its fractions.

12.3.3.3   Mass Spectra (MS)

By mass spectroscopy the molecular mass of a compound and its elemental composition can be easily determined, Further this method involves very little amount of the test sample, which will give molecular weight accurately.

12.3.3.4   HPTLC Fingerprinting

High Performance Thin Layer Chromatography (HPTLC) technique is most simple and fastest separation technique available today which gives better precision and accuracy with extreme flexibility for various steps [68]. The results showing number of peaks, maximum R_f value and total % area are given in Table 12.5 and the spectral range analyzed between 200 and 800 nm. The HPTLC fingerprinting of RSM and it fractions are shown in Figure 12.5. The RSM showed 6 peaks and the maximum percentage area covered was by peak No. 1, 2 and 3 (R_f value 0.00, 0.05, and 0.13, respectively) (Figure 12.5). The RSM I showed 5 peaks and the maximum percentage area covered was by peak No. 1, 2 and 3 (R_f value 0.01, 0.04 and 0.14, respectively) (Figure 12.5). The RSM II showed 5 peaks and the maximum percentage area covered was by peak No. 1, 4 and 12 (R_f value 0.02, 0.29 and 0.78, respectively) (Figure 12.5). HPTLC technique may be useful for both the identification and the quality-evaluation of preparations containing RSM.

12.3.4   ANTIMICROBIAL STUDY

MIC refers to the lowest concentration of the antimicrobial agent which is required for the inhibition of visible growth of the tested isolates [61, 77, 90]. MIC values were calculated using INT dye on a 96-well microtiter plate. The MBC is interpreted as the lowest concentration that can completely remove the microorganisms. In the present study, The MIC and MBC of the RSM, RSM I, RSM II, and standard antibiotics were tested for its antimicrobial activity at various concentrations against various Gram positive and Gram negative bacterial strains. The values are presented in Table 12.6.

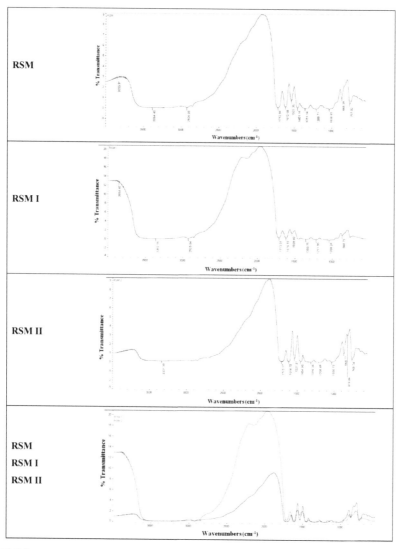

FIGURE 12.4   FTIR spectrum of RSM and its fractions.

The extracts exhibited concentration dependent inhibition of growth. For the Gram positive bacteria strains, MIC and MBC varied from 39 to > 1250 μg ml⁻¹ and 156 to >1250 μg ml⁻¹, respectively. *M. flavus* was found most susceptible bacterial pathogens to the RSM (MIC: 19.5 μg ml⁻¹).

TABLE 12.4   Interpretation of FTIR of RSM and Its Fractions

Wave numbers (cm⁻¹)	Main attributes	Functional group
RSM		
761, 868	Bending/strong	Alkene (=C-H)
1030	Stretch/strong	Alkyl Halide (C-F)
1200	Stretch/two bands or more	Ester (C-O)
1351, 1537	Stretch/strong, two bands	Nitro (N-O)
1451, 1612	Stretch/medium-weak, multiple bands	Aromatic (C=C)
1712	Stretch/strong	Acid (C=O)
2926	Stretch/strong	Alkane (C-H)
3384	Stretch/medium (primary amines have two bands; secondary have one band, often very weak)	Amine (N-H)
RSM I		
868	Bending/strong	Alkene (=C-H)
1038	Stretch/strong	Alkyl Halide (C-F)
1211	Stretch/strong	Acid (C-O)
1350, 1539	Stretch/strong, two bands	Nitro (N-O)
1616	Bending	Amide (N-H)
1713	Stretch/strong	Acid (C=O)
RSM II		
760, 819, 868	Bending/strong	Alkene (=C-H)
1030	Stretch/strong	Alkyl Halide (C-F)
1208	Stretch/medium-weak	Amine (C-N)
1316, 1537	Stretch/strong, two bands	Nitro (N-O)
1454	Stretch/medium-weak, multiple bands	Aromatic (C=C)
1614	Bending	Amine (N-H)
1713	Stretch/strong	Acid (C=O)
3331	Stretch/medium (primary amines have two bands; secondary have one band, often very weak)	Amine (N-H)

For the Gram negative bacterial strains, MIC and MBC ranged from 9.75 to 1250 μg ml⁻¹ and 78 to >1250 μg ml⁻¹, respectively. *P. morganii* was found most susceptible bacterial pathogens to the RSM II (MIC:

TABLE 12.5   HPTLC Spectral Analysis of RSM and Its Fractions

RSM			RSM I			RSM II		
Peak	Max R_f	Area (%)	Peak	Max R_f	Area (%)	Peak	Max R_f	Area (%)
1	0.00	41.56	1	0.01	36.77	1	0.01	35.50
2	0.05	26.39	2	0.04	31.84	2	0.06	26.28
3	0.13	28.47	3	0.14	29.86	3	0.15	35.46
4	0.28	0.98	4	0.29	0.72	4	0.29	1.10
5	0.35	1.82	5	0.35	0.81	5	0.37	1.67
6	0.86	0.77	—	—	—	—	—	—

9.75 µg ml^{-1}). For the Gram negative bacterial strains (*Pseudomonas* spp.), MIC and MBC ranged from 9.75 to >1250 µg ml^{-1} and 156 to > 1250 µg/ml, respectively. *P. pictorum* was found most susceptible bacterial pathogens to the RSM II (MIC: 19.5 µg ml^{-1}). For the standard antibiotics (CH and CF), MIC and MBC ranged from 1 to 32 µg ml^{-1} and 8 to >32 µg ml^{-1}, respectively.

Whilst Gibbons [30] suggests that isolated phytochemicals should have MIC < 1000 µg ml^{-1}. However, Madikizela et al. [47] also found antibacterial MIC values equal to or less than 1000 µg ml^{-1} for crude extracts that were considered active [53]. This study also supports these agreements. In this study RSM showed MIC values <1000 µg ml^{-1} except against some bacteria. China et al. [20] found that the petroleum ether extract of *S. grandiflora* flower showed MICs values ranging from 13–250 µg ml^{-1} against the tested pathogenic strains.

MBC/MIC ratio of value ≤ 4 is indicative of a bactericidal nature of the test sample [84]. In this study, bactericidal effect was shown by RSM against *S. epidermidis,* RSM I against two bacteria (*P. putida* and *P. morgani*), RSM II against three bacteria (*S. aureus 2, S. epidermidis* and *M. flavus*), the remaining showed bacteriostatic effects.

TABLE 12.6 MIC and MBC of Standard Antibiotics (Chloramphenicol and Ceftazidime), RSM and Its Fractions

Microorganisms		RSM		RSM I		RSM II		CH		CF	
		MIC	MBC	MIC	MBC	MIC	MBC	MIC	MBC	MIC	MBC
Gram positive	SA1	625	>1250	312	>1250	312	>1250	2	32	4	32
	SA2	1250	>1250	1250	>1250	156*	625*	2	>32	4	>32
	SE	312*	1250*	625	>1250	312*	1250*	4	>32	16	>32
	MF	19.5	156	78	625	39*	156*	4	16	16	32
	SAl	1250	>1250	625	>1250	625	>1250	4	32	2	16
	CR	1250	>1250	1250	>1250	1250	>1250	8	16	16	32
	LM	1250	>1250	156	1250	39	312	4	16	32	>32
	BC	1250	>1250	>1250	>1250	1250	>1250	4	>32	32	>32
	PMi	1250	>1250	39	1250	39	156	8	>32	32	>32
	ST	1250	>1250	1250	>1250	625	>1250	4	>32	1	8
	KP	1250	>1250	625	>1250	156	1250	4	32	4	>32
Gram negative	EA	1250	>1250	1250	>1250	625	>1250	8	>32	4	>32
	PMo	19.5	312	39*	156*	9.75	78	4	>32	16	>32
	CF	1250	>1250	1250	>1250	156	1250	8	>32	4	>32
	EC	1250	>1250	625	>1250	312	>1250	8	>32	4	>32
	KA	1250	>1250	1250	>1250	1250	>1250	2	>32	32	>32
Pseudomonas spp.	PA	156	1250	39	625	19.5	156	8	>32	32	>32
	PPu	1250	>1250	39*	156*	78	625	16	>32	32	>32
	PPi	1250	>1250	39	1250	9.75	156	32	>32	8	>32
	PS	625	>1250	1250	>1250	1250	>1250	16	>32	32	>32
	PT	625	>1250	1250	>1250	1250	>1250	32	>32	8	32

MIC and MBC: Values are expressed in μg ml^{-1}.

*: MBC/MIC ratio of value $\leq$ 4 is indicative of a bactericidal nature.

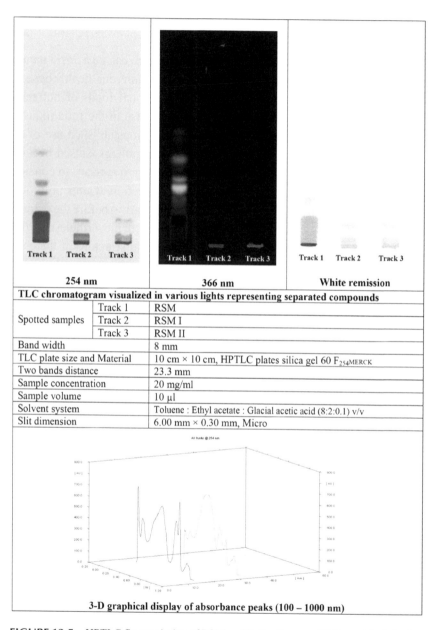

	Track 1	RSM
Spotted samples	Track 2	RSM I
	Track 3	RSM II
Band width		8 mm
TLC plate size and Material		10 cm × 10 cm, HPTLC plates silica gel 60 F$_{254}$MERCK
Two bands distance		23.3 mm
Sample concentration		20 mg/ml
Sample volume		10 µl
Solvent system		Toluene : Ethyl acetate : Glacial acetic acid (8:2:0.1) v/v
Slit dimension		6.00 mm × 0.30 mm, Micro

3-D graphical display of absorbance peaks (100 – 1000 nm)

FIGURE 12.5   HPTLC fingerprinting of RSM and its fractions; and 3-D graphical display of absorbance peaks (100–1000 nm).

12.4 CONCLUSIONS

The overall results indicated that ripe seed of Mango can be a novel natural source of antioxidants and antimicrobics with numerous health benefits that should be considered in the fields of functional foods or nutraceuticals. The extracts from fruit residues hold promise in the food industry as sources of bioactive compounds. In addition, an established use of the fruit residues will also help alleviate pollution problems caused because of the poor disposal of such residues. It is, however, necessary to consider both environmental (waste management and protection against pollution) aspects and economical aspects (extraction profitability) before the extracts from fruit residues could be commercially exploited. Nevertheless further studies are warranted for enlightening the chemical structure of the components present in ripe seed extracts responsible for biological potentiality.

12.5 SUMMARY

The present work was undertaken to examine the utilization of mango ripe seed, a waste material of the food industry, as a source of natural antioxidants and antimicrobics. Validation of the isolated fractions was determined by performing antioxidant and antimicrobial assays. The antioxidant activity was evaluated using 6 different antioxidant assays. Total phenol and flavonoid content was measured while RP-HPLC-DAD, UV, FTIR, MS and HPTLC techniques were also used for the detection of bioactive compounds. The evaluation of antimicrobial activity of the extracts was performed using a microwell dilution method against 21 disease causing bacteria. Methanol extract had the strongest and better antioxidant activity as compared to that of standards. The MIC and MBC were in the range of 9.75 to >1250 μg ml^{-1} and 156 to 1250 μg/ml, respectively. This study demonstrated that mango ripe seeds can serve as potential biological sources of natural antioxidants and antimicrobics for use in the preparation of dietary supplements or nutraceutical, food ingredients, pharmaceutical and cosmetic products.

KEYWORDS

- ABTS
- antimicrobics
- antioxidants
- antioxidants
- bactericidal
- bacteriostatic
- characterization
- cold percolation method
- DPPH
- fractionation
- FRAP
- FTIR
- HPTLC
- isolation
- kesar var.
- *mangifera indica*
- mango
- MBC
- methanol
- MIC
- MIC index
- MS
- OH
- pathogenic microbes
- purification

- RCA
- RP-HPLC-DAD
- seed
- SO
- TFC
- TPC
- UV
- validation

REFERENCES

1. Ademiluyi, A. O., & Oboh, G. (2011). Soybean phenolic-rich extracts inhibit key-enzymes linked to type 2 diabetes (α-amylase and α-glucosidase) and hypertension (angiotensin I converting enzyme) *in vitro*. *Experimental and Toxicologic Pathology, 65*, 305–309.
2. Akinyemi, K. O., Oladapo, O., Okwara, C. E., Ibe, C. C., & Fasure, K. A. (2005). Screening of crude extracts of six medicinal plants used in South-West Nigerian unorthodox medicine for antimethicillin resistant *Staphylococcus aureus* activity. *BMC Complementary and Alternative Medicine, 5*, 6–12.
3. Albanoa, S. M., & Miguelb, M. G. (2011). Biological activities of extracts of plants grown in Portugal. *Industrial Crops and Products, 33*, 338–343.
4. Al-Laith, A. A., Alkhuzai, J., & Freije, A. (2015). Assessment of antioxidant activities of three wild medicinal plants from Bahrain. *Arabian Journal of Chemistry*, http://dx.doi.org/10.1016/j.arabjc.2015.03.004.
5. Athukorala, Y., Kim, K. N., & Jeon, Y. J. (2006). Antiproliferative and antioxidant properties of an enzymatic hydrolysate from brown alga, *Ecklonia cava*. *Food and Chemical Toxicology, 44*, 1065–1074.
6. Babu, D. R., Pandey, M., & Rao, G. N. (2014). Antioxidant and electrochemical properties of cultivated *Pleurotus* spp. and their sporeless/low sporing mutants. *Journal of Food Science and Technology, 51*, 3317–3324.
7. Bekdeser, B., Ozyurek, M., Gucu, K., & Apak, R. (2012). Novel spectroscopic sensor for the hydroxyl radical scavenging activity measurement of biological samples. *Talanta, 99*, 689–696.
8. Benzie, I. F., & Strain, J. J. (1996). The ferric reducing ability of plasma (FRAP) as a measure of "antioxidant power": the FRAP assay. *Analytical Biochemistry, 239*, 70–76.
9. Bora, K. S., & Sharma, A. (2010). *In vitro* antioxidant and free radical scavenging potential of *Medicago sativa* Linn. *Journal of Pharmacy Research, 3*, 1206–1210.

10. Chanda, S., Amrutiya, N., & Rakholiya, K. (2013). Evaluation of antioxidant properties of some Indian vegetable and fruit peels by decoction extraction method. *American Journal of Food Technology, 8*, 173–182.

11. Chanda, S., Baravalia, Y., Kaneria, M., & Rakholiya, K. (2010). Fruit and vegetable peels-strong natural source of antimicrobics. In: *Current Research, Technology and Education Topics in Applied Microbiology and Microbial Biotechnology*, Mendez-Vilas, A. (Ed.). Formatex Research Center, Spain, pp. 444–450.

12. Chanda, S., & Dave, R. (2009). *In vitro* models for antioxidant activity evaluation and some medicinal plants possessing antioxidant properties: An overview. *African Journal of Microbiological Research, 3*, 981–996.

13. Chanda, S., Dave, R., & Kaneria, M. (2011). *In vitro* antioxidant property of some Indian medicinal plants. *Research Journal of Medicinal Plant, 5*, 169–179.

14. Chanda, S., Dudhatra, S., & Kaneria, M. (2010). Antioxidative and antibacterial effects of seeds and fruit rind of nutraceutical plants belonging to the family Fabaceae. *Food and Functions, 1*, 308–315.

15. Chanda, S., & Kaneria, M. (2012). Optimization of conditions for the extraction of antioxidants from leaves of *Syzygium cumini* L. using different solvents. *Food Analytical Methods, 5*, 332–338.

16. Chanda, S., Kaneria, M., & Baravalia, Y. (2012). Antioxidant and antimicrobial properties of various polar solvent extracts of stem and leaves of four *Cassia* species. *African Journal of Biotechnology, 11*, 2490–2503.

17. Chanda, S., Rakholiya, K., Dholakia, K., & Baravalia, Y. (2013). Antimicrobial, antioxidant, and synergistic properties of two nutraceutical plants: *Terminalia catappa* L., & *Colocasia esculenta* L. *Turkish Journal of Biology, 37*, 81–91.

18. Chanda, S., Rakholiya, K., & Parekh, J. (2013). Indian medicinal herb: Antimicrobial efficacy of *Mesua ferrea* L. seed extracted in different solvents against infection causing pathogenic strains. *Journal of Acute Disease, 2*, 277–281.

19. Chang, C. C., Yang, M. H., Wen, H. M., & Chern, J. C. (2002). Estimation of total flavonoid content in *Propolis* by two complementary colorimetric methods. *Journal of Food and Drug Analysis, 10*, 178–182.

20. China, R., Mukherjee, S., Sen, S., Bose, S., Datta, S., Koley, H., Ghosh, S., & Dhar, P. (2012). Antimicrobial activity of *Sesbania grandiflora* flower polyphenol extracts on some pathogenic bacteria and growth stimulatory effect on the probiotic organism *Lactobacillus acidophilus*. *Microbiological Research, 167*, 500–506.

21. Cho, J. Y., Park, S. C., Kim, T. W., Kim, K. S., Song, J. C., Lee, H. M., Sung, H. J., Rhee, M. H., Kim, S. K., Park, H. J., Song, Y. B., Yoo, E. S., & Lee, C. H. (2006). Radical scavenging and antiinflammatory activity of extracts from *Opuntia humifusa* Raf. *Journal of Pharmacy and Pharmacology, 58*, 113–119.

22. Christudas, S., & Savarimuthu, I. (2011). *In vitro* and *in vivo* antioxidant activity of *Symplocos cochinchinensis* S. Moore leaves containing phenolic compounds. *Food and Chemical Toxicology, 49*, 1604–1609.

23. Das, D. K., Dutta, H., & Mahanta, C. L. (2013). Development of a rice starch-based coating with antioxidant and microbe-barrier properties and study of its effect on tomatoes stored at room temperature. *LWT – Food Science and Technology, 50*, 272–278.

24. Daur, I. (2015). Chemical composition of selected Saudi medicinal plants. *Arabian Journal of Chemistry, 8*, 329–332.
25. Dolai, N., Karmakar, I., Kumar, R. B. S., Kar, B., Bala, A., & Haldar, P. K. (2012). Free radical scavenging activity of *Castanopsis indica* in mediating hepatoprotective activity of carbon tetrachloride intoxicated rats. *Asian Pacific Journal of Tropical Biomedicine, 2*, S242–S251.
26. Edziri, H., Ammar, S., Souad, L., Mahjoub, M. A., Mastouri, M., Aouni, M., Mighri, Z., & Verschaeve, L. (2012). *In vitro* evaluation of antimicrobial and antioxidant activities of some Tunisian vegetables. *South African Journal of Botany, 78*, 252–256.
27. Ercetin, T., Senol, F. S., Orhan, I. E., & Toker, G. (2012). Comparative assessment of anti-oxidant and cholinesterase inhibitory properties of the marigold extracts from *Calendula arvensis* L., & *Calendula officinalis* L. *Industrial Crops and Products, 36*, 203–208.
28. European Committee for Antimicrobial Susceptibility Testing (EUCAST) (2003). Determination of minimum inhibitory concentrations (MICs) of antibacterial agents by broth dilution. *Clinical Microbiology and Infection, 9*, 1–7.
29. Frey, F. M., & Meyers, R. (2010). Antibacterial activity of traditional medicinal plants used by Haudenosaunee peoples of New York State. *BMC Complementary and Alternative Medicine, 10*, 64–73.
30. Gibbons, S. (2005). Plants as a source of bacterial resistance modulators and antiin-fective agents. *Phytochemistry Reviews, 4*, 63–78.
31. Hseu, Y., Chang, W., Chen, C., Liao, J., Huang, C., Lu, F., Chia, Y., Hsu, H., Wu, J., & Yang, H. (2008). Antioxidant activities of *Toona sinensis* leaves extracts using different antioxidant models. *Food and Chemical Toxicology, 46*, 105–114.
32. Huang, W., Zhang, H., Liu, W., & Li, C. (2012). Survey of antioxidant capacity and phenolic composition of blueberry, blackberry, and strawberry in Nanjing. *Journal of Zhejiang University-Science B, 13*, 94–102.
33. Jayaprakasha, G. K., Girennavar, B., & Patil, B. S. (2008). Radical scavenging activity of red grapefruits and sour orange fruit extracts in different *in vitro* model systems. *Bioresource Technology, 99*, 4484–4494.
34. Kajdzanoska, M., Petreska, J., & Stefova, M. (2011). Comparison of different extrac-tion solvent mixtures for characterization of phenolic compounds in strawberries. *Journal of Agricultural Food Chemistry, 59*, 5272–5278.
35. Kaneria, M., Baravalia, Y., Vaghasiya, Y., & Chanda, S. (2009). Determination of antibacterial and antioxidant potential of some medicinal plants from Saurashtra region, India. *Indian Journal of Pharmaceutical Sciences, 71*, 406–412.
36. Kaneria, M., & Chanda, S. (2012). Evaluation of antioxidant and antimicrobial prop-erties of *Manilkara zapota* L. (chiku) leaves by sequential soxhlet extraction method. *Asian Pacific Journal of Tropical Biomedicine, 2*, S1526–S1533.
37. Kaneria, M., & Chanda, S. (2013). Evaluation of antioxidant and antimicrobial capacity of *Syzygium cumini* L. leaves extracted sequentially in different solvents. *Journal of Food Biochemistry, 37*, 168–176.
38. Kaneria, M., Kanani, B., & Chanda, S. (2012). Assessment of effect of hydroal-coholic and decoction methods on extraction of antioxidants from selected Indian medicinal plants. *Asian Pacific Journal of Tropical Biomedicine, 2*, 195–202.

39. Kaneria, M. J., Bapodara, M. B., & Chanda, S. V. (2012). Effect of extraction techniques and solvents on antioxidant activity of Pomegranate (*Punica granatum* L.) leaf and stem. *Food Analytical Methods, 5,* 369–404.

40. Kaneria, M. J., & Chanda, S. V. (2013). The effect of sequential fractionation technique on the various efficacies of pomegranate (*Punica granatum* L.). *Food Analytical Methods, 6,* 164–175.

41. Kaneria, M. J., Rakholiya, K. D., & Chanda, S. V. (2014). *In vitro* antimicrobial and antioxidant potency of two nutraceutical plants of Cucurbitaceae. In: *Traditional and Folk Herbal Medicine: Recent Researches–Vol. 2,* Gupta, V. K. (Ed.). Daya Publication House, India, pp. 221–247.

42. Kumar, M. S. Y., Dutta, R., Prasad, D., & Misra, K. (2011). Subcritical water extraction of antioxidant compounds from Sea buckthorn (*Hippophae rhamnoides*) leaves for the comparative evaluation of antioxidant activity. *Food Chemistry, 127,* 1309–1316.

43. Kunchandy, E., & Rao, M. N. A. (1990). Oxygen radical scavenging activity of curcumin. *International Journal of Pharmaceutics, 58,* 237–240.

44. Liu, Q., & Yao, H. (2007). Antioxidant activities of barley seeds extracts. *Food Chemistry, 102,* 732–737.

45. Loizzo, M. R., Tundis, R., Bonesi, M., Menichini, F., Mastellone, V., Avallone, V., & Menichini, F. (2012). Radical scavenging, antioxidant and metal chelating activities of *Annona cherimola* Mill. (cherimoya) peel and pulp in relation to their total phenolic and total flavonoid contents. *Journal of Food Composition and Analysis, 25,* 179–184.

46. Lou, H., Hu, Y., Zhang, L., Sun, P., & Lu, H. (2012). Nondestructive evaluation of the changes of total flavonoid, total phenols, ABTS and DPPH radical scavenging activities, and sugars during mulberry (*Morus alba* L.) fruits development by chlorophyll fluorescence and RGB intensity values. *LWT – Food Science and Technology, 47,* 19–24.

47. Madikizela, B., Ndhlala, A. R., Finnie, J. F., & Van-Staden, J. (2012). Ethnopharmacological study of plants from Pondoland used against diarrhea. *Journal of Ethnopharmacology, 141,* 61–71.

48. Maisuthisakul, P., & Gordon, M. H. (2014). Characterization and storage stability of the extract of Thai mango (*Mangifera indica* Linn. Cultivar Chok-Anan) seed kernels. *Journal of Food Science and Technology, 51,* 1453–1462.

49. Mc Cune, L. M., & Johns, T. (2002). Antioxidant activity in medicinal plants associated with the symptoms of diabetes mellitus used by the indigenous peoples of the North American boreal forest. *Journal of Ethnopharmacology, 82,* 197–205.

50. Mc Donald, S., Prenzler, P. D., Autolovich, M., & Robards, K. (2001). Phenolic content and antioxidant activity of olive extracts. *Food Chemistry, 73,* 73–84.

51. Mollick, M. M. R., Rana, D., Dash, S. K., Chattopadhyay, S., Bhowmick, B., Maity, D., Mondal, D., Pattanayak, S., Roy, S., Chakraborty, M., & Chattopadhyay, D. (2015). Studies on green synthesized silver nanoparticles using *Abelmoschus esculentus* (L.) pulp extract having anticancer (*in vitro*) and antimicrobial applications. *Arabian Journal of Chemistry,* http://dx.doi.org/10.1016/j.arabjc.2015.04.033.

52. Moyo, B., Oyedemi, S., Masika, P. J., & Muchenje, V. (2012). Polyphenolic content and antioxidant properties of *Moringa oleifera* leaf extracts and enzymatic activity

of liver from goats supplemented with *Moringa oleifera* leaves/sunflower seed cake. *Meat Science, 91*, 441–447.

53. Ndhlala, A. R., Stafford, G. I., Finnie, J. F., & Van-Staden, J. (2009). *In vitro* pharmacological effects of manufactured herbal concoctions used in KwaZulu-Natal South Africa. *Journal of Ethnopharmacology, 122*, 117–122.

54. Orhan, I., Kartal, M., Abu-Asaker, M., Senol, F. S., Yilmaz, G., & Sener, B. (2009). Free radical scavenging properties and phenolic characterization of some edible plants. *Food Chemistry, 114*, 276–281.

55. Parekh, J., & Chanda, S. (2007). *In vitro* antibacterial activity of the crude methanol extract of *Woodfordia fruticosa* kurz. flower (Lythraceae). *Brazilian Journal of Microbiology, 38*, 204–207.

56. Perez-Jimenez, J., Arranz, S., Tabernero, M., Diaz-Rubio, E., Serrano, J., Goni, I., & Saura-Calixto, F. (2008). Updated methodology to determine antioxidant capacity in plant foods, oils and beverages: Extraction, measurement and expression of results. *Food Research International, 41*, 274–285.

57. Petridis, A., Therios, I., Samouris, G., & Tananaki, C. (2012). Salinity-induced changes in phenolic compounds in leaves and roots of four olive cultivars (*Olea europaea* L.) and their relationship to antioxidant activity. *Environmental and Experimental Botany, 79*, 37–43.

58. Rakholiya, K., & Chanda, S. (2012). *In vitro* interaction of certain antimicrobial agents in combination with plant extracts against some pathogenic bacterial strains. *Asian Pacific Journal of Tropical Biomedicine, 2*, S876–S880.

59. Rakholiya, K., & Chanda, S. (2012). Pharmacognostic, physicochemical and phytochemical investigation of *Mangifera indica* L. var. Kesar leaf. *Asian Pacific Journal of Tropical Biomedicine, 2*, S680–S684.

60. Rakholiya, K., Kaneria, M., & Chanda, S. (2011). Vegetable and fruit peels as a novel source of antioxidants. *Journal of Medicinal Plants Research, 5*, 63–71.

61. Rakholiya, K., Kaneria, M., & Chanda, S. (2013). Medicinal plants as alternative sources of therapeutics against multidrug-resistant pathogenic microorganisms based on their antimicrobial potential and synergistic properties. In: *Fighting Multidrug Resistance with Herbal Extracts, Essential Oils and their Components*, Rai, M. K., & Kon, K. V. (Eds.). Elsevier, USA, pp. 165–179.

62. Rakholiya, K., Kaneria, M., & Chanda, S. (2014). Inhibition of microbial pathogens using fruit and vegetable peel extracts. *International Journal of Food Science and Nutrition, 65*, 733–739.

63. Rakholiya, K., Kaneria, M., & Chanda, S. (2015). *In vitro* assessment of novel antimicrobial from methanol extracts of matured seed kernel and leaf of *Mangifera indica* L. (Kesar Mango) for inhibition of *Pseudomonas* spp. and their synergistic potential. *American Journal of Drug Discovery and Development, 5*, 13–23.

64. Rakholiya, K., Kaneria, M., & Chanda, S. (2015). Antioxidant activity of some commonly consumed fruits and vegetables peels in India. In: *Recent Progress in Medicinal Plants, Volume 40*, J. N. Govil and Manohar Pathak (Eds.). Studium Press LLC; pp. 22–35.

65. Rakholiya, K., Kaneria, M., Desai, D., & Chanda, S. (2013). Antimicrobial activity of decoction extracts of residual parts (seed and peels) of *Mangifera indica* L. var.

Kesar against pathogenic and food spoilage microorganism. In: *Microbial Pathogens and Strategies for Combating them: Science, Technology and Education,* Mendez-Vilas, A. (Ed.). Formatex Research Center, Spain, pp. 850–856.

66. Rakholiya, K., Kaneria, M., Nagani, K., Patel, A., & Chanda, S. (2015). Comparative analysis and simultaneous quantification of antioxidant capacity of four *Terminalia* species using various photometric assays. *World Journal of Pharmaceutical Science, 4,* 1280–1296.

67. Rakholiya, K. D., Kaneria, M. J., & Chanda, S. V. (2014). Mango pulp: A potential source of natural antioxidant and antimicrobial agent. In: *Medicinal Plants: Phytochemistry, Pharmacology and Therapeutics,* Gupta, V. K. (Ed.). Daya Publishing House, India, pp. 253–284.

68. Rathee, D., Rathee, S., Rathee, P., Deep, A., Anandjiwala, S., & Rathee D. (2015). HPTLC densitometric quantification of stigmasterol and lupeol from *Ficus religiosa. Arabian Journal of Chemistry, 8,* 366–371.

69. Re, R., Pellegrini, N., Proteggentle, A., Pannala, A., Yang, M., & Rice-Evans, C. (1999). Antioxidant activity applying an improved ABTS radical cation decolorization assay. *Free Radical Biology and Medicine, 26,* 1231–1237.

70. Rice-Evans, C., Miller, N., & Paganga, G. (1996). Structure-antioxidant activity relationships of flavonoids and phenolic acids. *Free Radical Biology and Medicine, 20,* 933–956.

71. Rivas-Arreola, M. J., Rocha-Guzman, N. E., Gallegos-Infante, J. A., Gonzalez-Laredo, R. F., Rosales-Castro, M., Bacon, J. R., Cao, R., Proulx, A., & Intriago-Ortega, P. (2010). Antioxidant activity of oak (*Quercus*) leaves infusions against free radicals and their cardioprotective potential. *Pakistan Journal of Biological Sciences, 13,* 537–545.

72. Robak, J., & Gryglewski, R. J. (1988). Flavonoids are scavengers of superoxide anions. *Biochemical Pharmacology, 37,* 837–841.

73. Roby, M. H. H., Sarhana, M. A., Selima, K. A. H., & Khalela, K. I. (2013). Antioxidant and antimicrobial activities of essential oil and extracts of fennel (*Foeniculum vulgare* L.) and chamomile (*Matricaria chamomilla* L.). *Industrial Crops and Products, 44,* 437–445.

74. Rondevaldova, J., Novy, P., Urban, J., & Kokoska, L. (2015). Determination of antistaphylococcal activity of thymoquinone in combinations with antibiotics by checkerboard method using EVA capmat™ as a vapor barrier. *Arabian Journal of Chemistry,* http://dx.doi.org/10.1016/j.arabjc.2015.04.021.

75. Sawant, L., Prabhakar, B., & Pandita, N. (2010). Quantitative HPLC analysis of ascorbic acid and gallic acid in *Phyllanthus emblica. Journal of Analytical and Bioanalytical Techniques, 1,* 111–114.

76. Schieber, A. (2012). Functional foods and nutraceuticals. *Food Research International, 46,* 437.

77. Sharma, A., Gupta, S., Sarethy, I. P., Dang, S., & Gabrani, R. (2012). Green tea extract: Possible mechanism and antibacterial activity on skin pathogens. *Food Chemistry, 135,* 672–675.

78. Shen, J., Zhang, Z., Tian, B., & Hua, Y. (2012). Lipophilic phenols partially explain differences in the antioxidant activity of subfractions from methanol extract of camellia oil. *European Food Research Technology, 235*, 1071–1082.

79. Singh, R., Shushni, M. A. M., & Belkheir, A. (2015). Antibacterial and antioxidant activities of *Mentha piperita* L. *Arabian Journal of Chemistry, 8*, 322–328.

80. Sousa, A., Ferreira, I. C. F. R., Barros, L., Bento, A., & Pereira, J. A. (2008). Antioxidant potential of traditional stoned table olives "Alcaparras": influence of the solvent and temperature extraction conditions. *LWT – Food Science and Technology, 41*, 739–745.

81. Sowndhararajan, K., Joseph, J. M., Arunachalam, K., & Manian, S. (2010). Evaluation of *Merremia tridentata* (L.) Hallier f. for *in vitro* antioxidant activity. *Food Science Biotechnology, 19*, 663–669.

82. Sulaiman, S. F., Yusoff, N. A. M., Eldeen, I. M., Seow, E. M., Sajak, A. A. B., & Supriatno, K. L. (2011). Correlation between total phenolic and mineral contents with antioxidant activity of eight Malaysian bananas (*Musa* sp.). *Journal of Food Composition and Analysis, 24*, 1–10.

83. Tang, J., Meng, X., Liu, H., Zhao, J., Zhou, L., Qiu, M., Zhang, X., Yu, Z., & Yang, F. (2010). Antimicrobial activity of sphingolipids isolated from the stems of Cucumber (*Cucumis sativus* L.). *Molecules, 15*, 9288–9297.

84. Teke, G. N., Kuiate, J. R., Kuete, V., Teponno, R. B., Tapondjou, L. A., Tane, P., Giacinti, G., & Vilarem, G. (2011). Bio-guided isolation of potential antimicrobial and antioxidant agents from the stem bark of *Trilepisium madagascariense*. *South African Journal of Botany, 77*, 319–327.

85. Thiyagarajan, R., Sung-Won, K., Jong-Hwan, S., Seock-Yeon, H., Sang-Hyon, S., Sung-Kwang, Y., & Si-Kwan, K. (2012). Effect of fermented *Panax ginseng* extract (GINST) on oxidative stress and antioxidant activities in major organs of aged rats. *Experimental Gerontology, 47*, 77–84.

86. Thondre, P. S., Ryan, L., & Henry, C. J. K. (2011). Barley β-glucan extracts as rich sources of polyphenols and antioxidants. *Food Chemistry, 126*, 72–77.

87. Wang, J., Wang, Y., Liu, X., Yuan, Y., & Yue, T. (2013). Free radical scavenging and immunomodulatory activities of *Ganoderma lucidum* polysaccharides derivatives. *Carbohydrate Polymers, 9*, 33–38.

88. Wang, Y., Han, T., Zhu, Y., Zheng, C. J., Ming, Q. L., Rahman, K., & Qin, L. P. (2010). Antidepressant properties of bioactive fractions from the extract of *Crocus sativus* L. *Journal of Natural Medicine, 64*, 24–30.

89. Yang, Q. M., Pan, X., Kong, W., Yang, H., Su, Y., Zhang, L., Zhang, L., Zhang, Y., Yuling, Y., Ding, L., & Liu, G. (2010). Antioxidant activities of malt extract from barley (*Hordeum vulgare* L.) toward various oxidative stress *in vitro* and *in vivo*. *Food Chemistry, 118*, 84–89.

90. Ye, C., Dai, D., & Hu, W. (2013). Antimicrobial and antioxidant activities of the essential oil from onion (*Allium cepa* L.). *Food Control, 30*, 48–53.

91. Zarai, Z., Boujelbene, E., Salem, N. B., Gargouri, Y., & Sayari, A. (2013). Antioxidant and antimicrobial activities of various solvent extracts, piperine and piperic acid from *Piper nigrum*. *LWT – Food Science and Technology, 50*, 634–641.

92. Zhang, Q., Guo, Y., Shangguan, X., Zheng, G., & Wang, W. (2012). Antioxidant and antiproliferative activity of *Rhizoma Smilacis* Chinae extracts and main constituents. *Food Chemistry, 133*, 140–145.

93. Zhou, H. C., Lin, Y. M., Li, Y. Y., Li, M., Wei, S. D., Chai, W. M., & Tam, N. F. (2011). Antioxidant properties of polymeric proanthocyanidins from fruit stones and pericarps of *Litchi chinensis* Sonn. *Food Research International, 44*, 613–620.

CHAPTER 13

ISOLATION AND CHARACTERIZATION OF LYCOPENE FROM TOMATO AND ITS BIOLOGICAL ACTIVITY

YOGESH K. BARAVALIA, KOMAL V. POKAR, KHYATI C. BHOJANI, and SHAILESH B. GONDALIYA

CONTENTS

13.1   INTRODUCTION

Lycopene is a phytochemical, synthesized by plants and microorganisms but not by animals [16]. Lycopene is a pigment principally responsible for the characteristic deep-red color of ripe tomato fruits and product. The scientific name for the tomato is *Lycopersicon esculentum*, therefore it is commonly named lycopene.

Research has shown that lycopene can be absorbed more efficiently by the human body after it has been processed into juice, sauce, paste, or ketchup. In fresh fruit, lycopene is enclosed in the fruit tissue. Therefore, only a portion of the lycopene that is present in fresh fruit is absorbed. Processing of fruits makes the lycopene more bioavailable by increasing the surface area available for digestion. More significantly, the chemical form of lycopene is altered by the temperature changes involved in the processing. Also, because lycopene is fat-soluble (as are vitamins, A, D, E, and β-carotene), absorption into tissues is improved when oil is added to the diet. Although lycopene is available in supplement form, yet it is likely there is a synergistic effect when it is obtained from the whole fruit instead, where other components of the fruit enhance lycopene's effectiveness. Due to presence of unsaturated bonds in its molecular structure, lycopene is susceptible to oxidation and degrades easily when exposed to light and heat [12].

This chapter presents research study for the isolation and characterization of lycopene from tomato.

13.2 CHARACTERIZATION OF LYCOPENE

13.2.1 MECHANISM OF ACTION

The biological activities of carotenoids such as β-carotene are related to their ability to form vitamin A within the body. Since lycopene lacks the β-ionone ring structure, it cannot form vitamin A. Therefore, its biological effects in humans have been attributed to mechanisms other than vitamin A [1].

Lycopene is the most predominant carotenoid in human plasma, present naturally in greater amounts than β-carotene and other dietary carotenoids. This perhaps indicates its greater biological significance in the human defense system. Its level is affected by several biological and lifestyle factors. Because of its lipophilic nature, lycopene concentrates in low-density and very-low-density lipoprotein fractions of the serum. Lycopene is also found to concentrate in the adrenal, liver, testes and prostate. However, unlike other carotenoids, lycopene levels in serum or tissues do not correlate well with overall intake of fruits and vegetables.

13.2.2 SOURCES OF LYCOPENE

While the human body does not produce lycopene, it is readily available in variety of foods. Lycopene occurs naturally in many red colored foods such as tomatoes, watermelon, pink grape fruits, carrots, guava, red oranges and papaya. Cooking and processing tomatoes stimulates and concentrates the lycopene content, making tomato paste and sauces a rich source. This red colored pigment was first discovered in the tomato by Millardet in 1876 [10]. It was later named lycopene by Schunck. Tomatoes contribute over 85% of the lycopene intake by women. While there is currently no recommended daily intake of lycopene, suggestion ranges from about 5 to 35 mg/daily. This amount can be provided by one or two servings of tomato products each day [7].

13.2.3 STRUCTURE OF LYCOPENE

Isolation procedure for lycopene was first reported in 1910, and the structure of the molecule was determined in 1931.

Lycopene structure consists of a long chain of conjugated double bonds, with two open end rings. The structure lycopene is the longest of all carotenoids. Lycopene $[C_{40}H_{56}]$ with a molecular weight of 536.85 is an unsaturated hydrocarbon carotenoid containing 13 carbon–carbon double bonds, 11 of which are conjugated and arranged in a linear array (Figure 13.1). These conjugated double bonds are responsible for the vibrant red color of lycopene [2]. Lycopene is a lipophilic compound that is insoluble in water, but soluble in organic solvents.

Lycopene in fresh tomato fruits occurs essentially in the all trans-configurations. The main causes of tomato lycopene degradation during processing are isomerization and oxidation. Isomerization converts all transisomers to cis-isomers due to additional energy input and results in an unstable, energy-rich station. Determination of the degree of lycopene

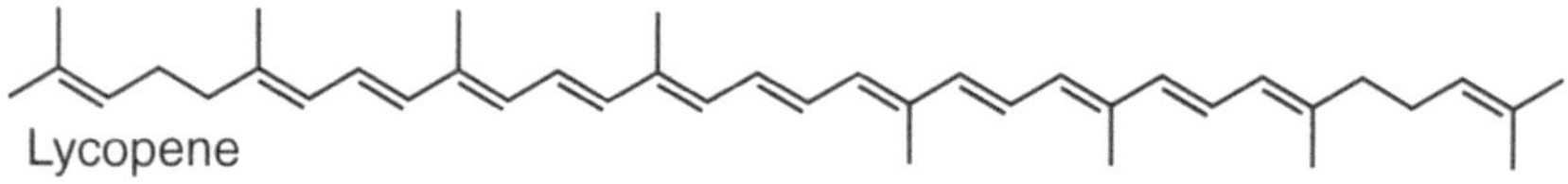

Lycopene

FIGURE 13.1 Structure of lycopene, $C_{40}H_{56}$.

isomerization during processing would provide a measure of the potential health benefits of tomato-based foods [3]. Table 13.1 shows physical properties of lycopene.

13.2.4 STRUCTURE OF TOMATO

Cross section of tomato is shown in Figure 13.2. Lycopene is synthesized in chromoplasts of fruit cells. Most of the cells in the pericarp near the epidermis synthesize higher lycopene levels than the inner tissue of tomatoes. In tomatoes, full ripening takes place 40–60 days after planting, during which chloroplasts change to chromoplasts upon synthesis of lycopene [12].

13.2.5 APPLICATION OF LYCOPENE

Tomato (*Solanum lycopersicum* L.) is one of the most important vegetables worldwide because of its high consumption, year round availability

TABLE 13.1 Physical Properties of Lycopene

Molecular formula	$C_{40}H_{56}$
Molecular weight	536.85 Da
Color	Powder form dark reddish-brown
Melting point	172–175°C
Sensitivity	Sensitive to light, oxygen, high temperature, acids, catalyst
Solubility	– Soluble in chloroform, hexane, benzene, carbon disulfide, acetone, petroleum ether and oil
	– Insoluble in water, ethanol and methanol

A   B   C

FIGURE 13.2 Hybrid tomato (A), desi tomato (B), and section of a tomato (C).

and large content of health related components. This section indicates applications of lycopene.

13.2.5.1 Antimicrobial Property

Infectious diseases caused by bacteria and fungi affect millions of people worldwide. Treating bacterial infections with antibiotics is beneficial, but their indiscriminate use has led to an alarming rate of resistance among microorganisms. Multiple drug resistance in microbial pathogens is an ongoing global problem. This results in loss of effective antibiotics and loss of budget for infectious disease treatment. There is an urgent and constant need for exploration and development of cheaper, effective, new plant-based drugs with better bioactive potential and the fewest possible side effects. Hence, attention has been directed toward biologically active extracts and compounds from plant species to fight against microbial diseases, as well as against degenerative diseases caused by free radicals.

This suggests that plants which manifest relatively high levels of antimicrobial action may be sources of compounds that can be used to inhibit the growth of pathogens. Bacterial cells could be killed by the rupture of cell walls and membranes and by the irregular disruption of the intracellular matrix when treated with plant extracts [6].

13.2.5.2 Anti-Cancer Property

Lycopene has a protective effect against stomach, colon, lung and skin cancers. Free radicals in the body can damage DNA and proteins in the cells and tissues, resulting in inflammation which may lead to cancer. Hence, the antioxidant properties of lycopene in eliminating free radicals may reduce the risk of cancer [2]. Research on breast, lung and endometrial cancers has shown that lycopene is even more effective than other bright vegetable carotenoids (α and β carotene) and delays the cell cycle progression from one growth phase to the next, thus inhibiting growth of tumor cells. Lycopene also plays a role in modulating intercellular communication by regulating irregular pathways that may be associated with cancer [14].

Studies show that taking high doses of lycopene can slow the progression of prostate cancer. The estimated intake of lycopene from various tomato products was inversely proportional to the risk of prostate cancer, a result not observed for any other carotenoid. Consuming ten or more servings of tomato products per week reduced the risk by almost 35%. The protective effects were highest for more advanced or aggressive prostate cancer [4].

13.2.5.3   Atherosclerosis and Heart Disease

Lycopene may be helpful in people with high cholesterol, atherosclerosis or coronary heart disease, possibly due to its antioxidant properties. Lycopene prevents oxidation of low-density lipoprotein (LDL) cholesterol and reduces the risk of arteries becoming thickened and blocked. Most published studies in this area used tomato juice as a treatment [1]. Drinking two to three (200 ml) glasses of a processed tomato juice would provide more than the 40 mg lycopene per day, recommended for reducing LDL cholesterol.

13.2.5.4   Cosmetics

Lycopene provides antioxidant protection from environmental damage, shielding skin from premature aging. In an article *"more beautiful skin with the 100 best anti-aging foods,"* author Allison Tannis points out that lycopene has the power to "reduce skin cell damage and redness" caused by environmental triggers such as the sun and pollution. Lycopene has the capability to improve cellular functions that are essential to keeping skin looking young. Lycopene strengthens skin by enhancing its ability to produce collagen and reducing the DNA damage that leads to wrinkles.

13.3   MATERIALS AND METHODS

Two tomato species (hybrid and desi tomato) were purchased from the local market of Rajkot, India and were used in this study.

13.3.1   *EXTRACTION OF LYCOPENE*

Liquid–liquid extraction (also known as solvent extraction) involves the separation of the constituents (solutes) of a liquid solution by contact with another insoluble liquid. Solutes are separated based on their different solubility in different liquids. Separation is achieved, when the substances constituting the original solution is transferred from the original solution to the other liquid solution. Extraction of lycopene was done by two methods.

13.3.1.1 Isolation of Lycopene by CCl₄ Method

The 100 grams of fresh tomatoes were cooked and then crushed by using mortar-pestle and prepared into a tomato paste. This paste was dehydrated by adding 130 ml methanol. This mixture was immediately shaken vigorously to prevent the formation of hard lumps. After 2 h, the thick suspension was filtered and the dark red cake was shaken for another 15 min with 150 ml mixture of equal volumes of methanol and carbon tetrachloride, and was separated by filtration (Figure 13.3A). The carbon tetrachloride was transferred to a separating funnel, water was added and the solution was shaken well. After phase separation, the carbon tetrachloride phase was evaporated and the residue was diluted with about 4 ml of benzene. Using a dropper, 2 ml of boiling methanol was added, then crystals of crude lycopene appeared immediately and the crystallization was completed by keeping the liquid at room temperature and ice

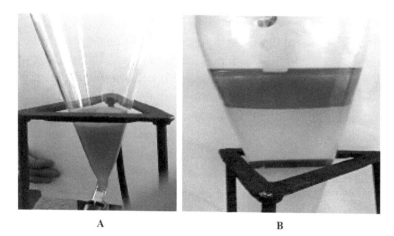

A   B

FIGURE 13.3 CCl₄ phase of separating mixture (Left: A); Acetone and petroleum ether phase of separating mixture (Right: B).

bath, respectively. The crystals were washed 10 times using benzene and boiling methanol [12].

13.3.1.2   Isolation of Lycopene by NaCl Method

The 100 grams of fresh tomatoes were cooked and was then crushed by using mortar-pestle and was prepared into a tomato paste. This paste was transferred into a 500-ml Erlenmeyer flask. Solid material was extracted by swirling the Erlenmeyer flask with a 250 ml mixture of equal volumes of acetone and low-boiling petroleum ether (Figure 13.3B). Any large clumps were broken with a spatula. Solution was filtered by Pasteur pipette and the filtrate was collected in a large test tube. It was again extracted with 50% mixture of acetone-petroleum ether and passed through the Pasteur pipette. Combined extracts were washed with 250 ml of saturated aqueous NaCl. Aqueous wash was removed with a Pasteur pipette, and then washed with 200 ml of 10% aqueous K_2CO_3. Finally, it was washed with another 250 ml of saturated aqueous NaCl and the aqueous wash was removed. This solution was kept in a beaker overnight for the evaporation of acetone and petroleum ether.

13.3.2   *PURIFICATION OF LYCOPENE*

13.3.2.1   Column Chromatography

Chromatographic column was arranged in vertical position on the stand. First column was washed with petroleum ether and solution was drained out, then was packed in a chromatographic column using silica (100–200 mesh size) slurry and allowed to settle down the slurry for 15 min. Column was saturated for about 2 h after adding 45 ml petroleum ether. Half volume of this solution was drained out and isolated lycopene sample was added to the top of the column. After some time, when the first yellow band starts to drain out of the column, mixture of 10:90:: acetone: petroleum ether was added to the top of the column and the eluent level was kept at constant level as before. When the lycopene layer (orange-red) begins to leave the column, orange-red layer was collected into the beaker. When the band was almost completely off the column, sample vial was removed [5].

13.3.2.2   Characterization of Lycopene

13.3.2.2.1   Thin layer chromatography

A thin *mark was made at the bottom of the plate with a pencil to iden-tify the sample spots.* During this time, mobile phase consisting of hexane:acetone:acetic acid (8:2:0.2) were allowed to saturate in the chamber for 30 min. Then the *solutions were applied on the spots marked on the line* at equal distances. The plate was immersed so that sample spot was well above the level of mobile phase but not immersed in the solvent for development. Sufficient time was allowed for development of spots. Then the plates were removed and allowed to dry. The sample spots were visualized in a suitable UV light chamber [13] and R_f value was calculated.

13.3.2.2.2   High performance thin layer chromatography (HPTLC)

Sample preparation
2 mg of each lycopene sample was dissolved in 1 ml of petroleum ether.

Mobile phase
Hexane:acetone:acetic acid (8:2:0.2) solution.

Sample application
The 10 μl of sample (1, 2, 3, and 4) was applied on 10×10 TLC plate (Aluchrosep silica gel 60/UV 254) by LINOMAT-5 sample applicator.

Mobile phase development
The plate was developed in twin trough chambers (10 × 10) up to 90 mm solvent front at room temperature.

Detection
CAMAG TLC scanner was used for scanning of lycopene sample at 256 nm.

13.3.2.2.3   Liquid chromatography–mass spectrometry (LC-MS)

The 0.2 mg of sample was diluted with HPLC methanol to makes up to one μg/ml. This mixture was vortexed for 15 seconds. The 0.5 ml of lycopene sample was added to the vial and was vortexed for 2 min. The sample was transferred into prelabeled auto-sampler vial that was placed on LC-MS system for analysis [11]. Table 13.2 indicates source dependent parameters.

13.3.3 ANTIOXIDANT ACTIVITY OF LYCOPENE

13.3.3.1 FRAP Assay

FRAP reagent (200 ml acetate buffer 300 mM, pH 3.6 + 20 ml TPTZ 10 mM + 20 ml $FeCl_3$ + 24 ml Distilled water) was warmed at 37°C for 20 min. The 3 ml of FRAP reagent was added into each test tube. Then 100 µl of sample was added in each test tube except the blank test tube (control). After every 4 min., absorbance was measured at 593 nm. The absorbance was taken until reading became stable [9].

13.3.4 ANTIMICROBIAL SUSCEPTIBILITY TEST

13.3.4.1 Test Microorganisms

For this study, four bacterial strains (viz. *Staphylococcus aureus* NCIM2901, *Bacillus cereus* NCIM2217, *Salmonella typhimurium* NCIM2501 and *Proteus mirabilis* NCIM2300) were obtained from National Chemical Laboratory, Pune

13.3.4.2 Antimicrobial Activity

The test organism was activated by inoculating a loop full of the strain in 25 ml of nutrient broth, which was kept overnight on a rotary shaker.

TABLE 13.2 Source Dependent Parameters

Parameters	Used
DL temperature	250°C
Drying gas flow	15.0 L/min
Heat block temperature	500°C
Interface temperature	300°C
Nebulizing gas flow	3.0 L/min

Mueller Hinton agar media was used for antibacterial activity. The assay was performed by the agar well diffusion method [15], with 200 µl of inoculum (1×10^8 cfu/mL) introduced into molten Mueller Hinton agar and poured into petri dishes, when the temperature reached 40–42°C. The media were solidified and wells were prepared in the seeded agar plates with the help of a cup borer (8.5 mm). Then, 100 µl of the test drug (10 mg/mL and 10 mg/mL in dimethyl sulfoxide [DMSO]) was introduced into the well and the plates were incubated at 37°C for 24 h. The DMSO was used as a negative control. All tests were triplicated under strict aseptic conditions. The microbial growth was determined by measuring the diameter of the zone of inhibition in mm.

13.4 RESULTS AND DISCUSSION

13.4.1 EXTRACTION OF LYCOPENE

The extracted yield of lycopene was greater in desi tomato compared to hybrid tomato. Compared to CCl_4 method, NaCl method gave more lycopene yield.

In a test tube containing 1 ml of extract, few drops of concentrated H_2SO_4 was added to conform the presence of lycopene in the extract. Test showed indigo blue color in the test tube indicating the presence of lycopene.

13.4.2 COLUMN CHROMATOGRAPHY

The Column chromatography was used to separate crude extract. After performing this column chromatography, it was observed that first band was yellow for β-carotene and second band was reddish orange for lycopene (Figure 13.4).

TABLE 13.3 Extracted yield of Lycopene

Tomato var.	Methods	Tomato sample, (gms)	Extract, mg
Hybrid	NaCl	100	25
Hybrid	CCl_4	100	27
Desi	NaCl	100	35.3
Desi	CCl_4	100	34.3

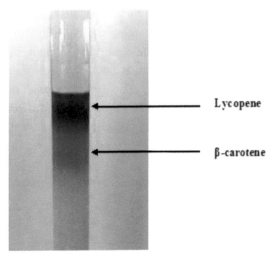

FIGURE 13.4   Column chromatography.

13.4.3   TLC AND HPTLC

R_f value of lycopene in TLC was 0.3 cm (Figure 13.5A). R_f value of lycopene in HPTLC was 0.3 cm (Figure 13.5B). Visualization was done by CAMAG Visualizer at 254 nm.

After performing TLC and HPTLC hypothetically, it was confirmed that lycopene was present in sample. As in other research studies, this research confirmed that R_f value of lycopene is 0.3 cm [8].

13.4.4   LIQUID CHROMATOGRAPHY–MASS SPECTROMETRY (LC–MS)

Molecular weight of lycopene is 536.85 Daltons. By performing LC-MS, Mass by charge (m/z) ratio of lycopene is 537.4, which was detected in ESI⁺ Ion Mode (Figures 13.6–13.9).

13.4.5   FRAP ASSAY

The *ferric reducing antioxidant power* (FRAP) of different extracts of lycopene gives almost the same results. The maximum antioxidant power was obtained in Desi tomato – NaCl method that was 1.039 Fe(III) mol/200 gm of tomato (Figure 13.10).

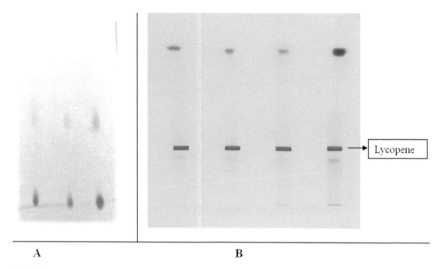

FIGURE 13.5   Separation of lycopene by TLC (Left – A); Separation of lycopene by HPTLC (Right – B).

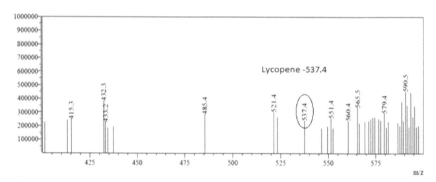

FIGURE 13.6   Characterization of lycopene from hybrid tomato – NaCl method by LC-MS.

13.4.6   ANTIMICROBIAL ACTIVITY

Moderate antimicrobial activity was observed in different isolated fractions of lycopene. Lycopene fractions from hybrid tomato by NaCl method were able to inhibit the growth of *S. aureus* and *B. cereus*. Whereas other lycopene fractions were not able to inhibit the growth of any of the bacterial strains under study.

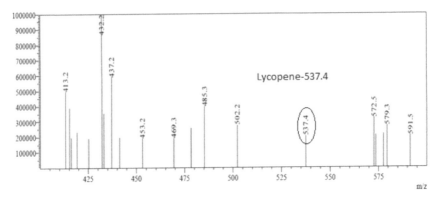

FIGURE 13.7 Characterization of lycopene from hybrid tomato – CCl₄ method by LC-MS.

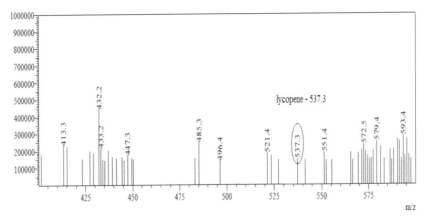

FIGURE 13.8 Characterization of lycopene from desi tomato – NaCl method by LC-MS.

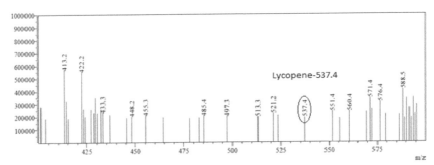

FIGURE 13.9 Characterization of lycopene from desi tomato – CCl₄ method by LC-MS.

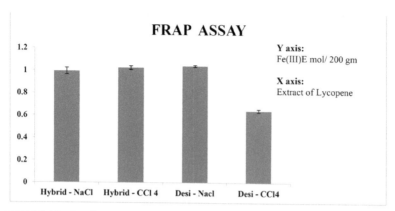

FIGURE 13.10   Ferric reducing antioxidant power (FRAP) of lycopene.

13.5   CONCLUSIONS

Conventional solvent extraction methods were used to extract lycopene from tomato paste. Liquid–liquid extraction method was used for this research because it requires less solvent and shorter extraction time than other methods and it is suitable for the extraction of lycopene from tomato paste. Between two methods (NaCl and CCl_4 methods), NaCl gave more yield of lycopene from desi tomato. From this study, it can be concluded that consumption of desi tomato is beneficial for human health.

13.6   SUMMARY

Lycopene is one of the 600 carotenoids found in nature and can be easily identified in tomatoes. Several epidemiological studies report that lycopene rich diets have beneficial effects on human health, showing strong correlations between the intake of carotenoids and a reduced risk of cancer, coronary and cardiovascular diseases.

This chapter presents research study for the isolation and characterization of lycopene from tomato. The purpose of this study was to develop a simple and effective method for solvent extraction of lycopene from tomato and study its antimicrobial activity. Two tomato species (hybrid and desi tomato) were used in this study, and the liquid–liquid extraction method was used for the extraction of lycopene by using NaCl and

CCl_4 methods followed by column chromatography. Compared to CC14 method, NaCl method yielded more lycopene from desi tomato compared to hybrid tomato. Further characterization was done by *Thin layer chromatography* (TLC), *High performance thin layer chromatography* (HPTLC) and *Liquid chromatography mass spectrometry* (LC-MS).

Antioxidant activity was assessed by FRAP. The maximum antioxidant power was obtained in desi tomato in NaCl method. The antimicrobial activities of lycopene were screened against pathogenic bacteria such as *Staphylococcus aereus, Bacillus cereus, Salmonella typhimurium, and Proteus mirabilis* by agar well diffusion method. From these results, it can be concluded that lycopene has great potential as antimicrobial compound against microorganisms and that they can be used in the treatment of infectious disease caused by resistant microorganisms.

KEYWORDS

- antimicrobial
- antimicrobial activity
- antioxidants
- carotenoid
- CCl4 method
- column chromatography
- FRAP
- high performance thin layer chromatography
- liquid chromatography mass spectrometry
- lycopene
- *lycopersicon esculentum*
- microorganism
- NaCl method
- pathogenic bacteria
- phytochemical
- thin layer chromatography
- tomato

REFERENCES

1. Agarwal, S., & Rao, A. V. (2000). Tomato lycopene and its role in human health and chronic diseases. *Canadian Medical Association Journal, 163*, 739–744.
2. Aghel, N., Ramezani, Z., & Amirfakhrian, S. (2011). Isolation and quantification of lycopene from tomato cultivated in dezfoul, Iran. *Jundishapur Journal of Natural Pharmaceutical Products, 6*, 9–15.
3. Chauhan, K., Sharma, S., Agarwal, N., & Chauhan, B. (2011). Lycopene of tomato fame: its role in health and disease. *International Journal of Pharmaceutical Sciences Review and Research, 10*, 99–115.
4. Choksi, P. M., & Joshi, V. Y. (2007). A review on lycopene – Extraction, purification, stability and applications. *International Journal of Food Properties, 10*, 289–298.
5. Church, W. H. (2005). Column chromatography analysis of brain tissue: An advanced laboratory exercise for neuroscience majors. *The Journal of Undergraduate Neuroscience Education, 3*, A36–A41.
6. Djifaby, S. P. A.E., Yacouba, C. A., Adama, H., Kiessoum, K., Marie-Hyacinthe, C. M., & Germaine, N. O. (2012). Carotenoids content and antibacterial activity from galls of *Guierasenegalensis* J.F. Gmel (*combretaceae*). *International Journal of Phytomedicine, 4*, 441–446.
7. Giovannucci, E., Ascherio, A., Rimm, E. B., Stampfer, M. J., Colditz, G. A., & Willett, W. C. (1995). Intake of carotenoids and retinol in relation to risk of prostate cancer. *Journal of the National Cancer Institute, 87*, 1767–1776.
8. Haroon, S. (2014). Extraction of Lycopene from Tomato Paste and its Immobilization for Controlled Release. A Thesis, The University of Waikato, New Zealand.
9. Kaushik, A., Jijta, C., Kaushik, J. J., Zeray, R., Ambesajir, A., & Beyene, L. (2012). FRAP (ferric reducing ability of plasma) assay and effect of *Diplaziumesculentum* (Retz) Sw. (a green vegetable of north India) on central nervous system. *Indian Journal of Natural Products and Resources, 3*, 228–231.
10. Kong, K. W., Khoo, H. E., Prasad, K. N., Ismail, A., Tan, C. P., & Rajab, N. F. (2010). Revealing the power of the natural red pigment lycopene. *Molecules, 15*, 959–987.
11. McIntosh, T. S., Davis, H. M., & Matthews, D. E. (2002). A liquid chromatography-mass spectrometry method to measure stable isotopic tracer enrichments of glycerol and glucose in human serum. *Analytical Biochemistry, 300*, 163–169.
12. Metkar, S., Saptarshi, S., & Kadam, A. (2014). Studies on extraction, isolation and applications of lycopene. *IndoAmerican Journal of Pharmaceutical Research, 4*, 5017–5028.
13. Mritunjay, K., Mondal, D. B., & Ananya, D. (2011). Quantification of Catechin and Lycopene in *Calendula officinalis* extracts using HPTLC method. *Asian Journal of Pharmaceutical and Clinical Research, 4*, 128–129.
14. Naz, A., Butt, M. S., Pasha, I., & Nawaz, H. (2013). Antioxidant indices of watermelon juice and lycopene extract. *Pakistan Journal of Nutrition, 12*, 255–260.
15. Perez, C., Paul, M., & Bazerque, P. (1990). An antibiotic assay by the agar well diffusion method. *ActaBiologiaeet Medicine Experimentalis, 15*, 113–115.
16. Rao, A. V., & Agarwal, S. (2000). Role of antioxidant lycopene in cancer and heart disease. *Journal of the American College of Nutrition, 19*, 563–569.

CHAPTER 14

FOOD PROCESSING USING MICROBIAL CONTROL SYSTEM: SHEA BUTTER

OFEORISTE D. ESIEGBUYA and F. I. OKUNGBOWA

CONTENTS

14.1   INTRODUCTION

Shea tree (*Vitellaria paradoxa* C.F. Gaertn.) belongs to the *Sapotaceae* family and was first named by the German botanist Carl Gaertner. It was later renamed *Butyrospermum parkii* (Kotschy) by the same author in

1961 [21]. Some authors considered *Butyrospermum parkii*as the synonym of *V. paradoxa*; and the name is generally accepted for the African Shea tree [32]. In West African region, it is classified as the subspecies *paradoxa* and as *nilotica* East Africa [24]. One of the major difference between the *V. paradoxa* and *V. nilotica* is that the butter from the latter is solid at ambient temperature while the former is liquid.

Shea tree grows wild in the noncoastal areas of dry savannas, forest and parklands of the Sudan zones of some African countries such as Senegal, Mali, Côte d Ivoire, Burkina Faso, Togo, Ghana, Benin, Nigeria, Niger, Cameroon, Uganda, Sudan and Ethiopia. These are sometimes refer to as the Shea belt [6, 16, 22]. Ghana and Burkina Faso are the main Shea nut exporters in Africa [17].

The oldest specimen of the Shea tree, as reported in literature was first collected by Mungo Park on May 26, 1797 [9, 15]. Shea tree grows up to a height ranging from 9–12 m (30–40 feet) and begins fruiting between 20 and 50 years of age and it reaches full maturity after 45 years, after which it can continuously produce its fruits for up to 200 years. Some authors believed that it is the harsh environmental conditions that the Shea tree is exposed to in the wild that makes it grow slowly within its first five years delaying its maturity and bearing of fruits [2, 31]. Fobil [15] reported that the harsh environmental conditions also contribute to the profuse branches, thick waxy and deeply fissured bark that makes it fire-resistant and live for up to 300 years [9, 15].

According to Ukers [34] and Steiger et al. [30], delay in the maturity of the Shea tree has discouraged commercialization of the Shea tree into plantation by farmers, when compared to some other economic tree like the coffee tree (*Coffee arabica*). In addition, quality deterioration of its main products is a problem as most Shea industries in African counties producing it depend on rural dwellers to pick the fruits from natural growing stands in their communities.

In Nigeria, the Nigerian Institute for Oil Palm Research (NIFOR) has recently added the Shea tree to its mandate crops, such as Oil, Coconut, *Raphia*, and Date palms. The institute mandate for the Shea tree focuses on research for the domestication of the Shea tree for cultivation by local farmers and export of its main products (Shea nuts and butter).

Rural dwellers in Nigeria believe that no one owns the Shea tree since it germinates and grows on its own without anyone caring for it. The Shea tree is found growing naturally especially in the Northern part of Nigeria and grows wild in States such as Niger, Nasarawa, Kebbi, Kwara Kogi, Adamawa, Benue, Edo, Katsina, Plateau, Sokoto, Zamfara, Taraba, Borno Federal Capital Territory and Oyo States. It has characteristic features of growing in a wide space and growing side by side. The products from the Shea tree are valued and highly exploited by the native inhabitants of the community [10].

The main product from the Shea tree is the Shea butter (Figure 14.1), which is called Shea in English, Karate in French, the Okwumain Igbos and Orioyo in Yorubas [10]. Figure 14.2 indicates that Shea butter is a triglyceride (fat) derived mainly from stearic and olein acid.

14.2 POSTHARVEST DISEASES OF SHEA NUTS AND KERNELS

Postharvest diseases are microbial diseases, which affect food crops after harvesting to storage. Some of the factors encouraging postharvest diseases caused by microorganisms include injury to fruits during harvesting, virulency of the pathogen, nutritional status of the fruits and environmental storage conditions [28]. The impact of postharvest diseases can result in the loss of food quantity and food quality along its supply chain.

14.2.1 MEASUREMENT OF POSTHARVEST FOOD LOSS

Postharvest food loss can be determined by either measuring the quantitative loss (reduction in the available quantity) or qualitative loss (reduction in economic or nutrient value). Qualitative loss results in changes, which lower the economic or nutrient value of the food, often requiring that it be discarded (quantitative loss). Qualitative food loss can be caused by both physiological changes and microbial infection.

Measuring postharvest food loss by quantitative approach involves quantification of diseases of the affected food part using various means of measurement of symptoms such as visual estimation and remote sensing.

Qualitative determination of postharvest food loss can be determined using analytical methods that can detect slight changes in the nutritional composition of the fruits. Esiegbuya et al. [13] qualitatively determined the postharvest loss of *Raphia hookeri* and Shea fruits, respectively using analytical methods described by Association of Official Analytical

FIGURE 14.1   Shea: Tree, nuts, butter, soap and cream.

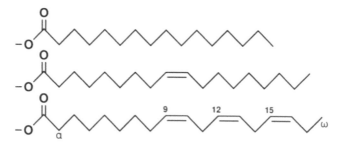

FIGURE 14.2 Shea butter (*Vitellaria paradoxa* C.F. Gaertn.) is a triglyceride (fat) derived mainly from stearic acid and olein acid.

Chemists (AOAC) and American Association of Cereal Chemists (AACC). These analytical methods can monitor and detect changes in the food component as caused by microbes.

Qualitative determination of postharvest loss by microbes have shown that microorganisms show differences in the pattern of nutrient utilization of food compound during postharvest deterioration, which results in the changes of the proximate and mineral composition of the fruit.

Esiegbuya et al. [13] observed the decrease in the carbohydrate, protein and fat components and increased moisture and fiber content of biodeteriorated Shea fruit. They also observed a decrease in all food components apart from the ash content of black rot infected *Raphia hookeri* fruits. Mba et al. [23] and Lawal et al. [19] have also observed differences in the pattern of nutrient utilization of postharvest deteriorated fruits by fungi. The research studies by these authors show that microbes have preference for different food components. The knowledge of nutrient utilization pattern for a particular microbe can be used to develop a control strategy for postharvest diseases. This is due to the ability of a host to resist diseases to depend on its ability to limit penetration, development and reproduction of the pathogen.

14.2.2 DISEASES OF SHEA NUTS AND KERNELS

Table 14.1 below highlights some of the diseases and its impact on Shea nuts and kernels. According to the authors apart from quality deterioration of the processed products, postharvest diseases can also result in loss of

TABLE 14.1   Diseases of Shea Nuts and Kernels

Diseases	Causal agents	Effect on the nuts/kernels
Nuts crack and holes	*Aspergillus niger, A. flavus*	Holes/cracks on the surface of kernels and insects
Nut discoloration	*Xyalria* sp, *A. persii* and *A. niger*	Color changes
Kernel discoloration	*A. niger, A. flavus, Mucor* and *Phoma* sp.	Color changes
Kernel deterioration	*A. niger, A. persii, A. flavus, Phoma* and *Fusarium* spp	Kernel damage

Source: Esiegbuya et al. [12].

viability and reduction in its market value. *A. flavus* and *niger* growing on Palm kernels are known to impact various colors to the seeds and also increase the free fatty acids content [12]. Their main impact on agriculture is in saprophytic degradation of products before and after harvesting and in the production of mycotoxins.

14.3   POSTHARVEST AND PROCESSING OPERATIONS ASSOCIATED WITH SHEA BUTTER PROCESSING

The three basic methods for processing Shea butter include: the traditional, semiimproved and mechanized methods. Carette et al. [7] reported that the traditional method is preferred by the processors in developing countries due to its profitability resulting from low fuel/maintenance cost, though it is less efficient when compared to mechanized Shea butter. The three methods of Shea butter processing begin with a process known as curing. The curing process involves the following stages:

1. Shea fruit harvesting.
2. De-pulping of fruits to get the Shea nuts.
3. Boiling of the nuts to prevent sprouting.
4. Nuts drying to ease separation of shell to get the kernels.
5. Cracking of shell to get kernels.
6. Sun drying of kernels to reduce moisture content and prevent fungal germination.
7. Shea kernels storage/Shea butter extraction.

The steps highlighted above are often exposed to poor postharvest handling, which affects the quality of the processed Shea butter as illustrated in Figure 14.3. In order to overcome the postharvest challenges, researchers in developing countries have contributed towards improvement of the curing stage to improve the quality of the processed products of Shea.

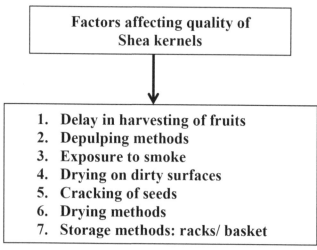

FIGURE 14.3   Factors affecting quality of Shea kernels.

Timely picking of Shea fruits and processing should be carried out without much delay, because heaping of Shea fruits for longer time encourages fungal deterioration and loss of food value, according to Shekarau et al. [29]. This challenge still linkers among local processors because of the nature of Shea fruit. Shea fruits are not harvested and must be allowed to drop by themselves and also the difficulty in walking a long distance to pick enough fruits for processing.

Depulping of Shea fruits should be carried out immediately so as to reduce fungal deterioration [14]. According to Shekarau et al. [29], depulped fruits should be washed several times in warm water so as to get rid of possible surface mold and oxidizing oil emitting from bad nuts. After which, the nuts are then sun dried to dehydrate them. Drying of Shea nuts in dirty surfaces should be avoided so as not to expose them to mites, which had the ability to borrow holes through the nuts to the kernels [11].

Germination of Shea fruits/nuts has been noticed to occur during delay in depulping and boiling of the nuts. The germination process can result to butter depletion, high free fatty acid, peroxide and iodine value of the processed Shea butter as a result of endogenous enzymes [1].

Standard operational procedure that controls the entire curing stages was proposed by Atehnkeng et al. [5]. The authors highlighted five critical points of microorganisms contamination of the Shea nuts and kernels in the production of Shea butter. The authors pointed out that there will be minimal contamination of the Shea products by afla-toxin and other polycyclic aromatic hydrocarbons with the following corrective measures:

1. Careful collection of fallen fruits.
2. Boiling of dried nuts.
3. Drying of both boiled and Shea nut/kernels.
4. Roasting of crushed kernels.

Preventive measures to exert control at these stages is to avoid delay in picking of fallen fruits, boiling of adequately dried fruits, avoid drying of boiled nuts and kernels on unhygienic surfaces and roasting of crushed kernels over naked flame. They also pointed out that aflatoxin contamination of the Shea nut/kernels was higher when compared to the obtained processed Shea butter. This was attributed to the heating of the Shea nut/kernels during processing, which has destroyed some of the heat labile aflatoxins. However, the authors did not mention the aflatoxin producing organisms.

Microorganisms reported to have potential in mycotoxin production have been reported by Esiegbuya et al. [11] to be associated with the postharvest fungal deterioration of Shea nuts and kernels. Some of these microorganisms include: *Aspergillus niger, A. flavus* and *A. persii.*

Obibuzor et al. [25–27] also highlighted the effect of processing of dete-riorated and microbes infested Shea kernels together with good Shea kernels on the quality of processed Shea butter. The authors found that the quality parameters such as free fatty acids and peroxide were higher in Shea butter processed from microbe infested Shea kernels when compared to deterio-rated and good Shea kernels. The authors then concluded that microbes play a role in the deterioration of Shea butter quality. Apart from these chal-lenges, Esiegbuya [12] also highlighted some processing challenges associ-ated with Shea butter processing in developing countries (Table 14.2).

TABLE 14.2   Processing Challenges of Shea Butter

Processing practices	Major challenges
Processing environment	Processing sites characterized by animal dung, unwashed processing materials and leaf litters.
Hygienic conditions of the processors	Processors were characterized by dirty clothing, hands and feet
Storage conditions of the pounded/ grinded kernels	Methods of storing pounded/grinded kernels expose it to domestic animals (fowl) and moisture buildup.
Method of cooling of the skimmed butter	Methods of cooling skimmed butter exposed it to cross-contamination by insects, rodents and quick rancidity.
Inconsistency (variation) in the use of processing materials	Use of aluminum or iron pots in heating stages (wet or dry) may result in contamination of the final product by toxic and prooxidative cations like Fe^{2+} and Cu^{2+}.
Processing water	Poor source of water serve as a source of introduction of coliform organisms and other chemical pollutants into the processed butter.
Use of stick for stirring boiling paste and processed butter	Wooden stick serve as a source attraction for dust, insects and microbe into the processed butter as a result of accumulation of butter.
Unhygienic method of keeping use and unused processing materials	The unhygienic manner of keeping used and unused materials encourages the presence of flies and other insect around the materials.
Poor method of waste disposal	The dumping of Shea waste (Shea cake) close to processing sites serve as a source cross-contamination of the processing and finished products

Source: Esiegbuya [12].

The identified poor processing practices were regarded as the major factors influencing the rate of direct and cross-contamination of the processed Shea butter during processing. The overall effect of these poor processing practices by local Shea butter processors (Figure 14.4) is the continuous cross-contamination of the processing equipments/utensils, raw materials and finished products by vectors such as flies, insects, rodents and domestic animals.

According to the author in order to address these challenges processing practices such as Good Processing Practices (GPP) and Sanitary and

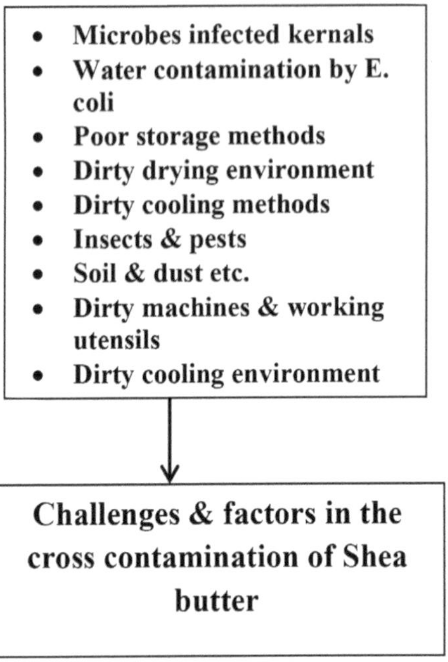

- Microbes infected kernals
- Water contamination by E. coli
- Poor storage methods
- Dirty drying environment
- Dirty cooling methods
- Insects & pests
- Soil & dust etc.
- Dirty machines & working utensils
- Dirty cooling environment

↓

Challenges & factors in the cross contamination of Shea butter

FIGURE 14.4 Continuous cross-contamination of the Shea butter processing line, *Source:* Esiegbuya.

Phytosanitory Standards (SPS) conditions should be adopted in order to place a barrier between these sources of contamination and the processed Shea products (Figures 14.4).

14.4 INNOVATIVE TECHNOLOGIES FOR IMPROVING THE QUALITY OF PROCESSED SHEA BUTTER

Developed countries, where Shea products are used, use state-of-the-art technologies for Shea butter extraction after the curing stages are employed. The technologies yield good quality butter of 42–50% from the kernels. After extraction, the Shea butter undergoes refining stages of de-gumming, neutralization, bleaching and deodorization. These refining processes take away some natural ingredients from the Shea butter. The de-gumming process helps to avoid color and taste reversion, while neutralization, bleaching and deodorizing processes help to remove phosphatides, free fatty acids, mineral and color bodies. The products resulting from these treatments undergo a process of fractionation (separating the stearin fraction from the olein fraction). This refining process tends to alter the nature of Shea butter making it sterile and lifeless (refined); this is what is available in the international market but buyers will opt for a high quality natural butter (unrefined) [35].

The major challenge in traditionally processed Shea butter is quality issues resulting from microbe contamination, high free fatty acids and peroxides, purity, etc. Studies have shown that microorganisms have the potential of causing quality deterioration of the processed Shea butter. According to Esiegbuya et al. [11], *Aspergillus* species possess the potential of causing color alteration, increased free fatty acids and peroxides and moisture content of the olein fraction of the Shea butter. This can also have possible effect on the stearin fraction.

In order to address the highlighted challenges in Table 14.2 associated with Shea butter processing, that causes microbial contamination of the different stages of processing and the processed Shea butter, an internal traceability system was developed by Esiegbuya [12] for the traditional and the semiimproved methods of Shea butter production using the schematic flowchart for processing Shea butter. The essence of the internal traceability system is to identify the points of microorganisms contamination of the Shea butter processing stages, detect and control their contamination of the processed Shea butter.

The efficiency of the traceability system for microbe detection was enhanced using the macroscopic, microscopic, biochemical and molecular characteristics of the identified microbes from each stage of processing to the processed Shea butter.

According to Esiegbuya [12] and Tella [32], food traceability system should have the ability to trace and follow a food through all stages of production and distribution. The major factors in a food traceability study include:

- The origin of the raw material to be processed.
- The processing methods/techniques.
- Source/introduction of external processing materials such as water.
- Hygienic conditions of the processors, environment and processing materials.
- Packaging of processed products.
- Products distribution.

Traceability system in food processing is important because it helps to determine the causes of food problem and taking of remedial actions, it can also help in the recalling of unsafe product [8, 33].

The internal traceability system developed for the traditional and semi-improved methods of Shea butter processing covers the microbial hazard analysis of each of the processing stages and also identifies the critical control points to be, the raw material (Shea kernels), sources of cross-contamination and the processing water.

If the traceability system is properly developed into a closed processing system with appropriative control feedback sensors, it will help to reduce the challenge of microbial contamination of the processing stages and cross-contamination of the processed Shea butter thus enhancing its quality.

14.5 EMERGING ENGINEERING TECHNOLOGIES FOR IMPROVING THE QUALITY OF PROCESSED SHEA BUTTER

14.5.1 FOOD PROCESSING CONTROL STRATEGY

Food process control is an important part of modern processing food industries. It is important because its helps to:

1. reduce variation in food product resulting from human errors;
2. achieve consistent production and yield;
3. ensure product safety;
4. reduce manpower and optimize energy;
5. waste reduction.

Inconsistency in the quality of Shea butter production/yield and safety still remains a major unaddressed challenge with Shea butter production.

Esiegbuya [12], Obubizor et al. [26] and Eneh [10] had previously highlighted that the inconsistency in Shea butter processing varied from village to village and from zone to zone and the quality of the processed butter only satisfied the local segment of the market thereby attracting low price. The authors also mentioned that the inconsistency in product was mainly attributed to variations at the different processing stage.

Reports have shown that the three methods of Shea butter processing (traditional, semiimprove and complete mechanized methods) yield less than 50% of the Shea butter from the Shea kernels [35]. This obviously shows that these methods of processing have not addressed the issue of extraction efficiency.

In order to address these challenges, there is the need to develop a cost effective closed system for Shea butter processing using on-line continuous sensors for monitoring rheology-related parameters in order to achieve higher quality, increased yields, reduced losses and microbial contamination.

Sensors have been developed that can be used to measure key processing parameters of processed food such as temperature, pressure, mass, material level in containers, flow rate, density, viscosity, moisture, fat content, protein content, pH, size, color, turbidity, etc. some of these sensors include penetrating, nonpenetrating and sampling sensors.

According to American Shea Butter Institute [3], the following tests are necessary in the determination of the quality of Shea butter:

1. Physical assessment (purity in terms of color, viscosity, moisture, etc.).
2. Chemical assessment (peroxide vale, fatty acid profile, etc.).
3. Heavy metal contamination.
4. Microbiological assessment (toxin and lipase producing organisms).
5. Solid fat profile.
6. NMR melting point.
7. Shelf life.

Two major issues necessitating these tests are genetic variations in Shea fruits and microbial contamination of the Shea nuts and processed Shea butter.

In order to overcome these challenges a closed system processing with enclosed online sensors devices that will aid in monitoring rheology-related parameters and quality assessments, is needed.

In detecting the presence of microbes in food, the heterogeneity of the food must be put into consideration since microbes can survive over a wide range of environmental parameters also in the design of a biosensor for food processing, the physical and chemical composition of the food must also be considered.

14.5.2   MICROBIAL MANAGEMENT OF SHEA BUTTER

Microbial management of Shea butter during processing is strategic for preventing contamination and for improving the product safety, quality and production hygiene. Florescent sensors have been designed that have the ability to detect saccharides of bacterial cell membranes in food [4]. In other application of genetic engineering principles, whole cell microorganisms can be used to detect chemical components such as environmental pesticides. *Pseudomonas syringae* has also been used as the biocatalyst to also measure BOD in water samples with a response time of 3–5 min. The biocatalyze was placed between cellulose and Teflon membranes [18]. *Pseudomonas putida* JS444 has also has been constructed to display organophosphorus hydrolase (OPH) activity on a dissolved oxygen electrode to detect synthetic organophosphate compounds (OP). In optimal condition it measured as low as 55 ppb for paraoxon, potent acetyl cholinesterase-inhibiting insecticides, without interference from other common pesticides [20].

With the knowledge of genetic engineering, biosensors can be designed and incorporated into a closed system processing that will aid the determination of the quality of processed butter across the different stages during processing.

14.6   SUMMARY

Shea tree is an important economic crop because of the heavy demand for its butter in the international market. It also contributes to the economies and livelihood improvement of processors in countries having it. The butter from the Shea tree is the major economic products from the Shea tree and it applicable in cosmetics, pharmaceutical industries and skin moisturizer due to its healing fraction. The major challenge of the Shea butter processing in African countries is mainly with postharvest and processing issues which affects the quality of the processed Shea butter making it to fall below international standards. Several authors and institutions have made contributions towards improving the quality of the processed Shea butter and this has yielded positive results. The aim of this book chapter is to make a review of the contributions made by authors and institutions and also to provide information on new and emerging technologies in the areas of technology innovations and engineering which can also be applicable to other food process control systems.

KEYWORDS

- bleaching
- close system processing
- cross-contamination
- curing
- de-gumming
- deodorizing
- depulping
- emerging technologies
- extraction of Shea butter
- food processing
- genetic engineering
- innovation
- neutralization
- postharvest
- postharvest diseases
- postharvest food loss
- processing operations
- refining Shea butter
- traceability system
- unrefined Shea butter

REFERENCES

1. Akueshi, T., Kojima, N., Kikuchi, T., Yasukawa, K., Tokuda, H., Masters, E., Manosroi, A., & Manosroi, J. (2010). Anti-inflammatory and chemo preventive effects of triterpene cinnamates and acetates from Shea fat. *Journal of Oleo Science, 59*(6), 273–280.
2. Alander, J. (2004). Shea butter: a multifunctional ingredient for food and cosmetics. *Lipid Technology, 16,* 202–205.
3. American Shea Butter Institute (ASBI), (2009). *Shea Butter.* 5th Edition, International Shea Butter Convention, Atlanta, Georgia.
4. Amin, R., & Elfeky, S. A. (2013). Fluorescent sensor for bacterial recognition. *Spectrochim. Acta Mol. Biomol. Spectros, 108,* 338–341.
5. Atehnkeng, J., Makun, H. A., Osibo, O., & Bandyopadhyay, R. (2013). *Quality production of Shea Butter in Nigeria. International Institute for Tropical Agriculture,* Ibadan, 52 pp.

6. Boffa, J. M., Yaméogo, G., Nikiéma, P., & Knudson, D. M. (1996). Shea nut (*Vitellaria paradoxa*) production and collection in agroforestry parklands of Burkina Faso. Department of Forestry and Natural Resources, Purdue University, West Lafayette, Indiana, USA. FAO document available at: http://www.fao.org/docrep/W3735e/w3735e17.htm.

7. Carette, C., Malotaux, M., Van Leewen, M., & Tolkamp, M. (2009). Shea nut and butter in Ghana, opportunities and constraints for local processing. Available at: http://www.resilience-foundation.nl/docs/Shea.pdf.

8. Denton, W. (2001). *Introduction to the Captured Fish Standard Traceability of Fish Products.* Tracefish EC-CA QLK1-2000-00164. Available at: http://www.tracefish.org.

9. Dogbevi, E. K. (2009). *Shea Nut has Economic and Environmental Values for Ghana.* Sekaf Ghana Ltd. Publication.

10. Eneh, M. C. C. (2010). An overview of Shea nut and Shea butter industry in Nigeria. *A Paper Presented at the National Seminar*, Hydro Motel Limited, Niger state 4th–5th August.

11. Esiegbuya, O. D., Okungbowa, F. I., & Obibuzor, J. U. Degradative effect of *Aspergillus* species on some physical and chemical parameters of olein fraction of Shea butter. *American Journal of Food Science and Nutrition.*

12. Esiegbuya, O. D. (2015). *Effect of Fungal and Bacterial Contamination on the Quality of Shea Butter and Identification of Sources of Contamination During Processing.* PhD thesis, University of Benin, Nigeria.

13. Esiegbuya, O. D., Okungbowa, F. I., Oruade-Dimaro, E. A., & Airede, C. E. (2013). Dry rot of *Raphia hookeri* fruits and its effect on the mineral and proximate composition. *Nigerian Journal of Biotechnology, 26*, 26–32.

14. FAO (2007). Corporate Document Depository. Minor oil crops. http://www.fao.org/docrep/X5043E/x5043E0b.htm.

15. Fobil, J. N. (2007). *Bole, Ghana: Research and Development of the Shea Tree and Its Products.* New Haven. CT: HORIZON Solutions International, May 8, 2007. Available at: http://www.solutionssite.org/artman/publish.

16. Goreja, W. G. (2004). Chapter 2. In: *Shea Butter: The Nourishing Properties of Africa's Best-Kept Natural Beauty.* Amazing Herbs Press. New York, NY. p. 5.

17. https://en.wikipedia.org/wiki/Shea_butter

18. Kara, S., Keskinler, B., & Erhan, E. (2009). A novel microbial BOD biosensor developed by the immobilization of *P. syringae* in microcellular polymers. *Journal of Chemical Technology and Biotechnology, 84*, 511–518.

19. Lawal, U. O., & Fagbohun, E. D. (2012). Nutritive composition and mycoflora of sundried millet seeds (*Panicummiliacieum*) during storage. *International Journal of Biosciences (IJB), 2*(2), 11–18.

20. Lei, Y., Mulchandani, P., Chen, W., & Mulchandani, A. (2004). Direct determination of *p*-nitrophenyl substituent organophosphorus nerve agents using a recombinant *Pseudomonas putida* JS444-modified clark oxygen electrode. *Journal of Agriculture and Food Chemistry, 53*, 524–527.

21. Maranz, S., & Wiesman, Z. (2003). Evidence for Indigenous Selection and Distribution of the Shea Tree, *Vitellaria paradoxa*, and its potential significance to prevailing parkland savanna tree patterns in sub-Saharan Africa north of the equator. *Journal of Biogeography, 30*, 1505–1516.

22. Masters, E. T., Yidana, J. A., & Lovett, P. N. (2004). Reinforcing Sound Management through Trade: Shea Tree Products in Africa. *Unasylva, 210,* 46–52.

23. Mba, M. C., & Akueshi, C. O. (2001). Some physicochemical changes induced by *Aspergillus flavus* and *A. niger* on Sesanumindicum. *African Journal of National Science, 4,* 94–97.

24. Mbaiguinam, M., Mbayhoudel, K., & Djekota, C. (2007). Physical and Chemical Characteristics of Fruits, Pulps, Kernels and Butter of Shea *Butyrospermumparkii* (Sapotaceae) from Mandoul, Southern Chad. *Asian Journal of Biochemistry, 2,* 101–110.

25. Obibuzor, J. U., Abigor, R. D., Omamor, I., Omoriyekemwen, V., Okogbenin, E. A., & Okunwaye, T. (2014). A two-year seasonal survey of the quality of Shea butter produced in Niger State of Nigeria. *African Journal of Food Science, 8*(2), 64–74.

26. Obubizor, J. U., Abigor, R. D., Omamor, I. B., Omoriyekemwen, V., Okunwaye, T., & Okogbenin, E. A. (2014). Evaluation and quantification of the contributions of damaged Shea kernels to the quality of Nigerian Shea butter. *Journal of Postharvest Technology and Innovations, 4*(1), 33–45.

27. Obubizor, J. U., Abigor, R. D., Omoriyekemwen, V., Okogbenin, E. A., & Okunwaye, T. (2013). Effect of processing germinated Shea kernels on the quality parameters of Shea (*Vitellariaparadoxa)* butter. *Journal of Cereals and Oilseeds, 4*(2), 26–33.

28. Okungbowa, F. I., & Esiebuya, D. O. (2015). Influence of Environmental Factors on the Prevalence of Postharvest Deterioration of Raphia and Shea Fruits in Nigeria. *Environmental Biotechnology.*

29. Shekarau, J., Oriaje, L., Kpelly, & Loveth, P. (2013). Current practices in production and processing of Shea nut. In: *Quality Production of Shea in Nigeria* by Atehnkeng, J., Makun, H. A., Osibo, O., & Bandyopadhyay, R. International institute for Tropical Agriculture, Ibadan, 52 pp.

30. Steiger, D. L., Nagai, C., Moore P. H., Morden, C. W., Osgood, R. V., & Ming, R. (2002). AFLP analysis of genetic diversity within and among *Coffea Arabica* cultivars. *Theoretical Applied Genetic, 105,* 209–215.

31. Tella, A. (1979). Preliminary studies on nasal decongestant activity from the seed of Shea butter tree, *Butyrospermumparkii. British Journal of Clinical Pharmacy, 7,* 495–497.

32. Tella, A. (2001). Traceability procedures based on FDA and CFIA regulations – an understanding. *Infofish International, 5,* 49–51.

33. Tracefish, (2001). *European Commission Concerted Action Project QLK1-2000-00164 Traceability of Fish Products.* Second Draft Information Standard for the Captured Fish Distribution Chains. http://www.tracefish.org.

34. Ukers, W. H. (1922). All about Coffee. Chapter 20, In: *The Tea and Coffee.* Trade Journal Company. New York, NY, USA.

35. USAID (2004). *Shea Butter Value Chain, Production Transformation and Marketing in West Africa.* WATH Technical Report 2.

INDEX

A

Absorbance, 128–130, 150, 151, 154, 155, 217–219, 223, 277–280, 287, 293, 301, 322

Absorption coefficient, 104, 105, 112

ABTS (2,2'-Azino-bis-(3-ethyl) benzo-thiazoline-6-sulfonic acid), 129, 132, 134–136, 139–141, 151–157, 215, 217–219, 221, 223–226, 275, 278, 286–288, 291, 292

  radical cation decolorization assay, 223

  radical cation scavenging activity/assay, 129, 134–136, 139, 151, 155, 217, 221, 278, 292, 288

    ferric reducing antioxidant power assay (FRAP), 129, 130

Acetate, 218, 237, 238, 247, 276, 322

Acetone, 148–157, 237–239, 241, 248, 247, 255, 267, 268, 291, 319–321

  extract, 148–152, 154–156, 291

Acetonitrile, 175

Acoustic, 116

Active packaging, 36

Acylglycerols, 164

Additive noise, 111

Adrenal, 314

Adulteration, 9, 13, 23 38

Aero dynamic, 100

Aerobic

  cells, 154

  respiration, 126

Aerogenes, 257, 259, 263, 264, 283

Aerva lanata, 140, 141

Aflatoxin, 8, 338

Agar

  disc diffusion method, 256, 267

  well diffusion method, 216, 226, 248, 323, 328

AgBF, 171, 172, 174

Agencies, 30, 34, 41, 44, 45, 48

Aggregation, 82

Air filtration systems (HVAC), 12

Albicans, 254

Algae, 37, 148, 149, 152, 154–157, 165

Alginate, 57, 58

Algorithms, 108–111, 113, 115

Alkaloids, 239, 266

Alkyl chain, 172

Alleviate pollution, 302

Aluminum chloride (AlCl), 128, 141, 215, 218, 275, 280

  colormetric, 128

Ambient temperature, 332

Amendments, 4, 57

American Association of Cereal Chemists (AACC), 335

Amino acids, 148

Amnesic shellfish poisoning (ASP), 8

Amorphous region, 72

Amylopectin, 71–73

Amylose, 71–73

Analyze hazards, 10

Anchovy (Engraulisringens), 198

Anesthesia, 40

Animal

  husbandry, 41

  research, 40

  welfare, 40

Anion radical scavenging activity, 150, 292

Annonaceae, 254, 255, 267, 268

Anti-aging foods, 318

Antiallergenic, 138

Antiartherogenic, 138

Antibacterial, 214–216, 222, 224–226, 235, 238, 256, 257, 259, 260, 264, 266, 267, 299, 323

T - #0816 - 101024 - C412 - 229/152/18 - PB - 9781774636855 - Gloss Lamination